6th International Conference on Traffic and Logistic Engineering (ICTLE 2018)

MATEC Web of Conferences Volume 259 (2019)

Bangkok, Thailand
3-5 August 2018

Editor:

Roberto Montemanni

ISBN: 978-1-5108-8069-6

Printed from e-media with permission by:

Curran Associates, Inc.
57 Morehouse Lane
Red Hook, NY 12571

Some format issues inherent in the e-media version may also appear in this print version.

No changes have been made to the content of these proceedings. There may be changes to pagination and minor adjustments for aesthetics.

Printed by Curran Associates, Inc. (2019)

For additional information, please contact EDP Sciences – Web of Conferences at the address below.

EDP Sciences – Web of Conferences
17, Avenue du Hoggar
Parc d'Activité de Courtabœuf
BP 112
F-91944 Les Ulis Cedex A
France

Phone: +33 (0) 1 69 18 75 75
Fax: +33 (0) 1 69 28 84 91

contact-edps@webofconferences.org

Additional copies of this publication are available from:

Curran Associates, Inc.
57 Morehouse Lane
Red Hook, NY 12571 USA
Phone: 845-758-0400
Fax: 845-758-2633
Email: curran@proceedings.com
Web: www.proceedings.com

TABLE OF CONTENTS

Preface

2018 6th International Conference on Traffic and Logistic Engineering (ICTLE 2018) provided a forum for accessing the most up-to-date and authoritative knowledge from both industrial and academic worlds, sharing best practices in this exciting field. ICTLE 2018 was held in Bangkok, Thailand, during the period August 3-5.

The event was held with presentations delivered by researchers and scholars from the international community, including keynote speeches and highly selective lectures.

The proceedings included papers which were from universities, research institutes and industries. All of the papers were subjected to peer-reviewing by conference committee members and international reviewers. The papers selected for the proceedings depended on their quality and their relevance to the conference. Studies presented in this volume cover the following topics: Transportation Safety, Transportation Planning, Emergency Respnse Systems and Technologies, Intelligent Transportation Systems, Energy Saving and Alternative Energy, Automotive Exhaust Treatment and Environment Protection Technologies, New Technology on Transportation Infrastructure (Pavement, Bridge), Highway Maintenance and Management, etc.

We would like to take this opportunity to thank many people. First and foremost we want to express our deep appreciation to keynote speaker, Prof. Roberto Montemanni from Dalle Molle Institute for Artificial Intelligence (IDSIA)&University of Applied Sciences of Southern Switzerland (SUPSI), Switerland; Prof. Kannapha Amaruchkul from National Institute of Development Administration (NIDA), Thailand; Prof. Khair S. Jadaan from University of Jordan, Jordan.

We would like to thank the committee members for their hard work in making sure a smooth running of this conference. Many thanks to the reviewers for their excellent job to maintain the academic quality of this conference. Special thanks is expressed to local chair, Prof. Pongchanun Luangpaiboon, Thammasat University, Thailand, who kindly gave us support.

Final thanks go to all authors and participants at ICTLE 2018 for helping to make it a successful event.

With our warmest regards,
ICTLE 2018 Organizing Committees
August 17, 2018

Statement of Peer review

In submitting conference proceedings to MATEC Web of Conferences, I certify to the Publisher that I adhere to the **Policy on Publishing Integrity** of the journal in order to safeguard good scientific practice in publishing.

1. All articles have been subjected to peer review administered by the proceedings editors.
2. Reviews have been conducted by expert referees, who have been requested to provide unbiased and constructive comments aimed, whenever possible, at improving the work.
3. Proceedings editors have taken all reasonable steps to ensure the quality of the materials they publish and their decision to accept or reject a paper for publication has been based only on the merits of the work and the relevance to the journal.

Title, date and place of the conference

August 3-5, 2018

Bangkok, Thailand

2018 6th International Conference on Traffic and Logistic Engineering (ICTLE 2018)

Proceedings editor(s):

Roberto Montemanni, Dalle Molle Institute for Artificial Intelligence (IDSIA)&
University of Applied Sciences of Southern Switzerland (SUPSI), Switerland

Date and editor's signature

Dec, 28. 2018

Building optimal routes for cargo delivery taking into account the dynamics of traffic flows

Viktor Danchuk, Olena Bakulich and Vitaliy Svatko

National Transport University, 01010 Kyiv, Ukraine

Abstract. A method for optimizing the route based on a modified ant algorithm is proposed, taking into account the unsteady dynamics of traffic flows in the urban road network. Within the framework of the method, simulation of the building of optimal routes was carried out using the example of Kyiv. It is shown the possibility of using the method for solving problems of effective control over the process of routing freight traffic in conditions of actual dynamics of traffic flows (synchronized traffic, wide moving jam, congestion, emergency situations, etc.)

1 Introduction

Accelerated motorization in the conditions of backlog of development of urban road networks (URN) leads to a sharp increase in the number of vehicles on highways, the intensification of traffic. This, in turn, significantly affects the workload of URN, contributes to the occurrence of the congestion and a decrease in the speed of traffic. Long congestion on the roads is increasingly causing not only temporary, but also economic losses. This issue is especially acute for enterprises that deliver their own, for example small-lot, goods to consumers without intermediaries, that is, using their own transport.

The main thing in urban transport logistics is the solution to routing problems using such methods of route optimization that would effectively take into account the actual dynamics of traffic flows on URN.

Thus, the relevance of the study is determined by the need to develop, implement and apply appropriate models, methods and software to optimize and manage the transport process in conditions of unsteady dynamics of traffic flows on the URN. The use of such developments will increase the efficiency of transportation of goods within large cities.

2 literature review and problem statement

Existing methods of discrete optimization of transport routes are not ideal and in most cases solve optimization problems for stationary states of the transport network [1-8].

In [1] the authors consider methods and models for assessing the efficiency of transport and various types of logistics operations in supply chains. Based on the analysis of the results it is shown that to date there is no single approach for obtaining adequate estimates of the effectiveness of the corresponding processes [1].

The model for solving the multi-purpose task of routing vehicles with time constraints was proposed in [2]. This problem is solved using a modified genetic algorithm. The optimal solution here is a trade-off between the cost of transporting the goods and the size of the car fleet necessary for customer service.

In [3], a new algorithm is proposed that allows an optimization procedure to be performed while simultaneously considering all paths of motion between graph nodes taking into account the throughput of each of its edges. This method is based on the representation of a graph in the form of an electrical network, each of whose sections has a certain resistance characterizing the corresponding capacity of the network sections.

It should be noted that the method of [3] is analogous to the method of discrete optimization, based on the use of the effects of self-organization of ant agents [4]. The classical method of self-organization of the ant colony [4] provides the possibility of finding the optimal path for the static graph. The various properties attributed to ant agents make it possible to solve a wide class of discrete optimization problems taking into account many characteristics of the investigated system. In particular, further numerous studies of various physical processes that were carried out using this method indicate that it is promising for solving discrete optimization problems of large dimension [5]. This concerns the routing of vehicles on URN and highways, information flows in computer networks, etc.

In addition, today a large number of modifications of the ant algorithm are presented. In [6], the authors proposed a modification of the classical algorithm with the introduction of "elite ants" Elitist Ant System into it. This modification allowed us to reduce the number of iterations of the algorithm and, as a result, the time of finding the optimal solution. The disadvantage of the proposed method is need to research of definition of the number of such elite ants.

The modification of the ant algorithm (Max-Min Ant System) presented in [7] offers limitations on the maximum and minimum concentration of pheromone. Here, the increase in the concentration of pheromone occurs on the best routes that have been covered by ants.

The proposed modifications make it possible to increase the efficiency of using the ant algorithm in comparison with the classical algorithm.

It should be noted that most of the modifications proposed in [6-8] are used to solve the static traveling salesman problem (TSP). One example of solving the dynamic traveling salesman problem (DTSP) by such modifications of the algorithm is the so-called problem of Cash Machines [9].

In our view, the application of the ant algorithm method can be promising for solving problems of optimization of processes occurring in open nonlinear dynamical non-stationary systems. Here, one must admit that in recent years there has been a rapid development of theoretical concepts, models, methods for describing nonlinear non-equilibrium dissipative processes and phenomena occurring in systems of various physical nature [10]. One of the main approaches used to describe such processes is a synergetic approach based on the principles of self-organization. For example, in [10], in the framework of a non-equilibrium nonlinear statistical-dynamic model, it was shown that when the temperature is raised, due to self-organization effects, it is observed a transition from a mono crystalline to a rotational crystalline phase in organic molecular crystals. Here, damping of molecular vibrational excitons occurs on orientation defects of various types. This leads to the appearance of a temperature dependence of the Davydov splitting of molecular vibrations in the infrared absorption spectra of these crystals. In [11], in the framework of the synergetic model of Lorentz, there were considered the phenomena of self-organization of enterprises' activity under the influence of external stochastic and deterministic (discrete, sharp, extreme) changes of the factors of a nonlinear market environment. This approach allows us to correctly describe and predict the time series of the evolution of the enterprise under the influence of such changes.

As noted above, the ant algorithm is also based on the use of self-organization effects of the relevant agents. This makes it possible to use it to describe the corresponding non-equilibrium, non-stationary processes and phenomena, which, in particular, are observed when considering the actual dynamics of high density transport flows.

Indeed, in [12], based on an analysis of the corresponding experimental data obtained on highways in Germany, England and the United States, B. Kerner proposed the theory of three phases of the dynamics of a transport stream (free flow, synchronized flow, wide moving jam or congestion). This theory allows us to explain and predict empirical spatial and temporal dependences of phase transitions from free to dense (synchronized) flow and between resultant structures in a dense transport stream. Within the framework of this theory, it is considered that the observed effects of phase transitions, hysteresis of velocity changes at these transitions are due to the manifestation of synergistic effects as a result of the non-equilibrium non-stationary dynamics of transport flows in the region of their high density [13].

In [14], it is proposed a method for optimizing the transportation route based on the ant algorithm, taking into account the change in the speed of traffic flows in certain sections of the URN. However, in [14], during the optimization, the dynamics of the change on the speed of movement on the sections of the URN were not taken into account. There, at the time of route optimization those sections of the road network, where speed decreased significantly, were excluded from consideration.

The purpose of this work is to develop a method for constructing an optimal route for the delivery of goods on the basis of a modified ant algorithm [14], taking into account the actual dynamics of traffic flows on the URN of large cities.

3 Route optimization method using modified ant algorithm

For the development of the method for constructing the optimal route for the delivery of goods, we represent URN in the form of a two-way oriented weighted graph. At the nodes of such a graph, the delivery points of the goods (warehouses, supermarkets, etc.) from the sender's warehouse using a vehicle moving on the URN are located. The edges connect each pair of nodes of the graph. The weight of the edges is determined depending on the nature of the problem being solved. This can be the distance between points, the average speed or the average time, the cost of delivery by the vehicle in the transport flow on certain section of the network. Optimization of routes of movement of a vehicle is carried out with the help of the ant algorithm in the framework of the traveling salesman problem. This problem is formulated as a task of finding a minimal closed route along all the nodes which salesman visits without the repetition on a certain weighted graph with the number of vertices m [4]. Here, the optimization parameters can be the length of the route, the average speed or the average time, extra expenses on delivery of goods. For convenience, we assume that the exit and return of the vehicle can occur from any node in the graph. A description of the classical method of the ant algorithm is given in [4].

The solution of the routing problem, taking into account the actual behavior of the traffic flow, will be realized with the help of the modified ant algorithm [14]. We consider the dynamic two-way oriented weighted graph. The weights of the edges in this graph are determined by the average travel time of the vehicle in the traffic flow at certain sections of the network between the points of delivery (graph nodes). The fundamental difference between the modification of the algorithm [14] and the modifications of the ant algorithm [6-8] is that the cyclic movement of the colony, where the ants move at constant speeds, is replaced by the asynchronous movement of each ant agent at a certain rate. In addition, it is possible to fixate the results of optimization of the partially covered path for calculating the further route when the edge length of the graph changes during movement. This allows simulation of route optimization in conditions of actual dynamics of traffic flows on URN, where ant agents, like vehicles analogues, move with corresponding speeds.

The application of such a modified method of the ant algorithm is performed under the following assumptions:

- the movement of a dedicated vehicle on all sections of the URN is carried out within the framework of a two-way traffic;
- there are always alternative routes in each set of n_j-sections of URN, which corresponds to the j-th edge of the graph;
- the change in the average travel time and average speed of movement is largely dependent on changes in the driving dynamics modes that do not include stops or delays due to traffic lights signals.

Each route between any two points within the metropolis has a number of alternatives. That is, it is possible to deliver goods from one point to another by several routes, which differ only in the length and average speed of the traffic flow on them. In our case, we assume that the average passage time along the n-th edge between the two nodes of the graph can be found as follows:

$$t_n = \frac{l_n}{\upsilon_n}, t \to \min \tag{1}$$

where t_n– the average time spent on passing the n-th edge, min;

l_n – the length of the n-th edge, km;

υ_n – average speed on the n-th edge, km/h.

Since the value of the actual traffic flow rate on each road network can vary, then, respectively, the value of the passage time of the corresponding sections of the URN is also a variable value. Thus, when constructing the optimal time route of cargo delivery, the data on the actual state of the URN at the time of route construction is taken into account. At the same time, it is clear that at certain limit values of the traffic flow velocity on certain sections of the URN it is possible to reconstruct the optimal time route of the traffic on the URN vehicle for the delivery of goods at destination points.

4 Case study: Application of the method by the example of urn of Kyiv. Results of simulation research and discussion

For approbation of the proposed method of constructing the optimal route of cargo delivery, data on the distance between the points of delivery in Kyiv city (Fig. 1) and the average traffic velocities at the corresponding sections of the URN were used [15, 16]. A fragment at the URN of Kyiv was presented as a two-way oriented weighted graph.

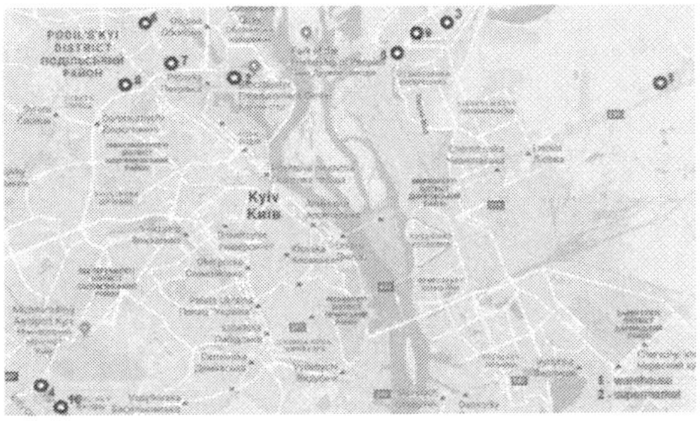

Figure 1. Scheme of the location of cargo delivery points at the URN of Kyiv. Point 1 – distribution center (warehouse of delivery of goods); points (2– 10) – points of delivery of goods (supermarkets, etc.)

The corresponding simulation researches on the construction of the optimal route were performed according to the proposed method, for the 10 nodes defining the points of delivery of the load. The average passage time for each section of the URN was based on the formula (1). The time taken to travel along the route and its length were selected for optimization parameters.

4.1 Construction of the optimal route for the delivery of cargo at the URN of Kyiv without taking into account the change in the speed of traffic flows

This section presents the results of simulation researches of the processes of constructing the optimal route provided that the average velocities are unchanged in the URN areas. At the same time, for calculations, data on the distance and average velocities in the regions of URN were used in accordance with [15, 16]. The results of these calculations are shown in Fig. 2 for 10 nodes respectively.

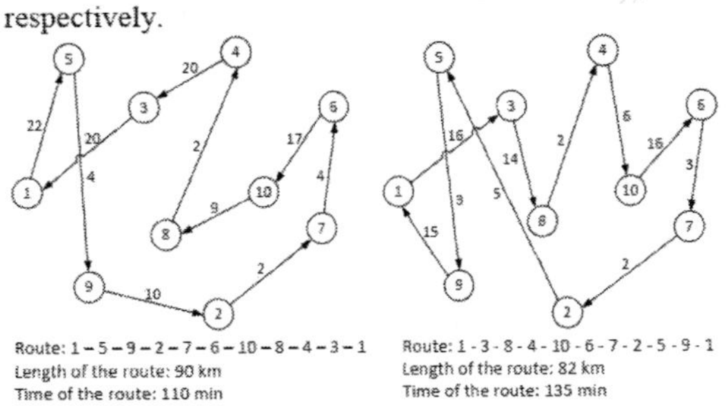

a b

Figure 2. Optimal delivery route for 10 points: a - by delivery time; b - the length of the route

As can be seen from Fig. 2, various optimal routes are obtained depending on the optimization parameter. In case of optimization of the route according to the time of delivery of goods (Fig. 2, a) the length of the found route is 90 km. Time of the route is 110 minutes. In the case of route optimization along the route length (Fig. 2, b), its total length is 82 km, which is less than the previous by 8 km. The time of delivery of the cargo is 135 minutes, which is more than in the previous case by 25 minutes.

From the obtained results it can be seen that, depending on what is the optimization parameter, for the same initial data, in general, we get different delivery routes with different characteristics. Consequently, this causes, when making a managerial decision on the delivery of goods to its destinations, the need for a correct choice of the optimization parameter, depending on the objective (minimum time or minimum delivery route).

In order to verify the correctness of the use of the modified ant algorithm to solve the problems of finding the optimal route, given in this work, the test of optimization of the route for the same network with 10 nodes was performed with the help of the exhaustive search method, as well as the method of branches and boundaries [14]. The obtained results of research fully coincide with those obtained with the help of the corresponding ant algorithm.

4.2 Construction of the optimal route for the delivery of cargo at the URN of Kyiv, taking into account the real dynamics of traffic flows

In this section the results of simulation modeling on optimization of the route of cargo delivery in conditions close to the real dynamics of transport flows at the URN of Kyiv.

According to [12], for a qualitative description of the states of the dynamics of a traffic flow, three phases are considered: free flow, synchronized flow and wide moving jam. Depending on certain limit values of the traffic intensity, density, and average speed of the vehicles forming the flow, all possible states of the traffic flow are divided according to [12] into certain modes of motion.

When conducting research as an input parameter, the optimal route was selected obtained by performing the time optimization procedure for unchanged initial velocities on the edges taken from [15], for a graph with 10 nodes (Fig. 2, a). In addition, for uniqueness, it was considered that the average speed of traffic flows due to changes in traffic conditions can simultaneously be changed in three sections of the original optimal route (Fig. 2, a). For this purpose, the selected sites were between nodes 2-7, 10-8 and 4-3 (Fig. 3a, red lines). In the initial optimal route, the average speeds and time of traffic flow in these areas of the URN were 53 km/h and 2 min, 49 km/h and 9 min, 49 km/h and 20 min respectively. The mean free-flow speed of the traffic flow at the 2-7 URN site was 60 km/h. That is, it was limited to the maximum allowed velocity value in Kyiv. For sections 10-8 and 4-3 URN in Kyiv, the average free-flow speed of the traffic flow was 50 km/h. That is, it is limited to the corresponding road signs and [17].

The results of the simulation researches are shown in Fig. 3. Here are shown the stages of the route-optimized route formation after each iteration associated with the decrease in speed at the specified areas of the URN.

As can be seen from Fig. 3a inbound route is optimal over time, when free traffic of vehicles is carried out in three sections of the WDM in accordance with [12]. After the second iteration of optimization over time, when on the site 2-7 already occurs according to [12] synchronized flow, and on the other two there is a free flow of cars, reducing the speed at the relevant sites leads to the emergence of two alternative optimal routes (T = 112 min). The first of which corresponds to the input route. And the second one is transformed so that it changes the area of URN 2-7, on which the mode of movement of cars has changed (Fig. 3, b, c). In this case, the length of the route for the second alternative route is greater than for the incoming (95 km and 90 km, respectively). Obviously, when the time of passing the routes is the same, and the length of the route is different, more suitable route can be considered, the length of which is less. From this it follows that for the adoption of correct management decisions it is necessary to have full quantitative information on the results of optimization. The importance of having, above all, quantitative information about the state of the URN is further illustrated by the results of the optimization carried out during the fourth iteration. Here, on three sections of the URN, where the speed changes in accordance with [15, 16], the same character of traffic flows associated with the synchronized flow of cars is observed qualitatively. However, the corresponding quantitative distribution of average velocities on all sections of the studied URN necessitates the restructuring of the optimal route after optimization (Fig. 3, d).

During the next, fifth iteration of optimization over time, the state of the URN will be answered, where at area 2-7 there is a movement of cars in the form of wide moving jam at low speed and with high probability of occurrence of a congestion [12], and on two other - the synchronized flow of cars. Here, such a reorganization of the route optimal is performed, when only one of three plots (area 10-8) remains in it. Here, in the framework of our model, the average speed of the traffic flow changes (Fig. 3, e, f).

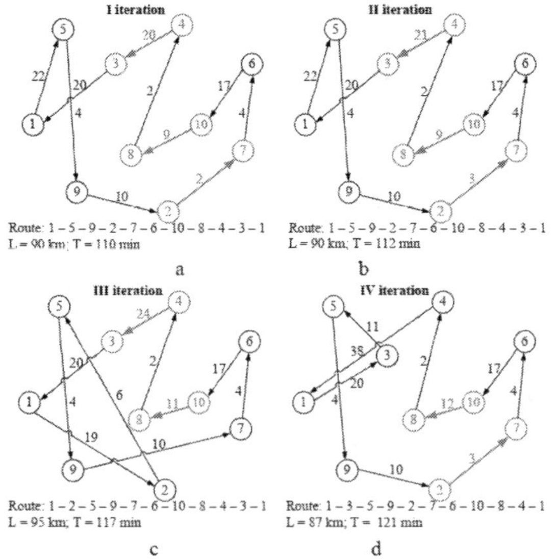

Route: 1 – 5 – 9 – 2 – 7 – 6 – 10 – 8 – 4 – 3 – 1
L = 90 km; T = 110 min

a

Route: 1 – 5 – 9 – 2 – 7 – 6 – 10 – 8 – 4 – 3 – 1
L = 90 km; T = 112 min

b

Route: 1 – 2 – 5 – 9 – 7 – 6 – 10 – 8 – 4 – 3 – 1
L = 95 km; T = 117 min

c

Route: 1 – 3 – 5 – 9 – 2 – 7 – 6 – 10 – 8 – 4 – 1
L = 87 km; T = 121 min

d

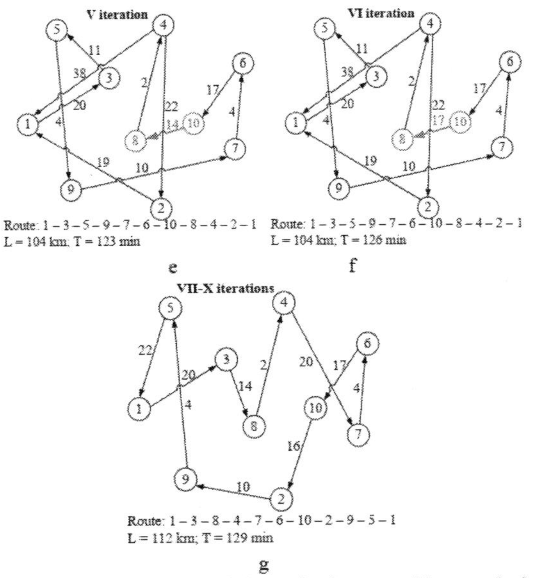

Figure 3. Construction of the optimal route with a gradual change in the speed of traffic at areas 2-7, 10-8 and 4-3 URN for 10 nodes. The red shows the areas of URN, which are changing the average speed of the traffic flow

Accordingly, since the seventh optimization iteration, the optimal route formed already does not include these areas of the URN (Fig. 3g). Areas where the reduction of average speeds of transport streams on the URN reach values that correspond to the mode of occurrence of traffic jam on them. It is clear that further reductions in average speeds on them no longer lead to the restructuring of the optimal route (Fig. 3g).

5 Conclusions

The method is proposed of optimization of cargo delivery routes in large cities in the conditions of non-stationary dynamics of transport flows in the regions of URN. In the framework of this method, the network is represented as a two-way oriented weighted graph. The method is based on the use of a modified ant algorithm. The main element of the modification is that is realized the possibility of asynchronous movement of ant colony agents at a certain speed. In addition, it is also possible to record the results of optimization of the partly traversed path to calculate the further route when changing the conditions of movement of these agents. This allows you to manage the process of optimizing the route, taking into account the dynamic state of the URN, which is associated with significant changes in the speed of vehicles in certain areas of this network. These changes in speed can be due to increased loading of sites on URN, the occurrence of congestion, emergency situations, etc. The method allows to carry out the process of optimization both in time and length of the route. In order to verify the correctness of the use of the modified ant algorithm, tests were conducted to optimize the route along the length using the exhaustive search method and the method of branches and boundaries. The obtained results of research fully coincide with those obtained with the help of the modified ant algorithm.

To approbation the proposed method, research simulations of route optimization processes were carried out within the framework of the traveling salesman problem for the example of the URN of Kyiv. During the simulation were detected a number of effects. These effects are related to the restructuring of the optimal routes with a decrease in the average speeds of vehicles on model parts of the URN to certain limit values, which correspond to certain modes of movement of traffic flows. The analysis of the revealed effects allowed formulating some methodical recommendations for making managerial decisions on issues of rational organization of cargo transportation.

The analysis of the results of simulation researches gives grounds to consider that the proposed method allows to adequately optimization the routes of cargo delivery in the conditions of unsteady traffic flow on the URN of large cities. In the future, he can find his application in solving the problems of operational management of the transportation process in metropolitan areas. For example, the formation in the online mode of the best routes, taking into account the real traffic on the URN.

The results obtained in this way can be used to improve existing intelligent transport systems and technologies for traffic management at URN of megacities.

References

1. V. Lukinskiy, V. Dobromirov. Transp. and Telecom. **17**, 55-59 (2016)
2. Y. Qingyou, Q. Zhang. Discr. Dyn. in Nat. and Soc. **2015**, 10-15(2015)
3. H. Knight. MIT News. **2014**, 21-26 (2014)
4. M. Dorigo. *Reader for CEU Sum. Univ. Course «Complex System»* (Budapest, Centr. Eur. Univ., 2001).
5. W. Zhaoyuan, X. Huanlai, L. Tianrui. IEEE Transact. on Evol. Comp. **20(3)**, 325-423 (2016)
6. R. Karmakar. MAYFEB J. of Comp. Sc. **1**, 8-15 (2016)
7. T. Stützle, H. Hoos. *IEEE Intern. Conf. on Evol.Comp.* 309-314 (1997)
8. M. Dorigo, L.M. Gambardella. IEEE Transact. on Evol. Comp. **1**, 53-66 (1997)
9. M. Mavrovouniotis, L. Changhe, Y. Shengxiang. Swarm and Evol. Comp. **33**, 1-17(2017)
10. G. Puchkovska, S. Makarenko, V. Danchuk. J. of Molec. Struc. **744-747**, 53-58 (2005)
11. M. Danchuk, A. Kravchuk. Act. Probl. of Econ. **148**, 207-213 (2013)
12. B.S. Kerner. *The Physics of Traffic*. Berlin, Heidelberg, New York: Springer. 683 p. (2004)
13. H. Haken. *Synergetics. Introduction and Advanced Topics*. Berlin, Heidelberg: Springer. 758 p. (2004)
14. V. Danchuk, O. Bakulich, V. Svatko. *Procedia Engineering*. **187**, 425-434 (2017)

15. Karta mista Kyieva dlia pobudovy marshrutiv po Ukraini. *Available at:* https://www.google.com.ua/maps/
16. Resurs sposterezhennia za chastynoiu dilianok VDM m. Kyieva. *Available at:* http://videoprobki.ua
17. Postanova Kabinetu Ministriv Ukrainy vid 10.11.2017 № 883 «Pro vnesennia zmin do Pravyl dorozhnoho rukhu». *Available at:* http://zakon3.rada.gov.ua/laws/show/883-2017-%D0%BF

Dynamic on-line multimodal route planning with least expected travel time in the stochastic and time-dependent bus network

Jinjian LI , Arnaud Doniec, Jacques Boonaert and Guillaume Lozenguez

IMT Lille Douai, Univ. Lille, Unité de Recherche Informatique, et Automatique, F-59000 Lille, France

Abstract. Route planning system in bus network is very important on providing passengers a better experience for public transportation services. This work presents a new on-line dynamic path planning algorithm in a stochastic and time-dependent bus network, with the objective of least expected travel time. Firstly, an initial optimal path is found when a passenger leaves from the starting location at time t for the destination location. Secondly, when the passenger reaches a bus stop, the real time traffic condition is checked to decide whether the optimal path should be modified and re-calculated. If so, the shortest path algorithm is re-applied to find the new path. The simulation scenario is executed to show the performance and efficiency of the proposed dynamic path planning system.

1 Introduction

Intelligent urban transportation services is one of the most important applications in smart cities and becomes more and more important for commuters. The main function for intelligent urban transportation services is to find an optimal routes in road and public transportation networks. The traditional route planners generally rely on fixed schedules, which is too ideal to be held in the realistic transportation network, because bus real arrival times at each stop mostly depend on actual traffic circumstances, which are extremely time-dependent and stochastic. Therefore, the path found by traditional route planners often are not the optimal solution in real world, leading to users' dissatisfaction. Then, there are many literatures that try to solve the drawbacks of traditional route planners in different point of views. Comparing with the static transportation network researched by traditional route planners, the travel time on each arc is random during the trip in the Stochastic and Time-Dependent (STD) network (e.g. a rail or bus network). On the one hand, the travel time is assumed to follow a discrete random distribution. Authors in [1-2] focus on the problem of searching the optimal path with the least expected travel time between two nodes in the STD network. On the other hand, the travel time can be estimated from the historical data without requiring the probability distribution of travel time. In [3], the approach only takes the range of uncertainty according to the historical data and experience of the decision-makers, and the route with the greatest robustness is found as the optimal path in STD network. In [4], the authors develop a dynamic route planner based on the stochastic time-dependent transportation network model, which is built by leveraging a large set of historical travel smart card data. However, the estimation of travel time in the above researches is only based on possibility distribution or historical data, and ignores the real traffic condition.

Another research direction in STD networks focus on finding an optimal routing policy, which is a hierarchical decision scheme specifying which bus to take next at each decision node. Works [5-7] research the traffic networks, where both link-wise and time-wise stochastic dependencies of link travel times are considered and online information is represented. In [8], the real-time information is incorporated into the routing policies in STD networks. However, the previous works only take into consideration of bus travel time and ignore waiting time. In reverse, works [9-10] considers the networks with stochastic waiting time, while the travel time is supposed to be deterministic. Therefore, the proposed work, named dynamic on-line route planning system, advances all the above proposals by taking into account waiting time, travel time and on-line traffic information in public transportation networks.

The main contributions of this paper are as follows: (i) introducing a dynamic on-line route planning structure in the STD network considering waiting time and travel time. (ii) processing the least expected travel time problem by applying a shortest path algorithm.

The paper is organized as follows: the first section surveys the background and literatures of the path planning in the STD network with the objective of least expected travel time. Section II describes the STD network to be researched. Section III presents the main methods, including shortest path algorithm and optimal path updating process. Section IV shows the simulation results for the validation of the proposed method. The last section concludes the current work and proposes some future works.

2 Problem description

Fig. 1 shows a hypothetical STD network having four rows and columns to be researched in this work. There are 5 bus stops and three bus lines. Passengers leave from the starting location at time t to the destination location. The problem is to dynamically find the optimal path with the least expected travel time during the trip based on the real time traffic information.

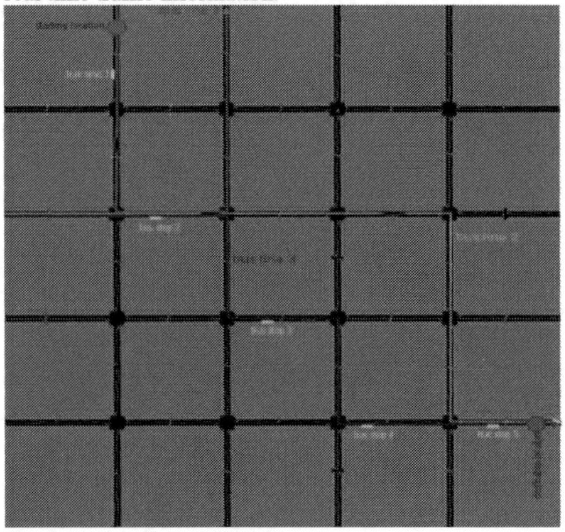

Figure 1. Hypothetical STD network.

The initial expected travel time for a passenger starting at time t to spend in traversing arc between bus stops i and j is denoted as $c_{ij}(t)$. That is to say, the earliest arrival time at bus stop j is $cij(t)+t$, if the passenger begins to cross an arc from bus stop i at time t. The time $cij(t)$ includes the waiting time wt_i in bus stop i and the routing time between two bus stops, which is estimated based on the timetable. In SUMO (Simulation of Urban Mobility, refers to section **IV**), the timetable of bus flow mainly consists of three elements: the final expected leaving time from the bus stop for the first bus in the flow (represented with variable *until* in SUMO), the *begin* and *end* time of bus flow, and the total numbers of bus during the flow (named by the parameter *number* in SUMO). Then the interval of two successive bus equals to the following value: *(end-begin)/number*. And the routing time between two successive bus stops equals to the gap of their leaving times. In the walking model, the travel time only depends on the passenger's walking distance and average speed.

$$c_{ij}(t) = \begin{cases} d_{ij}/v(t) & \text{walking mode} \\ wt_i(t) + until_j(t) - until_i(t) & \text{bus mode} \end{cases} \quad (1)$$

3 Dynamic on-line route planning algorithm

There are a great deal of route planning algorithms applied in road networks, and the main methods proposed in the literature can be classified into three categories based on the technique used to explore the solutions space [11], as the Fig.2 shows:

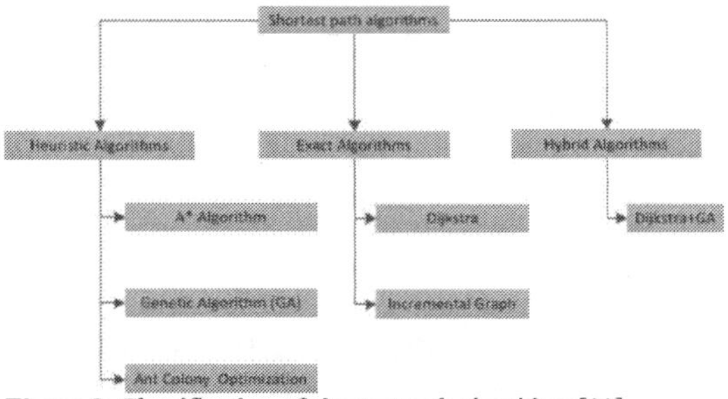

Figure 2. Classification of shortest path algorithms[11]

1) exact algorithms guarantee the global optimal path by the exploration of the whole set of possible solutions space.

2) heuristic based methods explore a subset of the available solutions space and usually find an approximate optimal path, whose qualities are close to those of the global optimal one.

3) hybrid based approaches combines the strengths of both the above two algorithms.

Among all the route planning algorithms, the Dijkstra is one of the most common optimal path algorithms. It can find the shortest path from one node to any other node in a transportation network. In other words, it includes the process of searching the path with the lowest cost (e.g. usually refers to the shortest path) from one node to another node in a road network. Its computation complexity is O(n2), where n means the number of nodes in the network [12].

Algorithm 1: Dijkstra's algorithm for shortest path with least travel time

Input: starting and destination location, starting time t, directed road network graph, estimated travel time on each arc.
Output: shortest path

for *each node i in directed graph* **do**
  $st_j(t) = infinity$;
  $pn_j(t) = NULL$;
end
$st_{starting\ location}(t) = 0$;
while *AN is not empty* **do**
  duplicate node γ=node n with the smallest $st_j(t)$ in AN;
  remove node n from AN;
  for *each node δ in AN connected directly with node γ* **do**
    temp=$st_\gamma(t) + m_{\gamma\delta}(t)$;
    if *temp*$< st_\delta(t)$ **then**
      $st_\delta(t) = temp$;
      $pn_\delta(t) = \gamma$;
    end
  end
end
extract and return the shortest path ;

The Dijkstra is also one of the labeling based shortest path algorithms. It is worthwhile to choose the Dijkstra algorithm for finding the shortest path from one-to-one problem, because this approach is terminated as soon as the destination node is labeled, which also means that the shortest path is found [13]. Therefore, in this work, the Dijkstra algorithm is applied to find the dynamic optimal path during the trip, as shown in the Alg.1. The notation $pn_j(t)$ represents the previous node at time t in optimal path from the starting location to the node j. And $st_j(t)$ means travel time from starting location to node j, from the time t. Symbol AN implies all nodes without finding optimal path.

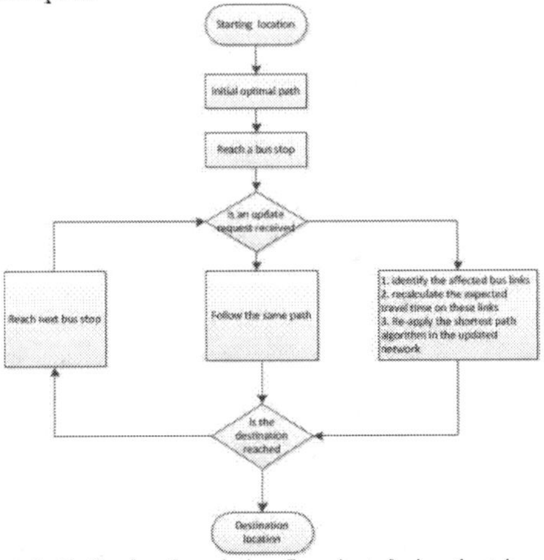

Figure 3. Optimal path updating flowchart during the trip

In Fig. 3, a flowchart describing the proposed dynamic path planning algorithm is drawed. This method should firstly interact with the dynamic changing transportation environment condition like traffic delay and temporary speed restriction, then adapts automatically the optimal path assigned to a passenger based on the real-time update received from the traffic management center. The proposed framework consisting of three main steps are explained as follows:

1) Calculation of an initial optimal path for passenger from starting location to destination location based on the static bus timetable with a shortest path algorithm (Dijkstra).

2) Recalculation of the optimal path due to an update in traffic condition. In other words, whenever a passenger reaches a bus stop, the traffic condition (i.e. expected travel time for bus) are checked for update. If there is an update impacting at least on the bus link included in the optimal path, the expected travel time for this bus linked is recalculated, and the shortest path algorithm is re-applied to get a new optimal path for passenger under the consideration of waiting time in bus stop. Otherwise, the passenger carries on his trip.

3) Judgement for the termination of dynamic optimal path process. If a passenger does not reach the destination location, the step 2 is repeated. Otherwise, the process is closed.

4 Performance evaluation with sumo

The professional transportation simulation tool named SUMO is chosen in this work to simulate the proposed algorithm, since it offers an open source package, which is deeply portable, applicable for the simulation, management and monitoring of each microscopic user (e.g. vehicle or passenger) in a transportation network [14]. Moreover, SUMO supports an interface en Python, named Traci, which allows users to modify travel strategies during runtime [15]. Therefore, a scenario with a specific road network is created, as shown in Fig.1. There are 5 bus stops and three bus lines. A passenger leaving the starting location at time t wants to find an optimal path to the destination location.

Table 1. Bus flow parameters in timetable

bus line	*begin* (s)	*end* (s)	*number*	*until* (s)				
				1	2	3	4	5
1	45	3600	12	8	25			
2	65	3620	12		13			50
3	75	3630	12		13	33	53	73

The timetable and flow configuration of bus is presented in Tab.I, where elements of *begin* and *end* mean the starting and ending time of bus flow, variable of *number* indicates the total numbers of bus during the interval between *begin* and *end*. And variable *until* presents the final leaving in each bus stop. For example, in the bus line 1, there are 12 buses during one hour, and the interval of two successive buses is 300s (5 minutes). The first bus leaves stop 1 at time 53s (8+45), stop 2 at time 70s (45+25). The second bus leaves the stop 1 at time 353s (300+45+8), stop 2 at time 370s (300+45+25). The estimated travel time between two bus stops equals to the difference of leaving time on bus stop (*until*), as Eq.(1) shows.

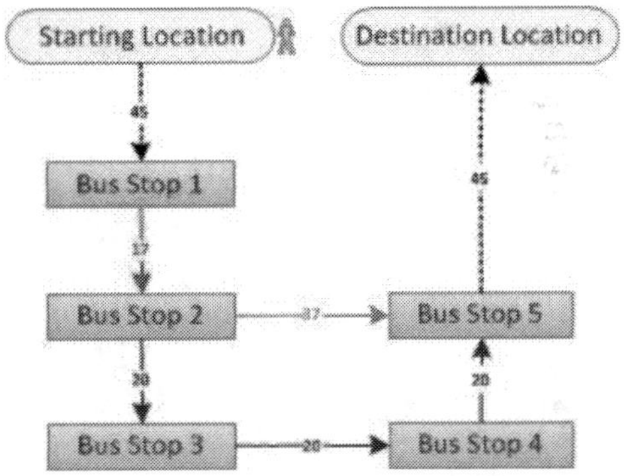

Figure 4. Graph model in Duarouter with initial travel time

For the reason of simplify, it is assumed that the walking time is 45s from the starting location to bus stop 1, or from bus stop 5 to the destination location. Therefore, for a passenger leaving from the starting location at time 0 to the destination location, the bus network topology with the initial travel time is shown in the Fig.4. Firstly, an initial route is found and assigned to the passenger. By applying the Dijkstra algorithm, the shortest path (bus lines from 1 to 2) is found with travel time 95s, instead of the path 2 (bus lines from 1 to 3) having travel time 118s.

When the passenger reaches an bus stop, the dynamic path planning system checks the events variable. If any road segment in the optimal path is blocked or closed because of some traffic condition, for example, incident or temporary speed limit, the Dijkstra algorithm is re-applied to re-calculate the optimal path from current bus stop of passenger to the destination location. Here, when the passenger arrive at the bus stop 2, an accident occurs on the edge between the bus stop 2 and 5 for bus line 2, and the travel time in this part takes longer time than expected, e.g. 237s, as Fig.5 shows. Then with the proposed process, the passenger change its path dynamically in bus stop 2 by taking the bus line 3 instead of staying at bus line 2 to reduce travel delay. This dynamical path planning process should be repeated until the passenger arrives at the destination location. To do so, the proposed dynamic path planning system applies the Traci interface to inform SUMO to modify the path already assigned to the passenger. As a result, the passenger's path can be updated during the runtime.

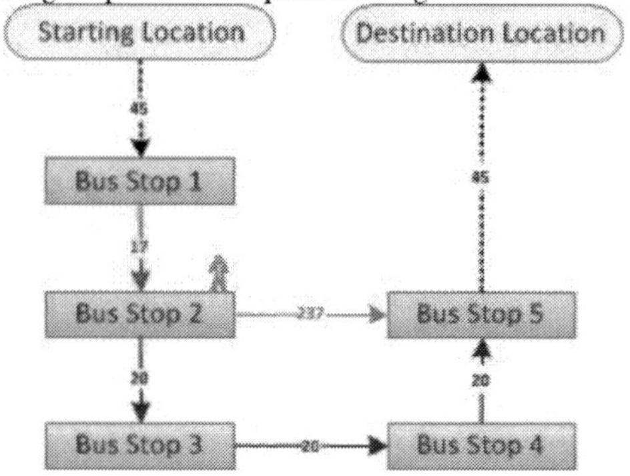

Figure 5. Graph model in Duarouter with dynamic travel time

5 Conclusion

In this paper, we propose a new on-line multimodal dynamic path planning method. The simulation is based on an open source tool named SUMO. An initial optimal path is assigned to passenger, and this path is updated dynamically during the trip according to the real traffic condition. The optimal objective is to minimize expected travel time by applying the Dijkstra algorithm. The simulation results show the efficiency of the proposed method. In the future, more traffic modes can be considered, such as boat, train and shared car. Other shortest path algorithms can also applied in this problem and the performances among different algorithms can be compared.

Acknowledgment

This work was supported by the project ORIO (Observing the peRformances of urban Infrastructures and mobility / preventing collisions with vulnerable people using Opportunistic radar). The project ORIO is within the framework of ELSAT2020, which is co-financed by the European Union with the European Regional Development Fund, the French state and the Hauts de France Region Council.

References

1. R. W. Hall. The fastest path through a network with random time-dependent travel times. *Transportation Science*,vol.20,no.3,pp.182–188, (1986)
2. E. D. Miller-Hooks and H. S. Mahmassani. Least expected time paths in stochastic, time-varying transportation networks, *Transportation Science*, vol. 34,no.2,pp.198–215, (2000)
3. S. Sun, Z. Duan, S. Sun, and D. Yang. How to find the optimal paths in stochastic time-dependent transportation networks?. *17th International IEEE Conference on Intelligent Transportation Systems*, Oct 2014, pp. 2348–2353.
4. P. Ni, H. T. Vo, D. Dahlmeier, W. Cai, J. Ivanchev, and H. Aydt. Depart: Dynamic route planning in stochastic time-dependent public transit networks. *2015 IEEE 18th International Conference on Intelligent Transportation Systems*, Sept 2015, pp. 1672–1677.
5. S. Gao. Optimal adaptive routing and traffic assignment in stochastic time-dependent networks. Ph.D. dissertation, Massachusetts Institute of Technology, (2005)
6. S. Gao and I. Chabini. Optimal routing policy problems in stochastic time-dependent networks. *Transportation Research Part B: Methodological*, vol. 40, no. 2, pp. 93–122, (2006).
7. S. Gao, E. Frejinger, and M. Ben-Akiva. Adaptive route choices in risky traffic networks: A prospect theory approach. *Transportation research part C: emerging technologies*, vol. 18, no. 5, pp. 727–740, (2010).
8. C. Wu, X. Zhang, and Y. Dong. Adaptive route guidance based on real-time information in stochastic time-dependent transportation networks. *17th International IEEE Conference on Intelligent Transportation Systems*, Oct 2014, pp. 2392–2397.
9. J. Boyan and M. Mitzenmacher. Improved results for route planning in stochastic transportation. *Proceedings of the twelfth annual ACM-SIAM symposium on Discrete algorithms*. pp. 895–902. (2001).
10. M. Datar and A. Ranade. Commuting with delay prone buses. *SODA*, pp. 22–29, (2000).
11. V. T. N. Nha, S. Djahel, and J. Murphy. A comparative study of vehicles' routing algorithms for route planning in smart cities. *2012 First International Workshop on Vehicular Traffic Management for Smart Cities (VTM)*, Nov 2012, pp. 1–6.
12. M. Fu, J. Li, and Z. Deng. A practical route planning algorithm for vehicle navigation system. *Fifth World Congress on Intelligent Control and Automation*, vol. 6, pp. 5326–5329, (2004)
13. F. B. Zhan and C. E. Noon. Shortest path algorithms: An evaluation using real road networks. *Transportation Science*, vol. 32, no. 1, pp. 65–73, (1998).
14. G. Kotusevski and K. Hawick. A review of traffic simulation software. CSTN, vol.13, no. 08, (2009).

15. A. Wegener, M. Piorkowski, M. Raya, H. Hellbruck, S. Fischer, and J.-P. Hubaux. Traci: An interface for coupling road traffic and network simulators. *11th Communications and Networking Simulation Symposium*, Apr 2008, pp. 155–163.

Hyperparameter search in periodic vehicle routing problem

Ekaterina Grakova[1] , Martin Golasowski[1], Roberto Montemanni[2], Kateřina Slaninová[1] , Jan Martinovič[1], Jafar Jamal[2], Kateřina Janurová[1] and Matteo Salani[2]

[1] *IT4Innovations National Supercomputing Center, VŠB - Technical University of Ostrava, 17. listopadu 15/2172, 708 33 Ostrava, Czech Republic*
[2] *Dalle Molle Institute for Artificial Intelligence (IDISA, USI/SUPSI), Galleria 2, 6928 Manno, Switzerland*

Abstract. The large number of real-world applications have shown that the use of computational method for distribution process planning produces substantial savings. Many of these applications lead to problem generally known as Vehicle Routing Problem. The real-world applications are highly computationally demanding for larger instances. This article aims to show the possibilities and benefits of using hyperparameter search for solving the Periodic Vehicle Routing Problem for exhausted oil collection by execution on the supercomputing infrastructure using HyperLoom platform. HyperLoom is an open source platform for defining and executing scientific pipelines in a distributed environment. This experiment was run on the supercomputer Salomon operated by IT4Innovations.

1 Introduction

The logistic industry heavily relies on optimal route planning as it represents a critical task. Indeed, transportation costs account for up to 20% of the overall logistic cost [1]. The Vehicle Routing Problem (VRP) was proposed more than 50 years ago [2] and it is a challenging combinatorial optimization problem. Variants of the VRP have been proposed to respond to the variety of operational constraints arising in the distribution industry.

One of such variants is the Periodic Vehicle Routing Problem (PVRP) in which routes are constructed over a period of time (e.g. days). The problem has been introduced by Christofides and Beasley [3] and has been solved mostly through heuristic approaches [4 - 6], [16].

In [7] we considered a variant of the PVRP in which exhausted oil must be collected from customers. Customers can be visited more than once during the planning period and the number of times they are visited is also a decision variable of the problem, furthermore all vehicles must return to the depot each day as wasted oil must be properly disposed. In the considered problem, exhausted oil accumulates at customers' location at a fixed ratio expressed in liters per week. From this perspective our problem shares features with that of waste collection problem [8 - 10].

High performance computing (HPC) architecture can be a solution to solve such large tasks, and to obtain the optimal solutions. The supercomputing infrastructure allows us to solve the complicated optimization problems and to use the optimization algorithms to find the optimal solutions. Many of these algorithms depend on configuration settings that are typically hand-tuned in the course of evaluating the algorithm for a particular data set. Such parameter tuning (hyperparameter search) is often presented as being incidental to the algorithm and correctly setting these parameters improves the quality of the algorithm results [11, 17].

HyperLoom has been successfully used at IT4Innovations for distributed hyperparameter search within machine learning applications for pharma industry [18]. This application aims to be another use case proving its usability beyond the pharmaceutic domain. Scientific pipelines such those in machine learning compose of multiple data processing tasks. This article aims to show the possibilities and benefits of using hyperparameter search for solving the Periodic Vehicle Routing Problem for exhausted oil collection by execution on the supercomputing infrastructure using HyperLoom platform. For the experiments, we created extended test case for exhausted oil collection. Experiments were run on the supercomputer Salomon.

2 Periodic VRP for oil collection

We formally describe the PVRP for exhausted oil collection. We are given a set N of customers and a central depot, denoted as node 0. The set $V = N \cup 0$ forms the node set of a graph G whereas a set of arcs A is linking every pair of nodes in N. For each arc $a_{i,j}, (i,j) \in N \times N$ let $t_{i,j}$ and $d_{i,j}$ the travelling time and travelling distance between nodes i and j, respectively. Let p_j be the expected exhausted oil accumulation ratio expressed in liters per week. We are also given a set K of identical vehicles of capacity Q located at the central depot 0 and a set M of planning periods. The problem asks to define up to $|K|$ routes per period, one for each vehicle, possibly returning to the depot more than once per period and visiting a subset of customers. The planning problem has two objectives: minimize the total travelled distance and balance the workload between the single routes. This last objective is considered computing the standard deviation in routes total travelling time. Routes must respect vehicle's capacity constraints and customers should not accumulate more exhausted oil than that transportable by a single vehicle in a single visit.

2.1 Test case for exhausted oil collection

In a collaboration with Caritas Suisse, we were asked to apply our heuristic algorithms to an exhausted oil collection problem in the area of Bali, Indonesia. We have been given a set of 308 different locations for the collection of the exhausted oil. For each location we know the expected amount of oil that accumulates within one week. We were asked to recommend one out of three potential locations for the exhausted oil processing plant. The travel times between any two locations are known. To collect the exhausted oil, cans of 25 liters capacity each are used. Every collection tour starts at the depot with a vehicle filled with empty cans. Empty cans are then exchanged at customer location with full ones. Finally, cans are delivered at the processing plant. As an additional requirement, the travel time for each route is not allowed to exceed a given bound on working hours corresponding to eight hours in our tests.

In order to produce fairly balanced routes, we introduce a weight parameter w which is used to balance total travelled distance versus the standard deviation of route duration. The parameter ranges from 0 to 1. When w is set to 0, total travel time is the only considered objective, when w is set to 1, the standard deviation of route durations is the only considered objective.

More formally, a solution S to the problem is a set of routes $r_{k,p}, k \in \{1,\ldots,|K|\}$, $p \in \{1,\ldots,10\}$, where k is the vehicle index and p is the period index, starting at the depot, finishing at the depot, visiting a customer at most once.

Let $t(r_{k,p})$ represent the total travel time of route $r_{k,p}$and $q_{v,j}$ represent the demand of customer $v \in V\setminus\{0\}$ accumulated until the customer gets visited at working day j. Let $s_v \in \{1,2,4\}$ be the decision variable representing the number of time customer v is visited in the planning period.

$$t(r_{k,p}) \leq 8 \; \forall \, k \in \{1,\ldots,|K|\}, p \in \{1,\ldots,10\}, \tag{1}$$

$$\textstyle\sum_{v \in r_{k,p}} \lfloor q_{v,p} / c \rfloor \leq Q \; \forall \, k \in \{1,\ldots,|K|\},\ p \in \{1,\ldots,10\}, \tag{2}$$

$$|\, r_{k,p} : k \in \{1,\ldots,|K|\},\ p \in \{1,\ldots,10\},\ v \in r_{k,p}\}| = s_v, \tag{3}$$

$$\left|\, r_{k,p} : k \in \{1,\ldots,|K|\}, \quad v \in r_{k,p}\right\}| \leq 1 \; \forall \, v \in V\{0\},$$
$$p \in \{1,\ldots,10\}, \tag{4}$$

Constraint (1) states that the total travel time for each route does not exceed the working time limit of eight hours and constraint (2) states that vehicle capacity is respected, here c is the amount of oil contained in one can and Q is expressed in number of cans. The third constraint relates the number of visits in the solution with variable s. Constraint (4) ensures that a customer is not visited twice per day. The objective is a combination of two objectives:

$$z_1(r) = \textstyle\sum_{k \in \{1,\ldots,|K|\}} \sum_{p \in \{1,\ldots,10\}} t(r_{k,p}), \tag{5}$$

$$z_2 = \sqrt{\frac{1}{10|K|} \textstyle\sum_{k \epsilon \{1\ldots|K|\}} \sum_{p \epsilon \{1\ldots10\}} t\left(\frac{(r_{k,p} - z_1(r))}{10|K|}\right)^2}\, . \tag{6}$$

The objective function can then be written as

$$z(r) = w\, z_1(r) + z_2(r) \tag{7}$$

We adopt a two level approach similar to the first assigned strategy used for PVRP [14]. The assignment of customers to vehicles and working days is managed on the first level of the heuristic and explored using local search. The second level optimizes the single routes of the different vehicles on different days. Nicely, the changes imposed by the first level local search algorithm affects a portion of the solution (two routes). Therefore, only two routes must be re-optimized and usually the starting solution is already of a good quality.

The heuristic starts with a random assignment of locations to vehicles and working days. It does so that the number of times a customer is visited in the planning period is compatible with its expected oil accumulation: none of the visit should occur after the customer accumulated more oil than the vehicle's capacity. For each vehicle and each working day an optimal route is computed. The obtained solution is the starting point for the local search algorithm. The local search exploits a shift operator and a swap operator. The shift operator takes a location assigned to a specific vehicle and a specific working day and shifts it to another vehicle and/or another working day. The swap operator considers two different locations and interchanges their assignments to vehicles and working days. In each iteration, the local search algorithm explores all the solutions in the local search neighborhoods induced by those two operators in a random order.

The second level optimization mechanism computes an optimal route by complete enumeration in the case where at most a given number of locations are assigned to a vehicle, otherwise a fast insertion heuristic approach is used.

If an improving solution is found, this solution replaces the current solution and a new iteration is started immediately. If there is no improving solution in the neighborhood of the current solution, the local search algorithm terminates, and the local optimal solution is returned by our approach as the final solution. Further details can be found in [7] and [15].

Histogram (Figure 1) reports the frequency of required running time for the algorithm to converge. Out of 75 runs, we observe that the algorithm convergence is almost always below 3 seconds. The proposed approach works very well for the given application anyway for the sake of challenging the HyperLoom platform a new and extended test case is necessary.

https://doi.org/10.1051/matecconf/201925901003

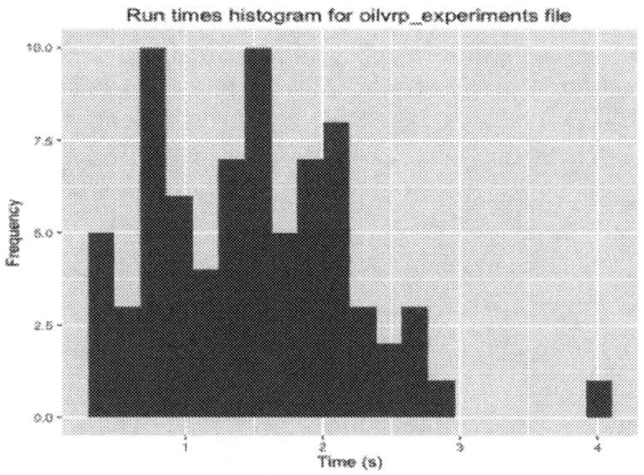

Figure 1. Run times histogram for test case

2.2 Extended test case for exhausted oil collection

We prepared an additional set of 1008 different locations for the oil collection problem. For each location we know the amount of oil. Every customer had a specific GPS location in Ostrava (Figure 2). We know the travel time between any two of the locations. Also, metric distances between the locations were created in Ostrava. We know the location of the central depot. The vehicles start at the central depot.

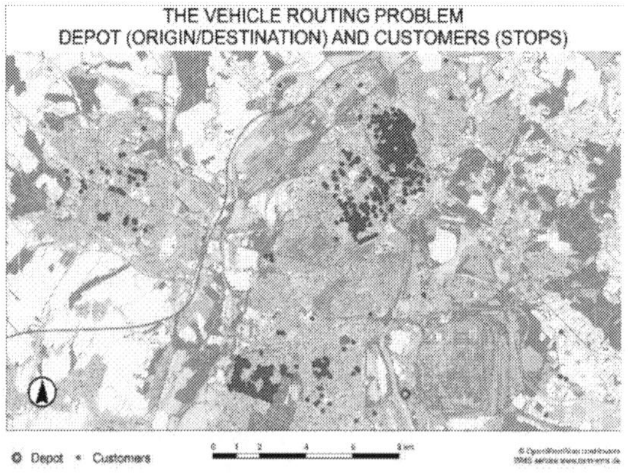

Figure 2. GPS locations in Ostrava

Planning time for algorithm testing was 10 working days.

3 HyperLoom for periodic VRP

HyperLoom is an open source platform for defining and executing scientific pipelines in a distributed environment developed at IT4Innovations. HyperLoom is able to efficiently execute millions of interconnected tasks on thousands of computational cores. In [12], the authors described a virtualization of HyperLoom – originally an HPC solution for defining and executing scientific workflows and compared the performance of the virtualized HyperLoom to the performance of the bare metal deployment.

We used HyperLoom to define and execute VRP hyperparameters sweep pipeline. HyperLoom enables to take an advantage of distributed systems by running multiple instances of VRP in parallel which promises an efficient usage of the provided resources. The quality of results for the heuristic algorithm depends on the adjustment of the configuration parameters of the algorithm. In this article, the following configuration parameters were used:

Table 1. Configuration parameters

Parameters name	Configuration values
Number_of_vehicles	15, 16, 17, 18, 19, 20
Depot_number	0
Weight	0.5, 0.05, 0.95
Random_Seed	1, 2, 3, 4
Horizon_days	10
Vehicles_capacity	34, 35, 36, 37, 38
Location_per_day	100, 101, 102, 103, 104, 105
Liters_per_day	200, 250, 300, 350

6000 combinations of configuration parameters were created for the algorithm. The experiments were performed on Salomon supercomputer, where each node has two – twelve – core Intel Xeon processors and 128 GB RAM (2xIntel Xeon E5 – 2680 v3, 2.5 GHz, 12 cores). We have carried out all the experiments on a 6, 12, 24, 48 identical physical computational nodes (Table 2). We have observed the dependence of the total execution time values on the number of nodes.

Table 2. Experiments results

Number of nodes	Total execution time, [s]	Strong scaling efficiency, [%]
6	5,128	98.52
12	2,841	88.92
24	1,443	87.53
48	839	75.27

Strong scaling efficiency was calculated for each experiment:

$$S = \frac{t_1}{(N t_N)}\ 100\%, \tag{8}$$

where:

t_1 - time to complete a work unit with 1 processing element, N - number of processing elements, t_N - time to complete the same unit of work with N processing elements.

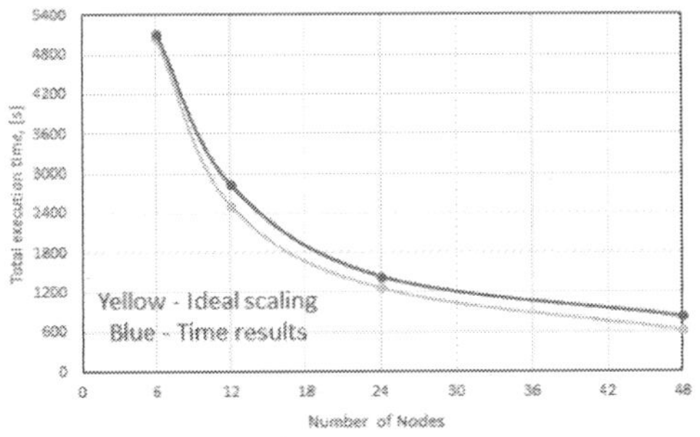

Figure 3. Strong Scaling

4 Algorithm results

The only result that was monitored during the experiments was the solution time of the algorithm which was between 62.999s and 201.839s for different configurations of input parameters described in Table 1. The histogram of solution time is given in Figure 4.

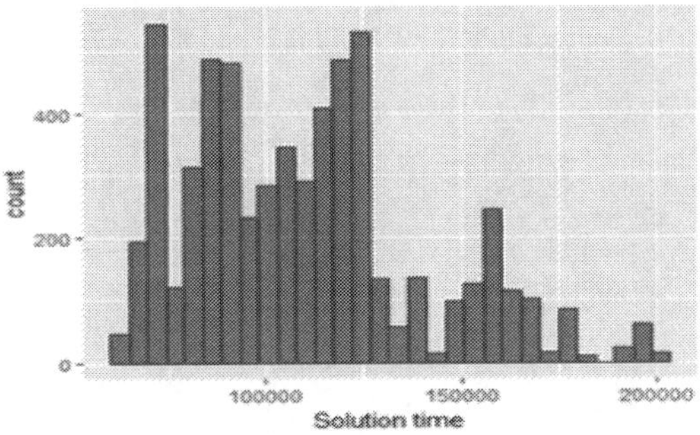

Figure 4. Histogram of solution time

We also assessed the dependence of solution time on the input parameter values and the importance of parameter values for total solution time.

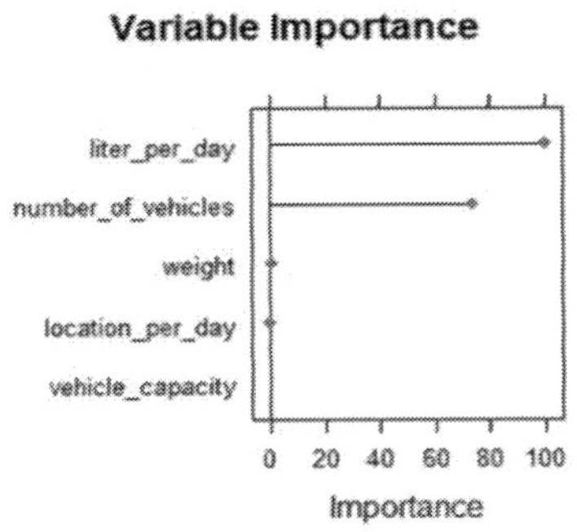

Figure 5. Parameters importance

The importance of the input parameters was determined as the proportion of the variance of the solution time that can be explained by the given input parameter (using the local regression model LOESS) [13]. This proportion was then recalculated as the relative measure where the importance of the input parameters is relative to the input parameter with the highest degree of importance (100%). The relative parameter importance is given in Figure 5 and together with correlation coefficients in Table 3.

As can be seen from the results, the total solution time is dependent only on two input parameters: liter_per_day and number_of_vehicles.

Table 3. Parameter importance and correlation coefficient

Parameter	Parameter Importance	Correlation Coefficient
liter_per_day	100%	-0.699
number_of_vehicles	74%	0.601
weight	0%	0.038
location_per_day	0%	0.002
vehicle_capacity	0%	0.000

The negative correlation coefficient for the liter_per_day input parameter means the fewer liter_per_day the higher solution time contrary to the number_of_vehicles, where the higher number_of_vehicles means the higher solution time.

5 Conclusion

In this article, we showed possibilities and benefits of using hyperparameter search for solving the Periodic Vehicle Routing Problem for the exhausted oil collection. HyperLoom platform was used for the definition of our hyperparameter search pipeline. We created an extended test case because the original instance of the problem was too small for test on the HPC infrastructures.

We have observed the dependence of the total execution time values on the number of nodes. For heuristic algorithm, we also assessed the dependence of solution time on the input parameter values and the importance of parameter values for total solution time. Using HyperLoom platform for our problem we showed that it is not necessary to parallelize used optimisation algorithm for Periodic Vehicle Routing Problem to use larger instances.

6 Acknowledgment

This work was supported by The Ministry of Education, Youth and Sports from the National Programme of Sustainability (NPS II) project 'IT4Innovations excellence in science - LQ1602', by the IT4Innovations infrastructure which is supported from the Large Infrastructures for Research, Experimental Development and Innovations project 'IT4Innovations National Supercomputing Center – LM2015070' and by the ExCAPE project - the European Union's Horizon 2020 research and innovation programme under grant agreement No. 671555.

References

1. P., Toth, D., Vigo. The vehicle routing problem. SIAM monographs on discrete mathematics and

applications, *Society for Industrial and Applied Mathematics*, (2002)

2. G,B., Dantzig, J.H., Ramser. The truck dispatching problem. *Management Science*, **6**, 80–9, (1959)
3. N., Christofides, J.E., Beasley. The period routing problem. *Networks*, **14**, 2, 237–256, (1984)
4. A., Angelelli, M.G., Speranza. The periodic vehicle routing problem with intermediate facilities. *European Journal of Operational Research*, **137**, 2, 233–247, (2002)
5. J.F., Cordeau, M., Gendreau, G., Laporte. A tabu search heuristic for periodic and multi-depot vehicle routing problems, *Networks*, **30**, 2, 105–119, (1997)
6. M., Gaudioso, G., Paletta. A heuristic for the periodic vehicle routing problem. *Transportation Science*, **26**, 2, 86–92, (1992)
7. D., Weyland, M., Salani, R., Montemanni, L.M., Gambardella. Vehicle routing for exhausted oil collection *Journal of Traffic and Logistics Engineering*, **1**, 1, 5–8, (2013)
8. B-I., Kim, S., Kim, S., Sahoo. Waste collection vehicle routing problem with time windows, *Computers & Operations Research*. **33**, 12, 3624-3642, (2006)
9. A.M., Benjamin, J.E., Beasley. Metaheuristics for the waste collection vehicle routing problem with time windows, driver rest period and multiple disposal facilities, *Computers & Operations Research*, **37**, 12, 2270-2280, (2010)
10. K., Buhrkal, A., Larsen, S., Ropke. The waste collection vehicle routing problem with time windows in a city logistics context, *Procedia - Social and Behavioral Sciences*, **39**, 241-254, (2012)
11. J., Bergstra, D., Yamins, D.D, Cox. Making a science of model search: hyperparameter optimization in hundreds of dimensions for vision architectures. *In: Proceedings of the 30th International Conference on International Conference on Machine Learning*, **28**, 115-123, (2013)
12. M., Kuhn. Caret package, *Journal of Statistical Software*, **28**, 5, (2008)
13. V., Cima, S., Bohm, J., Martinovič, J., Dvorský, J.T., Ashby, V., Chupakhin. HyperLoom Possibilities for Executing Scientific Workflows on the Cloud. *CISIS Complex, Intelligent and Software Intensive Systems*, 397-406, (2017)
14. S., Baptista, R.C., Oliveira, E., Zuquete. A period vehicle routing case study. *European Journal of Operational Research*, **139**, 2, 220–229, (2002)
15. Stochastic Vehicle Routing - From Theory to Practice, Dennis Weyland. *University of Lugano (Lugano, Switzerland)* (2013)
16. E., Grakova, K., Slaninová, J., Martinovič, J., Křenek, J., Hazelka. Waste collection vehicle routing problem on HPC Infrastructure. *CISIM Computer Information Systems and Industrial Management Applications*. (2018)
17. A., Fred, et al. (2016). Pattern Recognition: Applications and Methods: 4th International Conference, ICPRAM 2015, Lisbon, Portugal, January 10-12, 2015, Revised Selected Papers. Springer. Retrieved March 15, (2018)
18. V., Cima, S., Böhm, J., Martinovič, J., Dvorský J., Janurová, T., Vander Aa, J.T., Ashby V., Chupakhin. HyperLoom: A platform for defining and executing scientific pipelines in distributed environments. *In: Proceedings of the 9th Workshop and 7th Workshop on Parallel Programming and RunTime Management Techniques for Manycore Architectures and Design Tools and Architectures for Multicore Embedded Computing Platforms (PARMA-DITAM '18). ACM, New York, NY, USA*, 1-6. (2018)

The Possibility of CO_2 Pipeline Transport for Enhanced Oil Recovery Project in Poland

Paweł Wojnarowski[1], Robert Czarnota[1], Tomasz Włodek[1], Damian Janiga[1], Jerzy Stopa[1] and Piotr Kosowski[1]

[1] *AGH University of Science and Technology, 30-059 Krakow,* ***Poland****; Department of Drilling, Oil and Gas*

Abstract. Enhanced oil recovery schemes involve the transportation of large volumes of carbon dioxide from the capture source to the utilisation site. This research presents the possibilities of carbon dioxide transport using pipeline from the selected emission point to the oil reservoir located in Poland where greenhouse gas can be used as injecting fluid to improve oil production. In the first step, the different CO_2 thermodynamic states are analyzed. For the design purpose, length, operating pressures and flow rates are determined, then pipeline diameter is calculated. Furthermore, the pipeline transmission schemes for CO_2 transport are proposed. The study revealed, that the large amount of CO_2 produced at source power plant can be transported to the oilfield site more efficiently when CO_2 is converted into the liquid state. As a result, the smaller pipeline diameter can be applied for the transmission. Moreover, temperature decrease is observed when CO_2 is transported in gaseous state and temperature increase is present when CO_2 is in the liquefied state.

1 Introduction

Enhanced oil recovery (EOR) is a term used for a wide variety of techniques for increasing the amount of crude oil that can be extracted from an oil field. Carbon dioxide injection has been found to be commercially successful to improve oil production [1]. The process of CO_2-EOR project is divided into the stages of preparation, transport and injection of CO_2 into oil reservoir. Main methods of carbon dioxide land transportation are pipeline transmission. [2].

The most developed network of carbon dioxide pipelines is located in the United States. First CO_2 pipelines were constructed in the USA in the 70's of the twentieth century. Present, the overall length of CO_2 pipelines across the globe is above 6500 km. Million tons of CO_2 per year is transported using pipeline network [3].

Depending on thermodynamic properties, CO_2 can be transported in gaseous phase, in liquid state or in supercritical conditions. In this research, the possibilities of carbon dioxide transportation by buried pipeline from anthropogenic emission source to oil reservoir for CO_2-EOR project in Poland are presented.

2 Analysis of CO_2 thermodynamic states

2.1 CO_2 pipeline transmission in gas phase

The thermodynamic conditions of CO_2 gaseous phase are limited for pipeline transport. The first problem is the maximum operating pressure in the pipeline [4]. In Poland the ground temperature during Spring/Autumn at the depth of 1 m below ground (the pipeline level) does not exceed 10°C. As result, the maximum operating pressure in the pipeline can be equal of 4 MPa. During Summer time, ground temperature of 20°C at the depth of pipeline placement allow to increase the maximum operating pressure to 4.8 MPa. During Winter, it is often that at a depth of one meter temperature is close to 0°C. In these conditions the maximum allowable operating pressure drops below 3 MPa. Such restrictive values of maximum operating pressure for CO_2 transmission are related to thermodynamic boundaries. Increase of pressure above saturation pressure at given temperature causes phase transition into liquid phase, what pose a risk to the transmission equipment. For these reasons, CO_2 pipeline transport in gas phase should be considered for short distance [5].

2.2 CO_2 pipeline transmission in liquid phase

Carbon dioxide in the liquid phase has higher density than in the supercritical state or in gaseous state, this fact together with slight compressibility of the liquid CO_2 lead to the smaller pressure drops along the pipeline. Consequently, it is possible to transport liquid CO_2 at relatively long distances using smaller pipe diameters comparing to other ways of transport. As a result the much larger quantities of CO_2 can be transported efficiently [6]. However, viscosity pose a problem for the transportation of CO_2 in the liquid phase because its value is greater than in the gas phase and in the supercritical state. The great threat in liquid CO_2

transportation is uncontrolled pressure drop in the pipeline what may result in liquid-gas two-phase system flow [7].

In Poland the possible operating range of the liquid pipeline CO_2 transport is within the pressure of 3.5-5.8 MPa for ambient temperature from 0 to 20°C. Taking into consideration the temperature fluctuation during the year, carbon dioxide should be transported in pressure above 5.8 MPa. Despite the challenges associated with liquid CO_2, the pipeline transmission is the most advantageous way transportation due to the favorable thermodynamic parameters.

2.3 CO_2 pipeline transmission in supercritical phase

Supercritical state of carbon dioxide can be obtained by exceeding the critical parameters: pressure 7.4 MPa and temperatures 31.1°C. Supercritical carbon dioxide has a density comparable to the liquid state and its viscosity and compressibility are comparable the gas phase. These supercritical CO_2 characteristics determine that the thermodynamic transport of CO_2 in the supercritical state as advantageous solution [8]. Due to the high critical temperature, additional CO_2 heating stations are required. Moreover, in case of temperature or pressure drop below the critical value, the occurrence of a two-phase system can be observed. This fact will result in the CO_2 density decrease and pressure drop is observed [9]. Energy consumption and additional costs make transport CO_2 in supercritical state as unprofitable in Poland.

3 CO_2 pipeline design

3.1. Pipeline diameter calculation

Pipeline diameter has a decisive influence on the flow characteristics and the energy loss. In order to analyse the most important factors such as mass flow rate, pipeline operating pressure range and land elevation change on pipeline diameter, the following formula can be used:

$$D = \sqrt[5]{\frac{16 \cdot \lambda \cdot z^2 \cdot R^2 \cdot T^2 \cdot L \cdot M^2}{\pi^2 \cdot [z \cdot R \cdot T \cdot (P_1^2 - P_2^2) - 2 \cdot g \cdot P_{av}^2 \cdot \Delta h]}} \quad (1)$$

Where: λ-friction factor, z-compressibility factor, R-gas constant, T-temperature, L-length, M-mass flow rate, P_1-ilet pressure, P_2-outlet pressure, g-gravity acceleration, P_{av}-average pressure, Δh-elevation difference.

Design calculations and cost comparative analysis allow the investors to choose a pipeline transport of carbon dioxide using larger diameters and less pumping stations or smaller diameters with more pumping stations. For this reason, the design of the CO_2 transport pipeline should be customized to the efficient transfer of contracted CO_2 quantities, while maintaining the lowest capital expenditures and the operating expenses.

3.2. Pipeline route

The proposed pipeline length from CO_2 source (EDF Krakow power plant) to Grobla oilfield is equal to 34.8 km. The pipeline will be located in the densely populated and industrial areas of Krakow. There are numerous collisions with the linear infrastructure (public roads, power lines, water pipelines, gas pipelines, etc.). Moreover, it will be necessary to create a crossing under the Vistula River e.g. by using trenchless technology. The route of the planned pipeline is shown in the Figure 1.

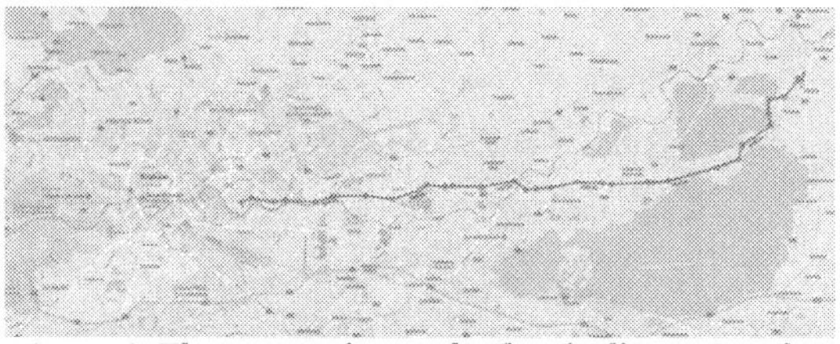

Figure. 1. The proposed route for the pipeline connecting EDF Krakow (CO_2 emitter) with Grobla oilfield.

2.4 Pipeline operational range

Maximum operating pressure for carbon dioxide in gas phase should not exceed 3 MPa in Winter and 4 MPa during Summer. Climate conditions in Poland determine these limits. Dew point for carbon dioxide at temperature of 0°C is equal to 3.4 MPa and for temperature of 20°C about 5.6 MPa. In case of necessity to transport relatively large amount of captured CO_2 from the emission source to the Grobla injection site, the transmission of CO_2 in the liquid phase will also be considered. Worldwide experience with liquid CO_2 transmission indicates the minimum operating pressure above 8 MPa. The CO_2 transmission flow rates for injection into Grobla oil reservoir are assumed as:

a) minimum flow rate equal to 9 750 Nm^3/hr,
b) maximum flow rate equal to 107 650 Nm^3/hr,
f) variant for 60% and 30% of maximum carbon dioxide source capacity,
e) above flow rates of gaseous CO_2 transferred in the liquefied state.

3.3. CO_2 pipeline transmission

2.1.1 Gas phase

For the CO_2 transmission in gaseous state, the double compression system is proposed. At first, CO_2 from the capture installation for pipeline transport is pressurized for the purpose of transmission and prior to injection, CO_2 is compressed the second time at oilfield site. In Figure 2 the schematic sketch for CO_2 pipeline transportation in gaseous state from source to oilfield injection site is presented.

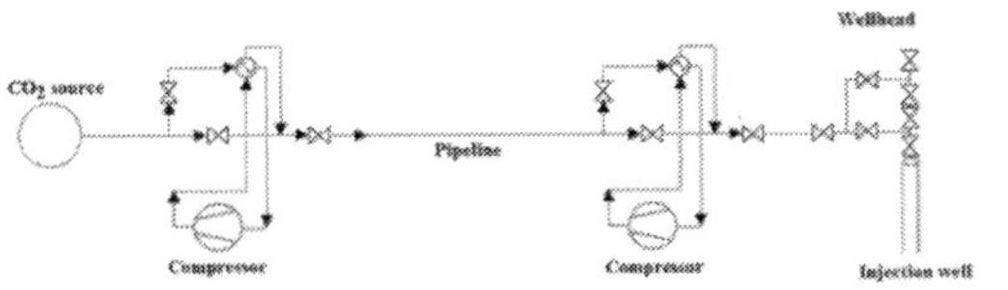

Figure. 2. Schematic sketch for CO_2 pipeline transportation in gaseous state from source to oilfield injection site.

2.1.1 Liquid phase

In the second transmission method, CO_2 is transferred in liquid state what allows for higher operating pressures. For this purpose, after capturing CO_2, gas is compressed to 10 MPa, then cooled and converted into the liquid phase. In the next stage carbon dioxide is sent by pipeline using pumps. Depending on the required injection pressure, it is possible to place the additional CO_2 injection pump at the end of pipeline - directly before injection well. This method eliminates using large pipeline diameters at high flow rates because the pressure drop for liquid flow is lower comparing to gas flow. However, the construction of pumping station, heat exchangers and condensers to convert CO_2 into liquid phase, which also significantly increases the investment costs. The CO_2 booster pump demands less input energy than the compressors. Scheme of second proposed method is presented in Figure 3.

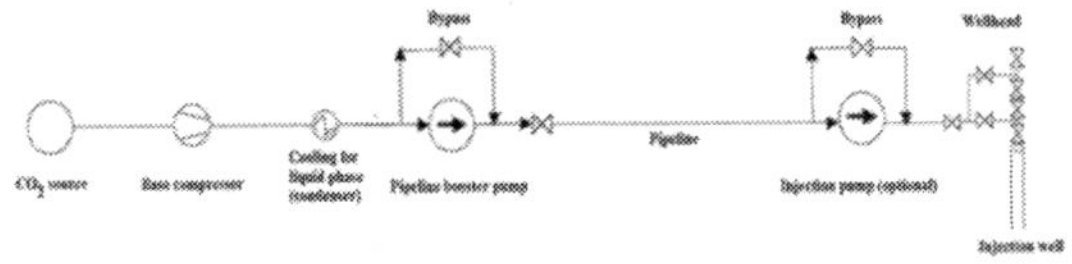

Figure. 3. Schematic sketch for CO_2 pipeline transportation in liquid state from source to oilfield injection site.

4 Results and discussion

The most influential parameters on determining the pipeline diameter are: length, flow rate and operating pressure. It was assumed that the inlet pressure for gaseous carbon dioxide pipeline transmission is equal to 3 MPa for Winter and 4 MPa for Summer. For liquid phase, the minimum operating pressure is set above 8 MPa. For the pipeline diameter determination, several variants of flow rate were taken into consideration which are described in the previous section. The calculated diameters for the designed pipeline as function of pressure are presented in Figures 4 - 8 for the assumed variants.

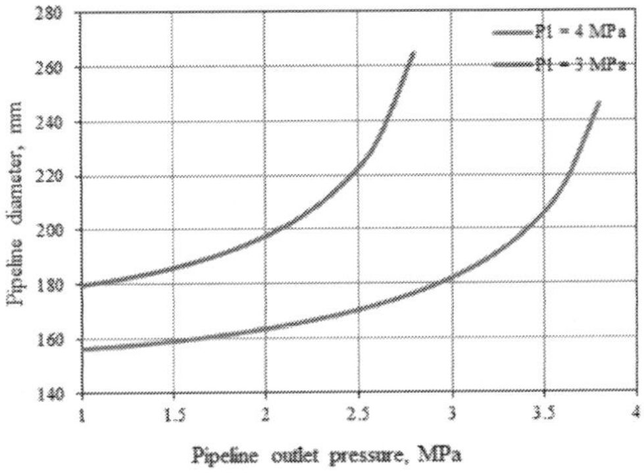

Figure. 4. Diameter estimation for pipeline from EDK Krakow source for minimum flow rate variant and two inlet pressures as a function of outlet pressure

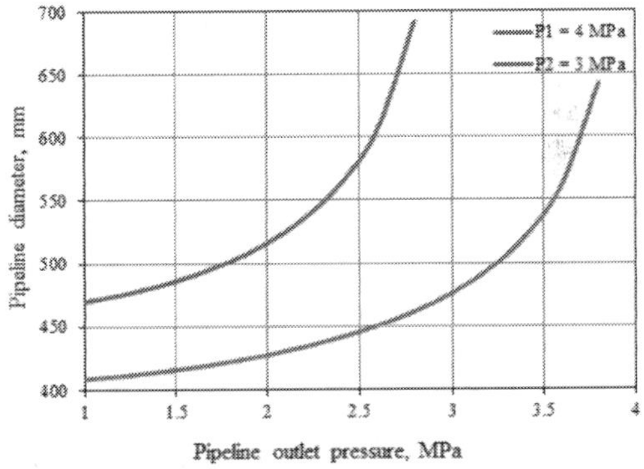

Figure. 5. Diameter estimation for pipeline from EDK Krakow source for maximum flow rate variant and two inlet pressures as a function of outlet pressure

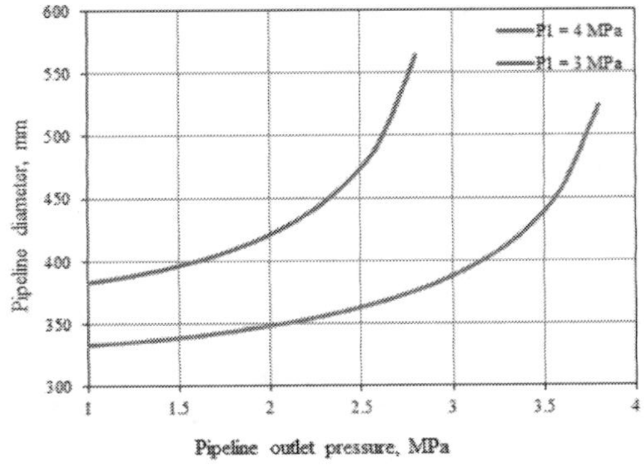

Figure. 6. Diameter estimation for pipeline from EDK Krakow source for 60% of maximum flow rate variant and two inlet pressures as a function of outlet pressure

 https://doi.org/10.1051/matecconf/201925901004

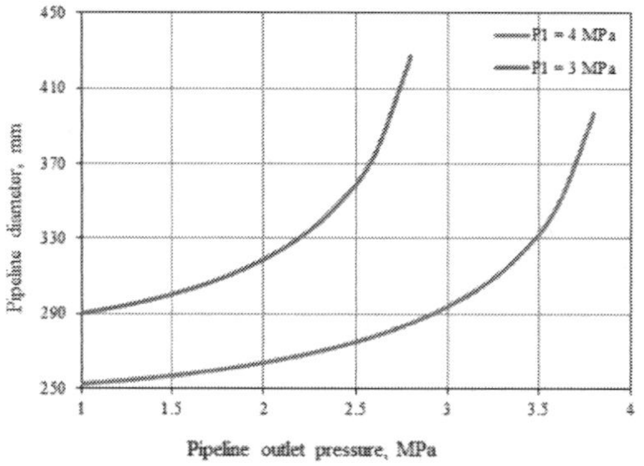

Figure. 7. Diameter estimation for pipeline from EDK Krakow source for 30% of maximum flow rate variant and two inlet pressures as a function of outlet pressure

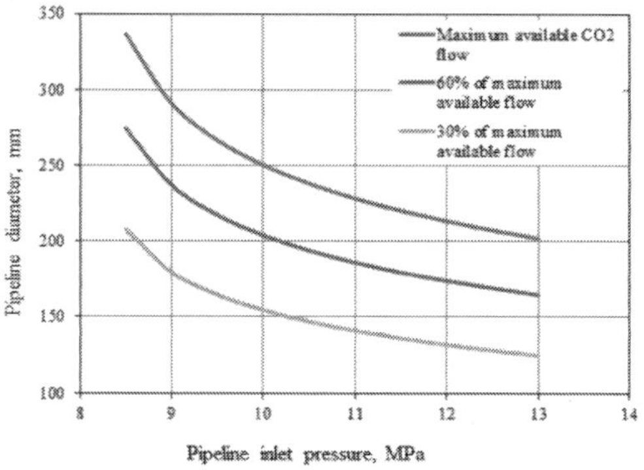

Figure. 8. Diameter estimation for liquid CO2 pipeline from EDK Krakow source for three variants of flow rate as a function of pipeline inlet pressure.

Figures 4-8 show the pipeline diameters for pipeline from EDF Krakow CO_2 emission source to the Grobla oil reservoir for the different flow rate variants as function of the pipeline pressure. For the projected pipeline, the optimum diameter for the minimum gaseous CO_2 flow rate variant is 200 mm. As a result, the outlet pressures will be 3.4 MPa for higher inlet pressure and 2.2 MPa for lower inlet pressure, respectively. For the maximum flow rate variant (gaseous CO_2), the optimum pipe diameter is equal to 500 mm and outlet pressures will be 1.75 MPa for lower inlet pressure and 3.2 MPa for higher inlet pressure. For lower inlet pressure the pipe diameter of 600 mm should be considered to obtain outlet pressure of 2.6 MPa. For gaseous state, two more variants of flow rates were analyzed - 60% and 30% of maximum declared CO_2 flow rate. In the first case, the proposed pipeline diameter should be 400 mm and in the second case 350 mm. For liquid carbon dioxide transportation calculated diameters are lower than for gaseous phase. The proposed diameter for maximum variant of flow rate is 250 mm (for assumed 2 MPa pressure drop in the pipeline). The diameters for liquefied CO_2 transfer in quantity of 60% and 30% of the maximum flow rate are equal to 200 mm and 150 mm, respectively.

In Figures 9-14 pressure and temperature profiles as function of pipeline length are presented for the maximum assumed flow rate variant and determined pipeline diameters. The surrounding ground temperatures are related to the ambient temperatures for Summer and Winter.

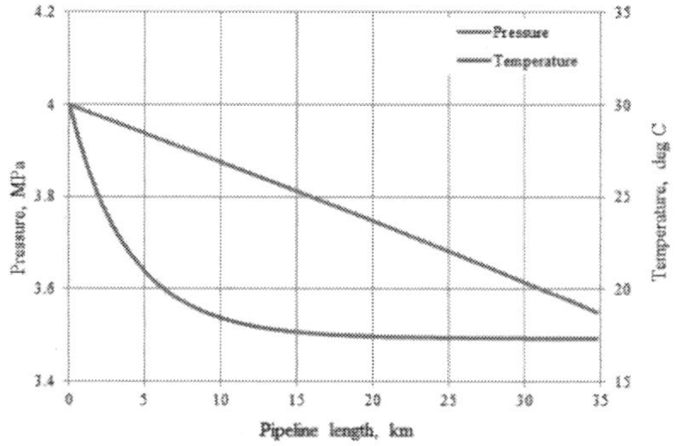

Figure. 11. Pressure and temperature profiles of gaseous carbon dioxide as a function of pipeline length for 500 mm, minimum available flow rate, inlet pressure 4 MPa, inlet temperature 30°C, the surrounding ground temperature 18°C.

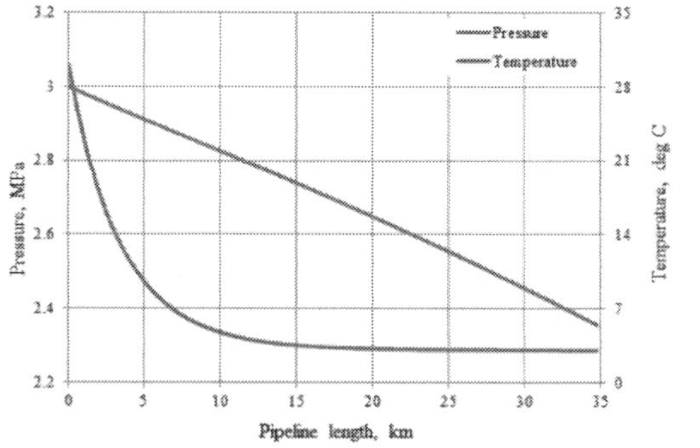

Figure. 12. Pressure and temperature profiles of gaseous carbon dioxide as a function of pipeline length for 500 mm, minimum available flow rate, inlet pressure 3 MPa, inlet temperature 30°C, the surrounding ground temperature 4°C.

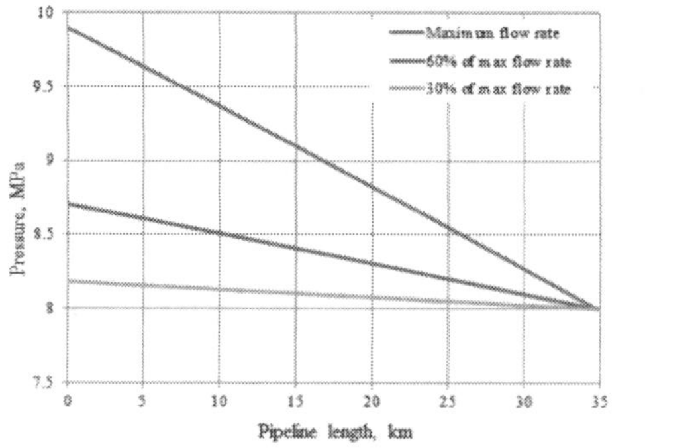

Figure. 13. Pressure profiles of liquid carbon dioxide as a

function of pipeline length for 200 mm, the three flow rate variants, outlet pressure 8 MPa, inlet temperature 10°C, the surrounding ground temperature 18°C.

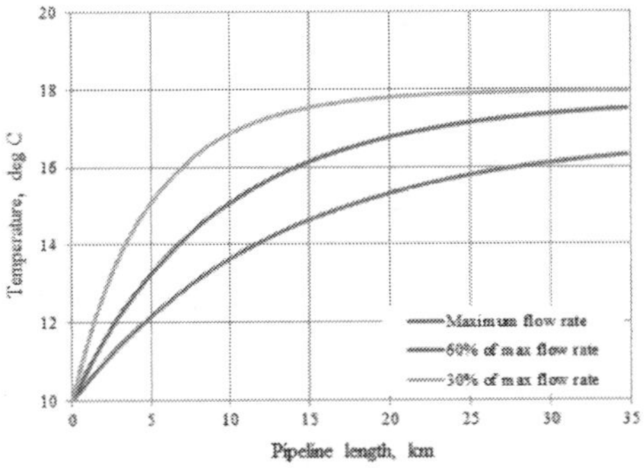

Figure. 14. Temperature profiles of liquid carbon dioxide as a function of pipeline length for 200 mm, three variants of flow rate, outlet pressure 8 MPa, inlet temperature 10°C, the surrounding ground temperature 18°C.

For gas and liquid cases of the transport, the pressure and the temperature profiles along the 34.8 km long pipeline were determined. The linear pressure drop is observed in Winter and Summer. In terms of temperature drop for gaseous CO_2 transport, carbon dioxide reach the temperature of the surrounding ground after 20 km distance for both summer and winter periods. For the liquid CO_2, temperature increase of 8°C is observed at oilfield site when 30% of maximum carbon dioxide source capacity is transported. For the maximum flow rate variant, more than 6°C increase is registered. Temperature increase is caused by the surrounding temperature of pipeline, since it is higher than the CO_2 fluid temperature at inlet point. The pressure and temperature drops obtained in this study are in line with the literature [10].

4 Conclusions

In this study, two methods were used to transport CO_2 over long distance: either as a gaseous state or as a subcooled liquid. Gas phase transport is disadvantaged by the low density and consequently large pipe diameter. Carbon dioxide in the gas phase is relatively heavy and its density causes high pressure drop. This is the reason that carbon dioxide is transmitted in the gas phase mainly for short distances with small flow rates. Gaseous CO_2 transport is additionally limited by the maximum pressure. For the winter conditions in the case of low temperature, the maximum allowable operating pressure cannot exceed 3 MPa, above which carbon dioxide may condense in the pipeline. In the analyzed cases, pipeline transport for the CO_2 gas phase is possible, however, for the maximum flow rate variant, the required pipeline diameters is equal to 500 mm. The application of large pipeline diameter significantly raises the project investment costs. Therefore, in the case of large quantities of carbon dioxide, the more efficient way of transport is to convert CO_2 into the liquid phase. This solution requires heat exchanger installation, however, the pump booster application allows for a significant operating cost reduction comparing to the gas compressors.

Acknowledgement

Work performed within the statutory research program of AGH UST No. 11.11.190.555 and grant No. 15.11.190.723

References

1. R. Czarnota, E. Knapik, P. Wojnarowski, J. Stopa, D. Janiga, P. Kosowski, The chemical composition of crude oil produced by supercritical CO_2 injection into oil reservoirs. Przemysl Chemiczny, **96(5)**, 964-967, (2017).
2. Z. Zhang, G. Wang, P. Massarotto, V. Rudolph, Optimization of pipeline transport for CO_2 sequestration. Energy Conversion and Management, **47(6)**, 702-715, (2006).
3. Noothout P., Wiersma F., Hurtado O., D. Macdonald , J. Kemper, K. van Alphen, *CO_2 Pipeline infrastructure–lessons learnt*. Energy Procedia, **63**, 2481-2492, (2014).
4. S. Sim, I. Cole, Y. Choi, N. Birbilis, A review of the protection strategies against internal corrosion for the safe transport of supercritical CO_2 via steel pipelines for CCS purposes. International Journal of Greenhouse Gas Control, **29**, 185-199, (2014).
5. V. Onyebuchi, A. Kolios, D. Hanak, C. Biliyok, V. Manovic, A systematic review of key challenges of CO_2 transport via pipelines. Renewable and Sustainable Energy Reviews, **81(2)**, 2563-2583, (2018).
6. T. Włodek, Wybrane aspekty techniczne rurociągowego transportu dwutlenku węgla. AGH Drilling, Oil, Gas, **29**, 323-335, (2012).
7. J. Jakobsen, S. Roussanaly, R. Anantharaman, A techno-economic case study of CO_2 capture, transport and storage chain from a cement plant in Norway. Journal of cleaner production, **144**, 523-539, (2017).
8. S. McCoy, *The economics of CO_2 transport by pipeline and storage in saline aquifers and oil reservoirs*, (2009).
9. M. Mohitpour, H. Golshan, A. Murray, *Pipeline Design & Construction: A Practical Approach*. American Society of Mechanical Engineers Press, (2000).
10. C. The, A. Barifcani, D. Pack, M. Tade, The importance of ground temperature to a liquid carbon dioxide pipeline. International Journal of Greenhouse Gas Control, **39**, 463-469, (2015).

Traffic speed prediction using ensemble kalman filter and differential evolution

Lukáš Rapant

IT4Innovations, VŠB – Technical University of Ostrava, Ostrava-Poruba, Czech Republic

Abstract. Importance of traffic state prediction steadily increases with growing volume of traffic. Ability to predict traffic speed in short to medium horizon (i.e. up to one hour) is one of the main tasks of every newly developed Intelligent Transportation System. There are two possible approaches to this prediction. The first is to utilize physical properties of the traffic flow to construct an exact or approximate numerical model. This approach is, however, almost impossible to implement on a larger scale given the difficulty to obtain enough traffic data to describe the starting and boundary conditions of the model. The other option is to use historical traffic data and relate information and patterns they contain to the current traffic state by application of some form of statistical or machine learning approach. We propose to use combination of Ensemble Kalman filter and Cell Transmission Model for this task. These models combine properties of physical model with ability to incorporate uncertainty of the traffic data.

1 Introduction

The objective of this study is to propose a macroscopic traffic speed prediction model, which can be based both on stationary and dynamic data sources. Most straightforward solution to the traffic speed prediction problem is to solve it by application of some form of kinematic wave model or car following model. They accurately describe behavior of traffic flow by application of physical models. Examples of this approach are articles written by Tang et. al. [1] and [2] and Costesqua et. al. [3]. However, the absence of traffic intensity or density information in FCD and sparsity of other sensors rules out most of these traffic flow models, which are based these quantities. Even though there are some models which circumvent this problem (for example Cell transmission model - velocity (CTM-v), I still have insufficient data to exactly describe their initial and boundary conditions. While their results may seem realistic (as a dissolution of traffic jam from my CTM-v model shown in the Figure 1, their quality is quickly eroding with each iteration. Therefore, other prediction scheme for this problem must be proposed.

Other possible approach to traffic state prediction is to use historical data about traffic and relate them to the current state of the traffic by application of some statistical or machine learning approach. This can be described as a task of finding traffic state time series prediction

$$P_l = \max_{s \in S} p(\, s \mid x_1, \dots, x_n \,),$$

where P_l is prediction of length l, S is set of possible predictions of certain length l, s is member of S and $x_1, \dots, x_n$ are n last measurement of speed.

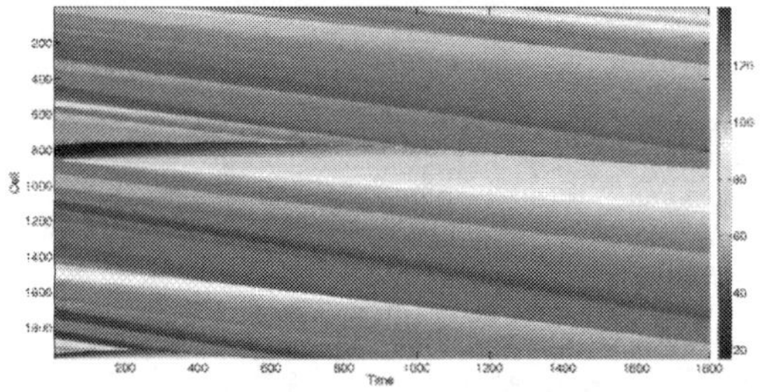

Figure 1. Result from simulation of D1 by CTM-v model (time is in seconds and color represents speed in km/h)

Examples of this approach are described by Calvert et. al. [4] or Huang [5] and some other authors. This task is, however, also not a simple one. It is complicated by complex nature of the traffic data. They can, depending on type of their source, be inaccurate or not available in required quantity. Despite this fact, this approach is preferable in case of traffic state prediction on Czech Republic highways due to the unavailability of boundary condition data (i.e. there are not measurements on all ramps leading to and out of motorway).

One of the possible approaches is Ensemble Kalman filter (EnKF). EnKF model was, amongst other reasons, chosen because its ability to work with coarse physical model and to represent nonlinearity of the traffic. For the physical model, CTM-v was chosen because it is only physical model that requires only speed for the computation (and FCD contain only information about speed, not density or intensity). However, its application would require finding heuristics for boundary conditions (beginning and end of the modelled road and ramps). This heuristic will be based on analysis of historic data. Then this model will be implemented into Ensemble Kalman filter and resulting model will be tested and optimized on the historical data. EnKF is an algorithm that uses a series of measurements observed over time, containing noise and other uncertainties. It produces estimates of unknown variables that tend to be more precise than those based on a single measurement alone.

2 Theoretical background

2.1 Kalman filter

The Kalman filter is used to solve the general problem of estimating the state of a discrete-time controlled process that is defined by the linear stochastic difference equation

$$x_k = Ax_{k-1} + Bu_k + w_{k-1}, \quad (1)$$

with a measurement $z \in R^m$ that is

$$z_k = Hx_k + v_k. \quad (2)$$

The matrix A in the difference Equation 1 relates the state at the previous time step to the state at the current step. The matrix B relates the optional control input to the state x. The matrix H in the Equation 1 relates the state to the measurement z_k. The random variables w and v represent the process and measurement noise. They are assumed to be independent and with normal probability distributions.

$$P(w) \sim N(0, Q) \quad (3)$$
$$P(v) \sim N(0, R) \quad (4)$$

Q and R are the process noise covariance, respectively measurement noise covariance matrices.

The Kalman filter estimates a process by using a form of feedback control: the filter estimates the process state at some time and then obtains feedback in the form of (noisy) measurements. As such, the equations for the Kalman filter fall into two groups: time update equations and measurement update equations. The time update equations are responsible for advancing the current state and error covariance estimates in time to obtain the a priori estimates for the next time step. The measurement update equations are then responsible for the feedback. Their task is to incorporate a new measurement into the a priori estimate to obtain an improved a posteriori estimate. The time update equations can be considered as prediction equations, while the measurement update equations can be considered as correction equations.

The specific equations for the time updates are following:

$$\widehat{x_k^-} = A\widehat{x_{k-1}} + Bu_k$$
$$P_k^- = AP_{k-1}A^T + Q$$

These time update equations project the state and covariance estimates forward from time step (k-1) to step k. A is from Equation 1 and Q is from Equation 3. The specific equations for the measurement updates are:

$$K_k = P_k^- H^T (HP_k^- H^T + R)^{-1},$$
$$\widehat{x_k} = \widehat{x_k^-} + K_k(z_k - H\widehat{x_k^-}),$$
$$P_k = (I - K_k H)P_k^-.$$

H is from Equation 2 and R is from equation 4. The first task during the measurement update is to compute the Kalman gain K_k. The measurement is incorporated after this step to obtain z_k, and then to generate a posteriori state estimate. The final step is to obtain a posteriori error covariance estimate.

After each time and measurement update cycle, the process is repeated with the previous a posteriori estimates used to predict the new a priori estimates. This recursive nature is one of the very appealing features of the Kalman filter because it makes practical implementations much more feasible.

2.1.1 Ensemble kalman filter

The ensemble KF (EnKF) was introduced as an alternative to the extended KF. Extended KF performance is poor when the state transition function is highly nonlinear. The EnKF belongs to a group of suboptimal estimators that use Monte Carlo or ensemble integration. The EnKF uses a collection of state vectors (called the ensemble members of system states) to propagate the state in time and to compute the mean and covariance needed for the correction step. The covariance estimated in this way is used to compute the Kalman gain. The correction step equation stays the same as in the traditional KF. The EnKF algorithm can be described by the following steps:

1. Generate N ensemble members of system states x_0^n by drawing N samples from a Gaussian distribution ($n = 1,2, \ldots, N$).
2. Make a time update $\widehat{x_t^n} = A(\widehat{x_{t-1}^n})$.
3. Compute the mean of the ensemble:

$$\mu_t = \frac{1}{N}\sum_{n=1}^{N} \widehat{x_t^n}$$

4. Compute the covariance of the predicted state:

$$P_t = \frac{1}{N-1} E_t (E_t)^T$$

where $E_t = \left[\widehat{x_t^1} - \mu_t, \dots, \widehat{x_t^N} - \mu_t\right]$.

5. Calculate the Kalman gain:

$$K_k = P_t H^T (H P_t H^T + R)^{-1}$$

6. Obtain a new ensemble from the measurements:

$$x_t^n = \widehat{x_t^n} + K_t\left[z_t - H\widehat{x_t^n}\right]$$

7. Return to step 2.

2.2 Differential evolution

Differential evolution has proven to be an efficient method for optimizing real-valued functions. DE is trying to breed as good population of individuals as possible in loops called generations [6].

The very first step is defining constants affecting behavior of evolution algorithm. Those constants are crossover probability (CR) which influences probability of choosing a parameter of the mutated individual instead of the original one, mutation constant (F) which determines volume of mutation of the mutated individual, population size (NP) and finally trial specimen. Now I have a specimen and I can create initial population vector X.

When I have initial population prepared (called generation), loop begins. In each generation cycle, nested loop is executed. This loop is called evolution cycle. Each individual from current generation is bred to have better characteristic using mutations in evolution cycle. Algorithm finishes either if a solution is good enough (individual) or after defined number of generation loops had been executed.

For my purposes, the DE/rand/1 algorithm mutation is the best choice. The notation DE/rand/1 specifies that the vector v to be perturbed is randomly chosen and the perturbation consists of one weighted vector.

$$v = x_{r1,j}^G + F \cdot \left(x_{r2,j}^G - x_{r3,j}^G\right),$$

where G is number of generation. Due to this mutation, new individuals are not affected by the temporary best individual from generation and space of possible solutions is searched through uniformly. More detailed description of Differential Evolution can be found in [6]. Differential evolution can be described by following pseudocode.

DE{X,f_cost,NP,F,CR,Specimen}

1. X: initial population (vector)
2. f_cost: function returning fitness of current solution
3. NP: population size
4. F: mutational constant
5. CR: Crossover threshold
6. Specimen: trial specimen
7. For I < Generation do begin
8. For J < NP do begin
9. Select J individual
10. Select three random individuals from population
11. Compute the new individual fitness
12. Choose a better one from both individuals to new population
13. EndFor
14. EndFor

2.3 Cell transmission model – velocity

CTM-v is based upon classical Lighthill-Whitham-Richards (LWR) partial differential equation. LWR PDE for modelling traffic on highways is:

$$\frac{\partial \rho}{\partial t} + \frac{\partial q}{\partial x} = 0 \tag{5}$$

where $q(x,t)$ is flow and $\rho(x,t)$ is density of the vehicles at location x and time t. This equation is derived from hydrodynamics theory and expresses the conservation of mass for a fluid of density ρ and of flux q . Empirical relation called fundamental diagram $q(x,t) = Q(\rho(x,t))$ is used for the expression of q as a function of ρ. Q is flux function and is independent of variables. To transform LWR PDE to velocity LWR (LWR-v) PDE, I will use the specific flux function called Greenshields:

$$v = v_{max}\left(1 - \frac{\rho}{\rho_{max}}\right)$$

where v_{max} and ρ_{max} denote the maximal velocity respectively the maximal density allowed by the model. This function expresses relation between density and velocity. Greenshields flux function can be easily inverted to express ρ as a function of v. Now I can substitute this expression into Equation 5 to obtain LWR-v PDE:

$$\frac{\partial v}{\partial t} + \frac{\partial R(v)}{\partial x} = 0 \tag{6}$$

where $R(v) = v^2 - v_{max}v$. Variable change $v = v - \frac{v_{max}}{2}$ transforms Equation 6 into its final form:

$$\frac{\partial v}{\partial t} + \frac{\partial v^2}{\partial x} = 0$$

on the domain $(x,t) \in \langle a,b\rangle \times \langle 0,T\rangle$. The initial and boundary conditions are too long to be presented here and can be found for example in [7]. For practical implementation, the LWR-v PDE is discretized using a Godunov numerical scheme [8]. By application of Godunov scheme, I obtain CTM-v. Let $n \in N$ be number of time steps of length $\delta t = \frac{T}{N}$, $M \in N$ number of space cells of length $\delta x = \frac{b-a}{M}$. Then v_i^n is called discrete value of v at time step n in cell i ($0 \leq n \leq N$ and $0 \leq i \leq M$). At each time step v_i^{n+1} is computed from the previous time step. This is done by formula:

$$v_i^{n+1} = v_i^n - \frac{\Delta t}{\Delta x}\left(g(v_i^n, v_{i+1}^n) - g(v_{i-1}^n, v_i^n)\right) \quad (7)$$

where the numerical flow function g is:

$$g(v_1, v_2) \begin{cases} R(v_2) & if\ v_1 \leq v_2 \leq v_c \\ R(v_c) & if\ v_1 \leq v_c \leq v_2 \\ R(v_1) & if\ v_c \leq v_1 \leq v_2 \\ max(R(v_1), R(v_2)) & if\ v_1 \leq v_2 \leq v_c \end{cases}$$

where $v_c = \frac{v_{max}}{2}$, $R(v) = v^2 - v_{max}v$ and v_{max} is maximum speed allowed in the model. Note that Δt and Δx must satisfy Courant-Friedrichs-Lewy (CFL) condition:

$$\left|\frac{\Delta t}{\Delta x} v_{max}\right| \leq 1$$

For implementation of boundary conditions, I use ghost cells placed at each side of the domain defined by the boundary conditions to be satisfied.

3 Algorithm

Ensemble Kalman filter based algorithm is presented in this part. Parts of this approach were presented in [9] and [10]. Once again, this algorithm utilizes both historical and current data from both my data sources. Main difference between this algorithm and the other ones is, that this one predicts the speed on the entire motorway and not only parts of it. Also, this algorithm is not only based on the historical data but also on a physical model. As a standalone, physical model (in my case CTM-v) would be unable, due to the missing data regarding its boundary conditions, cope with this task. However, thanks to the smart combination of historical data and physical model it is possible to utilize it in the prediction. Moreover, thanks to the integration of CTM-v into Ensemble Kalman filter, efficient assimilation of current situation data is guaranteed. Ensemble Kalman filter was chosen, because the problem of the traffic modelling is generally considered to be non-linear and not suitable for general Kalman filter. Due to the assimilation step, this algorithm is more useful for relatively short predictions until next traffic state data are available (5 minutes in my case because of ASIMs). General progress of this algorithm can be summarized into following points (note that steps 3 and 4 are performed repetitively while the prediction is being performed and current data supplied):

1. Learn the boundary conditions from the historical data.
2. Initiate an ensemble.
3. Perform an ensemble of predictions based on the learned physical model.
4. Assimilate current state data by EnKF.

Let us discuss each of these steps in detail. In the first step, learning of the boundary conditions is performed. It can be likened to the inverse modelling task of finding a set of boundary conditions B, such that

$$\mathbf{B} = \underset{b}{argmin}\ |\mathbf{m}(\mathbf{x\_0}, \mathbf{b}) - \mathbf{x}|, \quad (8)$$

where b are possible boundary conditions, x are available data and function m represents my physical model (i.e. I am minimizing the difference between the physical model and data given the boundary conditions). Given that I need to define boundary condition for each ramp on the motorway and that this condition should change in time, this optimization is not a simple one. It can be expected, that minimizing Equation 8 would be very difficult due to the high number of local minimums. Solution of this problem was found in the field of evolutionary computations and is named Differential evolution. This optimization technique is well proven to handle such difficult optimization problems as for example is proven in [9]. Optimized variable b is a matrix where rows represent individual ramps and columns its progress in time (this forms my specimen). Crossbreeding between specimen is then performed between rows of the b. In an optimal situation, model should be optimized for a week-long period, as the week is larges strictly periodical time interval in traffic. These specimens are then inserted into a CTM-v model and their results are compared to the original data producing the measure of quality. Result of this optimization is utilized in the step 3.

The second step is easy and straightforward. **N** members of ensemble are initialized by using free flow speed $\mathbf{v_{ff}}$ randomized by variance $\boldsymbol{\sigma^2}$ obtained from the historical data for each cell

$$\mathbf{v_i} = \mathbf{v_{ff}} + \mathbf{N}(\mathbf{0}, \boldsymbol{\sigma^2}),$$

where $\mathbf{v_i}$ is initial speed in the cell.

The next phase is a prediction phase. This phase is based on CTM-v and is realized by a modification of Equation 7.

$$\mathbf{v_i^{n+1}} = \mathbf{v_i^n} - \frac{\Delta \mathbf{t}}{\Delta \mathbf{x}}\left(\mathbf{g}(\mathbf{v_i^n}, \mathbf{v_{i+1}^n}) - \mathbf{g}(\mathbf{v_{i-1}^n}, \mathbf{v_i^n}) \pm \mathbf{B_i^n}\right),$$

where $\mathbf{B_i^n}$ is control variable governing the boundary conditions and learned in previous step. Addition or deduction depend whether it is in-ramp (-) or off-ramp (+). Length of each cell $\mathbf{x}$ and time step $\mathbf{t}$ must be set to adhere to Courant-Friedrichs-Lewy conditions. This means that if I want to have a suitably small cell, each model step must be reasonably short, and it is necessary to advance the model many times each prediction phase by calculating Equation 7 (for example, 5 minute prediction requires 300 prediction steps without any further data). This results in problem of finding balance

 https://doi.org/10.1051/matecconf/201925902001

between detail and accuracy and will be topic for the future research. Because this algorithm is using EnKF, this prediction is not only done once but for each **N** members of the ensemble. Resulting prediction from the entire ensemble can be obtained by calculating an interval of confidence for mean speed of the ensemble (as usually ensemble has more than a few members).

The last step of my algorithm is current state data assimilation. This step is performed once per minute or per 5 minutes depending whether ASIM sensor data are used. It is also quite straightforward as this step consist only of calculating ensemble prediction covariance $\mathbf{P_t}$, calculating Kalman gain K_k and updating the ensemble by combining the new measurements with current ensemble. Measurements covariance **R** is calculated from historical data. Measurements used for this step come mostly from the FCD (TNC segments are divided into cells with the same measured value) and also can come from ASIM, where they are related to the cell with the gate. This provides us with more accurate measurement but also lengthens my prediction step. The entire algorithm can be summarized into following pseudo code.

EKF predict{**X**, **Xc**, **N**, **R**, **evol**}

1. **X**: historical FCD and ASIM values
2. **Xc**: current speed time series
3. **N**: ensemble size
4. **R**: measurements covariance
5. **evol**: vector with settings for DE
6. **B**=LearnDE(**X**, **evol**);
   \ *Learning of boundary conditions by DE*
7. **En**=InitializeEnsemble(**N**, **X**);
   \ *Ensemble initialization*
8. While {$\mathbf{x_c}$}
9. **EnsemP**=PredictEnsemble(**Ensem**, **B**);
   \Prediction step
10. **Ensem**=Assimilate(**Xc**, **EnsemP**, **R**);
    \Assimilation of current measurements
11. EndWhile

4 Experimental results

Experiments with this algorithm were performed on data from D1 (roughly from Humpolec to Brno) coming from 21.4.2014. Unfortunately, quantity of the data is insufficient to perform thorough testing and evaluation, however I do not possess larger data set at the moment. These results should be therefore considered as demonstration of usability. This problem would be solved soon with access to new datasets.

Experiment was performed only on FCD data, however, I kept 5-minute prediction step. I used 10 hours from 6:00 to 16:00 for learning and measurement covariance estimate. I used 50 generations of 75 individuals and set crossbreeding threshold to 0.7 and mutation constant to 1.2. After learning the model, I performed 20 prediction and assimilation steps. Spatial discretization for CTM-v model was chosen to be 100 m and temporal 1 second (by the CFL conditions). Results from prediction steps 6, 12 and 20 can be seen in Figures 2,3 and 4. These Figures show true traffic speed on entire section of the D1 motorway in the given step and prediction confidence interval.

RMSEs of these predictions are 8.65, 10.10 and 12.31 respectively. While these numbers may not seem to too high, few glances at the Figures 2, 3 and 4 show that results are roughly capable of capturing the magnitude of speed but rarely the trend. This fact seriously degrades value of the results. However, these problems are probably caused by small training set and may be solved by introduction of the new and larger data sets.

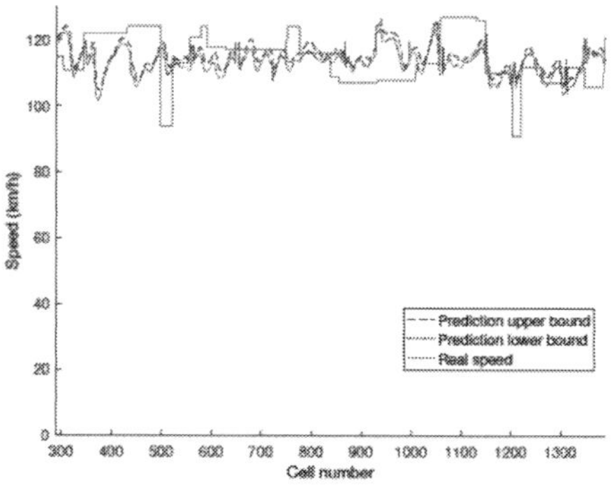

Figure 2 Traffic speed on segment of D1 after prediction step 6

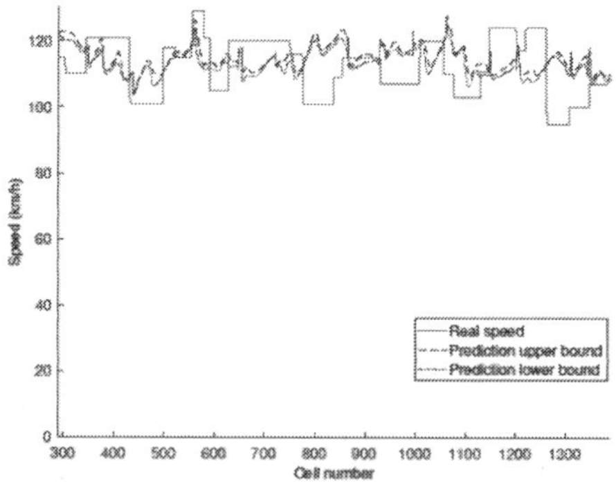

Figure 3. Traffic speed on segment of D1 after prediction step 12

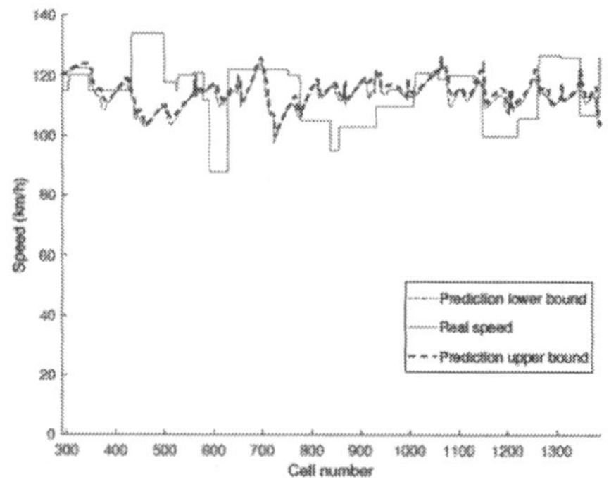

Figure 4. Traffic speed on segment of D1 after prediction step 20

For the comparison to the contemporary works, I have chosen articles written by Coric et. al. [11], Jiang et. al. [12], Castillo et. al. [13] and Xie et. al [14]. Coric et. al.

[11] used Mean Absolute Error (MAE) which are generally smaller than RMSE for equivalent data. These works were chosen because they present similar solutions and also have comparable results. Resulting errors of their algorithms are shown in Table 1.

Table1. Comparison of errors (RMSE or MAE) of contemporary works

Author	Jiang et. al. [12]	Castillo et. al. [13]	Coric et. al. [11]	Xie et. Al [14]
Average RMSE	9	8.7	6.2 (MAE)	9.1

From the table, it is evident that my algorithms have similar error results, even with small data sets. HMM algorithm even performs better than the other ones. Also, advantage of my algorithms is, that they were tested on real data with all of their problems and uncertainties. This is for example not true in case of Jiang et. al. [35] who used artificial data from SUMO traffic simulator.

4 Conclusion

From the scientific perspective, this work contains utilizes some interesting and original solutions. While proposed algorithms are generally based on existing methods, some subtasks required development of specific approaches. The most important innovation of EnKF algorithm is utilization of differential evolution for inverse modelling of boundary conditions. Perhaps more important achievement of this work is from the perspective of its application. Experimental results of all proposed algorithms indicate their usefulness in real intelligent traffic systems. From the perspective of future work, developed algorithms can still utilize some improvements to overcome their minor issues (other than enlarging the learning data sets). When trained on the suitably large data set, EnKF algorithm will be improved by thorough calibration. The purpose of this calibration is to find the best spatial and temporal discretization and also the optimal prediction window. The algorithm will also be extended for modelling of other traffic quantities, namely traffic density and intensity. Combination of these attributes can be utilized in our models to provide higher level of prediction accuracy. This combination could improve the prediction because of known interaction of these attributes, which simplifies identification of different traffic situations and reduces uncertainty.

5 Acknowledgment

This work was supported by The Ministry of Education, Youth and Sports from the National Programme of Sustainability (NPU II) project „IT4Innovations excellence in science - LQ1602“.

References

1. T.-Q. Tang, L. Caccetta, Y.-H. Wu, H.-J Huan, and X.-B. Yang. A macro model for traffic flow on road networks with varying road conditions. *Journal of Advanced transportation* **48**, pp. 304–317 (2014).
2. T. Q. Tang, J. G. Li, H. J. Huang, and X. B. Yang. A car-following model with real-time road conditions and numerical tests. *Measurement* **48** (2014).
3. G. Costesequea and J. P. Lebacque. A variational formulation for higher order macroscopic traffic flow models: Numerical investigation. *Transportation Research Part B: Methodological* **70** (2014).
4. S. C. Calvert, T. Henk, M. Snelder, and S. P. Hoogendoorn. Application of advanced sampling for efficient probabilistic traffic modelling. *Transportation Research Part C: Emerging Technologies* **49** (2014).
5. M.-L. Huang. Intersection traffic flow forecasting based on v-gsvr with a new hybrid evolutionaryalgorithm. *Neurocomputing* **147**, pp. 343–349 (2015).
6. K. Price. An introduction to differential evolution. *New Ideas in Optimization*, pp. 79–108 (1999).
7. D. Work, O. Tossavainen, S. Blandin, A. Bayen, T. Iwuchukwu, and K. Tracton. An ensemble kalman filtering approach to highway traffic estimation using gps enabled mobile devices. *In Proceedings of the 47th IEEE Conference on Decision and Control* (2008).
8. J. P. Lebacque. The godunov scheme and what it means for first order traffic flow models. In *Proceedings of 13th International Symposium on Transportation and Traffic Theory* (1996).
9. P. Dubec, J. Plucar, and L. Rapant. Use of the bio-inspired algorithms to find global minimum in force directed layout algorithms. *In Proceedings of the 6th International Conference on Multimedia Communications, Services and Security, Communications in Computer and Information Science* **368**, pp. 194–203. Springer Verlag (2013).
10. L. Rapant. Traffic flow modeling based on sparse data. *In Proceedings of WOFEX 2014*, pp. 526–531. VSB-TUO (2014).
11. S. Vucetic V. Coric, N. Djuric. Traffic state estimation from aggregated measurements with signal reconstruction techniques. *Transportation Research Record: Journal of the Transportation Research Board*, **2315** (2012).
12. B. Jiang and Y. Fei. Traffic and vehicle speed prediction with neural network and hidden markov model in vehicular networks. *In Proceedings of 2015 IEEE Intelligent Vehicles Symposium* (2015).
13. E. Castillo. Stochastic demand dynamic traffic models using generalized beta-gaussian bayesian networks. *IEEE Transactions on Intelligent Transportation Systems* **13** (2012).

14. Y. Xie, Y. Zhang, and Z. Ye. Short-term traffic volume forecasting using kalman filter with discrete wavelet decomposition. *Computer-Aided Civil and Infrastructure Engineering* **22**, pp. 326–334 (2007).

Preliminary establishment and analysis of automated driving benchmark in China

Hui Xiong, Yinghan Wang, Qiang Ge, Ziqing Gu, Mingyang Cui, Jianqiang Wang, Keqiang Li and Qing Xu

State Key Laboratory of Automotive Safety and Energy, Tsinghua University, Tsinghua Park N0. 1, Beijing, China

Abstract. In order to promote the localization of Automated Driving (AD) in China, it is necessary to collect large-scale traffic scene data with Chinese characteristic for future analysis. In this paper, we propose the methodologies and rules of establishing AD benchmark involving how to configure sensors, how to design the collection schema to show Chinese traffic characteristics and the rules of elaborating distinctive scenes and routes, what to label, and it is also demonstrated that the benchmark can support the future application of extended AD research. Data collection lasted about one month covering diverse scene data such as campus, highway, park, etc. from three representative Chinese cities and driving data from 30 different drivers. Moreover, some statistical results and analyses are produced in accordance with the designed methodologies as instances of potential application. Up to now, the dataset contains about 7,000 labelled image frames and corresponding LiDAR, GPS and Controller Area Network (CAN) data. Labels cover scene type, road user, traffic sign, traffic light, and lane marker. This benchmark can help researchers better understand Chinese traffic situation in aspects of environmental perception, driving behavior analysis, risk assessment, automated vehicle decision and control.

1 Introduction

In the research on the adaptability of AD by identifying and analysing Chinese unique complex traffic environment, also in order to find more key research topics in the field of AD, firstly researchers need to establish benchmark standards and build the collection platform. Afterwards a road environmental survey on scene features, road structures, and road user behaviours is launched. This paper makes a statistical analysis, and based on which, a conclusion on scene characteristics and road parameters such as the radius of road curvature, lane width, vehicle-following time gap, which could demonstrate the potential application of the dataset. To find out the behaviors of road users and drivers, we also focus on the relative distance, velocity, intention of other road users, which could deduce the native driving style for the automated vehicle to learn. We hope that data collected from real drivers would guide the research on automated driving.

Inspired by sustained progress in naturalistic driving benchmark [1], [2], [3], a preliminary establishment and analysis of automated driving research platform for specified scenarios in China have been carried out. Its primary goal is to provide a precise instruction on developing relevant algorithms for well-rounded research fields of AD including the perception, fusion, decision, and control parts. Specifically, the original image data from the monocular camera, detection output from the functional camera, and the 3D information of two full-view LiDARs could be used for environmental perception [4] and sensor fusion[5]. Besides, the recorded different driving behaviors and the vehicle dynamics from CAN assisted by Inertial Measurement Unit (IMU) and Global Positioning System (GPS) would be useful for automated decision and control research [6], [7].

In the rest of this paper, the overview of approach architecture for AD benchmark and the vehicle platform are introduced in Section 2. The basis of scene design is explained, as well as the covered scenario illustration and route exhibition in Section 3. Next section introduces the details of data acquisition and labelling. The potential application analysis is discussed in Section 5. Finally, we sum up our work and propose the next research steps.

2 System architecture

2.1 Approach architecture

As shown in Figure 1, firstly we determine dataset category so that the sensor configuration can be confirmed. Then scene collection schema is designed to show Chinese characteristics. To cover these scenes, it is economical to select some representative cities, especially in different regions of China. Route design should consider the sizes of cities and hardware foundation. In labelling stage, it is necessary to make sure that we can cover all scenes' types; we also need to consider what and how to label to support AD research.

Finally, we can make analyses on the labelled data to demonstrate the potential application.

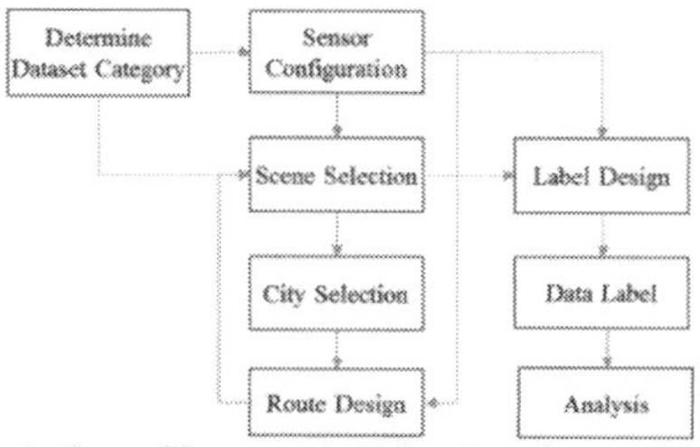

Figure 1. The architecture of AD benchmark establishment.

2.2 Acquisition platform

Our testbed is established based on a modified electric NISSAN Leaf. The sensor configuration for our AD benchmark in the test vehicle includes two LiDARs, GPS, IMU and three types of camera--monocular camera, stereo camera and functional Mobileye camera in Figure 2. The information of vehicle dynamics and driver's operation can also be acquired from vehicle's CAN. The basic rule of sensor configuration is to support full-view perception and comprehensive understanding of decision/control parts with effective cost.

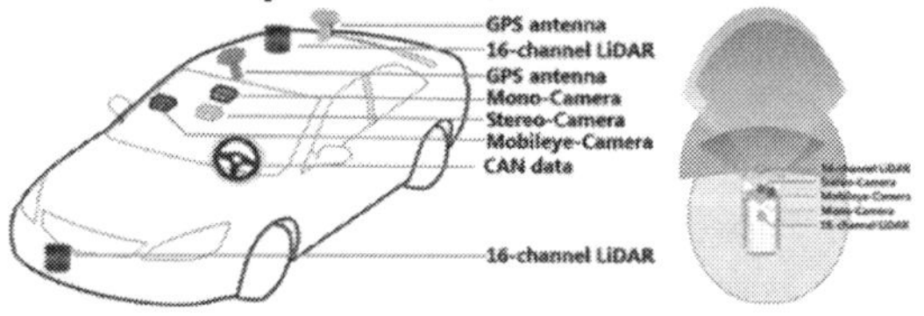

Figure 2. Main sensors installed on our testbed.

Here list the details of each sensor:

- Mono-Camera: 1×IDS UI-5250CP-C-HQ, 1,920,000 pixels, 35.6fps, global shutter, CMOS.
- Stereo-Camera: 1×Stereolabs ZED stereo camera, max to 1080x720, field of view: 110°, up to 100fps.
- Mobileye-Camera: Mobileye 5-Series, the output of lane detection, velocity, position of target.
- LiDAR: 2×Velodyne VLP-16, 16 Channel, range: 100m, FOV: 360° (horizontal), ±15° (vertical), accuracy: ±3cm.

We achieve the camera calibration and LiDAR calibration using sensor drivers in Robot Operating System (ROS) [8] individually. The open source tool of joint calibration [9] is used, as presented in Figure 3. The camera and LiDAR are calibrated to the same coordinate system in the head of vehicle.

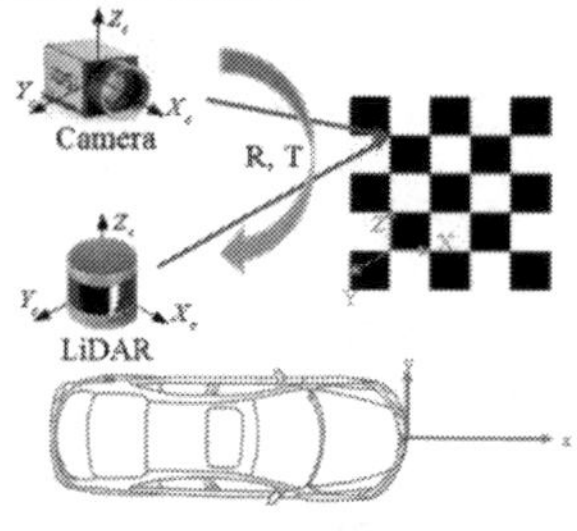

Figure 3. Joint calibration of camera and LiDAR sensor.

3 Scene design

Vast area and unbalanced developments in both infrastructure and economy bring China great challenges to promote AD. Scenes vary in all aspects: modern/traditional blocks; structured/unstructured road; main road users; truck and car on highway. Varied scenes then breed more behaviors on the road, mostly violating traffic rules, such as jaywalking, converse driving and illegal parking. Based on these features, this section elucidates design basis, describes covered scenarios, illustrates concrete route exhibition.

3.1 Design basis

A large scale of high-quality natural driving dataset is indispensable for AD development and testing. Therefore, it is necessary to design concise and representative scenarios and routes to build the dataset that meets the needs. We divide the currently general driving datasets into two categories: one is the universal dataset proposed by the field of pure computer vision, only including the element of VRUs like cyclists [14]; the other is AD dataset [10], [11], including IMU, GPS and other information which can be acquired from the driving simulator or the real-natural driving which is our concern.

We investigate the features of typical datasets under the real-natural driving scene finding that the dataset concerning the AD is mostly used in environmental perception and related fields in computer vision, ignoring the integration of drivers' behavior sets. Correspondingly, our proposal is to assist the development and testing of the overall AD process. Compared with current existing AD datasets (Table 1), our work covers more kinds of traffic scene (shown in next part), which varies with three Chinese cities, whereas previous papers focus more on climate variable [11] or limited type of scene [12], [13].

Table 1. Features of typical AD dataset.

Dataset	Features
KITTI[10]	In a mid-size city; In rural areas and on highways; real-world computer vision benchmarks
Oxford RobotCar[11]	One route; Over one year; In all weather conditions, including heavy rain, night, direct sunlight and snow
Cityscape[12]	Cover 50 cities; Several months (spring, summer, fall); Daytime; Good/Medium weather conditions; urban roads
Comma.ai[13]	7.25 hours of driving data of highway

3.2 Covered scenarios

Our dataset mainly contains ten natural scenarios (Figure 4) in three cities that are commonly used in the AD development, including *low-speed urban road, urban expressway, urban overpass, urban undercrossing tunnel, main road downtown, industrial park without signal light, original urban road, highway, campus* and *community*. They can reflect different traffic conditions in different types of cities in China under the same scenario and guide the adaptive development of AD research.

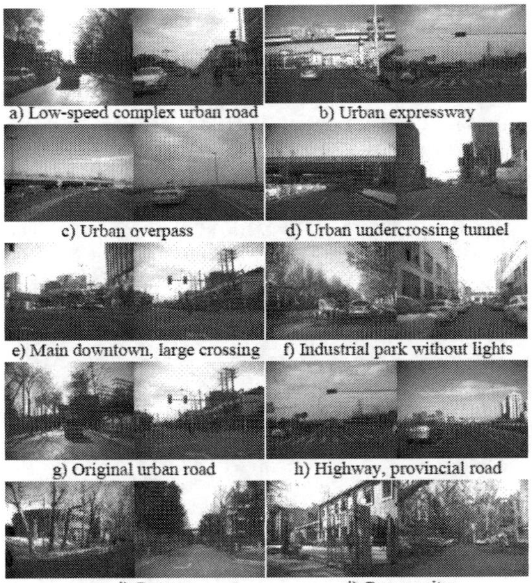
a) Low-speed complex urban road b) Urban expressway
c) Urban overpass d) Urban undercrossing tunnel
e) Main downtown, large crossing f) Industrial park without lights
g) Original urban road h) Highway, provincial road
i) Campus j) Community

Figure 4. Normal scenarios.

Route in Beijing can be taken as a good example, which contains: Tsinghua University (*campus*), Wudaokou (*congested intersection*, *from low-speed urban road complex conditions*), Wanquanhe Bridge (*urban overpass*), Xiaojiahe Bridge (*urban expressway*), G7 (*highway*), Software Park (*industrial park without signal light*), Xi'erqi (*original urban road*), Zhongguancun (*commercial district, from main road downtown, large crossing*) and Nanlou (*community*).

a) b) c) d)
e) f) g) h)
i) j) k) l)
m) n) o) p)

Figure 5. Special scenarios.

a) T-shape crossing	b) Y-shape crossing
c) Right-turning lane split from straight lane	d) Complicated scene, easy to get crowded
e) illegal road occupancy	f) Vogue lane marks
g) spotted lane mark	h) Patched road surface
i) Road under construction with lane width changed	j) Old-fashioned block and market, one-way lane
k) Mixed traffic flow on narrow road	l) Narrow road outside, few people or vehicles
m)Main road with unique left-turning lane occupancy	n) Urban road contains roundabout
o) Interior roads in hospital with narrow one-way lanes, and many parked cars aside	p) Undercrossing tunnel downtown, white balance problem occurs when entering/exiting tunnel

Furthermore, there are sixteen special scenarios as typical representations in Figure 5, which can well reflect features of the road with Chinese characteristics. For example, no matter in the large city or small city, illegal road occupancies are very common. The T-shape or Y-shape crossing is unique leading to more complex traffic flow. Also, there are some roads under construction with lane width changed. These unconventional datasets with local characteristics of traffic scenes can better guide the AD technology to land in China.

3.3 Selected routes

We choose three relatively developed and representative cities located in the northern, central and southeast region of China -- Beijing, Luoyang and Suzhou (see Figure 6). As the capital of China, Beijing is the most representative first-tier city. Compared with the national average level, the management of pedestrians and non-motor vehicles is more standardized. Luoyang is a second-tier city located in central China, the reconstruction of the Old City and construction of the New District are underway, which produces the mixture of structured road and unstructured road scenes. Suzhou, located in the southeast, different from the two cities mentioned before. Many township roads, bridges, newly-developed commercial areas, high-speed entrances and exits could effectively supplement previous scenes.

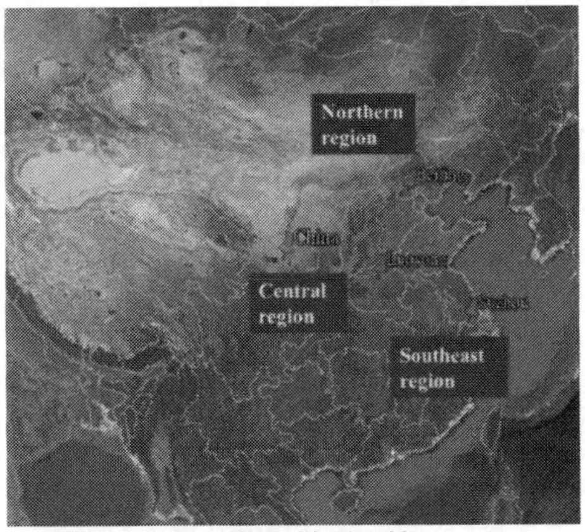

Figure 6. Three representative cities in China.

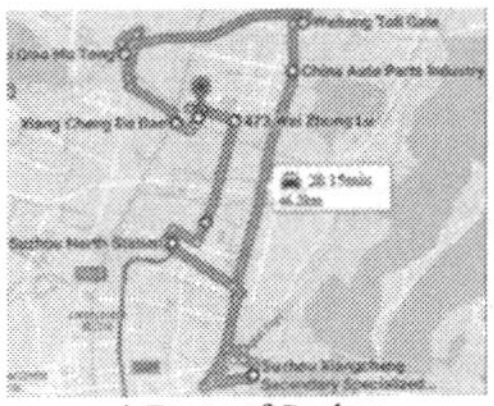
a) Route of Suzhou

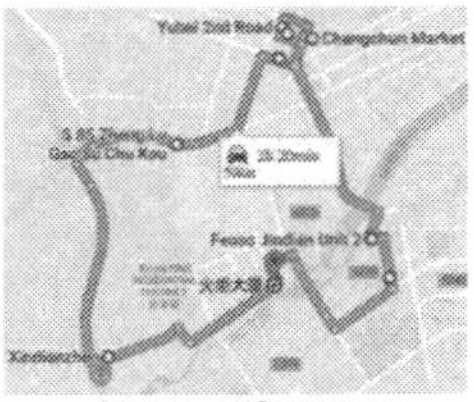
b) Route of Luoyang

https://doi.org/10.1051/matecconf/201925902002

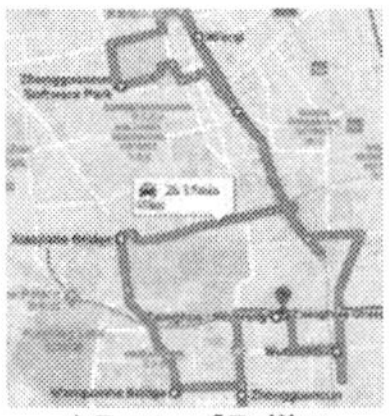

c) Route of Beijing

Figure 7. Designed routes in Suzhou, Luoyang and Beijing.

Due to limited storage and cruising mileage, design principle of the route is to cover all scenarios around the city center, and as many local features of traffics as possible in 2~2.5 hours' journey covering 45~50 km, so that we can drive to collect twice a day each city, as depicted in Figure 7.

4 Data acquisition and labelling

4.1 Data acquisition and extraction

To achieve the data acquisition function of independent sensors, we use the top-down design idea to design the data-collecting program in the ROS platform, and all the procedures apply hierarchical modular design to improve the portability and scalability of the software in Figure 8.

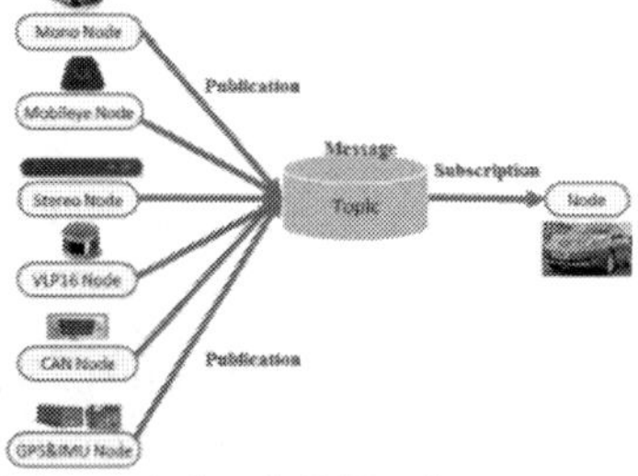

Figure 8. Software platform in ROS nodes.

Because the collected sensor data is saved in a unified ROS bag format, we need to extract the image file, point cloud file, CAN information, IMU/GPS information and Mobileye-camera results from the generated bag files. As far as the timestamp synchronization is considered, the integrity of the data is guaranteed using combined shell and python script by loading the complete message information from the bag file, and the concrete sensor information of the real traffic environment is extracted, as shown in Table 2.

Table 2. Extracted sensor data.

Item	Format	Description
Mono-Camera	PNG	Provide the frontal environment information
Mobileye Camera	CSV	Provide the detection information
VLP16 LiDAR	PCL	Provide 360-degree perception
CAN	CSV	Provide vehicle dynamics information
GPS&IMU Data	CSV	Provide the position and vehicle posture
Scene Type	CSV	Provide different scene types

4.2 Labelling contents and rules

4.2.1 Labelling content design

For subsequent AD research regarding to the environmental perception such as multiple object detection, scene segmentation [15], tracking [16], even intention prediction [17] and risk assessment [18], we design to label all the road users, all the traffic signs and lights, as well as any road markers and road users' abnormal behaviors. In addition to common park-highway-city AD testing scene roadmap from closed to open road, we also consider cluster campus scene as a semi-closed road. Moreover, considering varied movement model, air resistance and illumination condition, the 2nd level scenes -- uphill, downhill, on or under the bridge, and tunnel scene are designed. The labelling contents are summarized as follows:

Table 3. Labelling content design of scenes and objects.

Scenes	**1st level**	Campus	Park	City	Highway
	2nd level	Uphill	Downhill	On or Under Bridge	Tunnel
Objects	Road user		Traffic sign	Traffic light	Road marker

To understand the scene type, here shows some 1st scenes:

a) Campus scene b) City scene

c) Highway scene d) Park scene

Figure 9. The example of the 1st scene types.

Table 4. Labelling content design of abnormal behaviors.

Abnormal behaviors	Description
abparking	Illegal parking in wrong place
abLanechange	Illegal lane change
abovertaking	Illegal overtaking
jaywalk	Jaywalk
abDrive	Driving between two lanes
accident	Exist accident
converse	Driving in the oppsite direction
disobey	Drive through red light

Non-motor examples of road user types namely Vulnerable Road Users (VRUs) are shown:

Figure 10. The non-motor example of road users.

4.2.2 Labelling rules

In order to better refine the workflow of the labelling, there are some specific regulations listed below, according to the requirements of future AD research:

For scene labelling, each image could have one and only one 1st level scene type, and one or more 2nd level scene labels.

For object labelling, line segments are used for labelling lane and the boundary of the road, while bounding boxes are used for road users, traffic signs and traffic lights. The lane marking should be marked from near" to far", and other lanes should be marked in the same order. Lane markers less than two pixels or of which starting points and endpoints are nearly connected should be ignored. With regard to the traffic signs, indicate signs and guide signs are distinguished, so is the background color of them.

Besides, in order to cover special situations in China, we also categorize abnormal behaviors [22] of motor and non-motor vehicles labelled with bounding boxes. These behaviors involve abnormal parking, abnormal lane changing, abnormal overtaking especially for the motorcycle, and jaywalking or crossing the road for non-motor vehicles. Also, some common abnormal behaviors may exist for both of them, such as abnormal driving, converse, disobey and accident.

5 Dataset application analysis

In this section, some potential applications of our dataset will be discussed. There are various types of data in our dataset which we hope can support the automated driving research in the future.

5.1 Road condition analysis in China

5.1.1 Radius of the road curvature

Road curvature can be estimated by the lane or the boundary of the road in the image, which labelled as several segments defined as [(*x1*, *y1*), (*x2*, *y2*)] in the image coordinate system [19]. All segments are classified based on whether they belong to the same lane. After warping the image to the birds-eye-view image, it is necessary to interpolate some points between (*x1*, *y1*) and (*x2*, *y2*) to avoid under-fitting in the second order polynomial fitting. Finally, road curvature radius can be calculated by fitted formula.

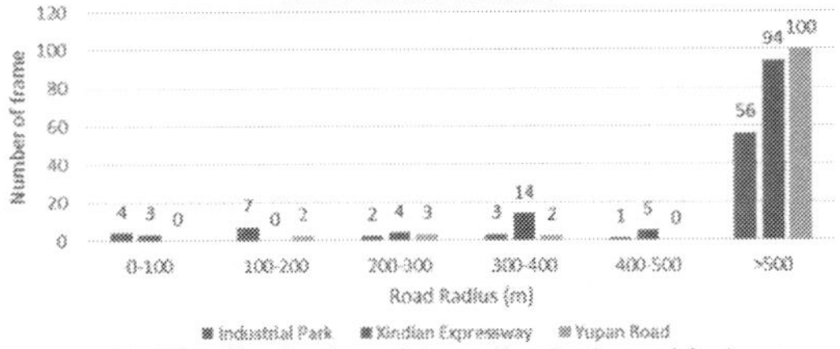

Figure 11. The distribution of the radius in three cities' scenes.

Three different scenes are selected to show road radius. Comparing the distribution of radius, Suzhou Yupan road is the straightest followed by Luoyang Xindian expressway and Beijing industrial park, as shown in Figure 11.

This example proposes a simple method to fit road lane and calculate road curvature, which shows that our dataset can support the research on road environment perception for AD, such as the research of lane detection model.

5.1.2 Lane width statistics

In the collected images, we label a straight line in the lane from the left to the right. The width of lane W can be estimated by $W=wd/f$, where w is the length of the labelled line, d is the distance from the camera to the position of labelled line and f is the focal length. The distance d can be estimated by $d=Hf/l$, where H is the height of the camera and l is the vertical distance between the labelled line and the vanishing point.

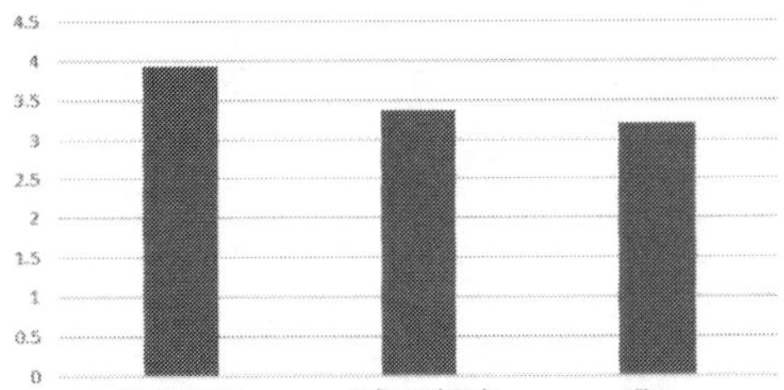

Figure 12. The average width of lane in three scenes.

Typically, the width of lane in China is between 3.25m and 3.75m. But it may have some changes in different cities, different scenes or different types of roads. In our statistics, the lane width of expressway is the largest, as given in Figure 12.

Lane width is an important factor, as varied as the scenes, in the lateral control as the lane helps drivers operate vehicles closer to the center of the lane [20]. This example implies the collected data can support the research on lateral control and decision for AD.

5.2 Vehicle – following distance

Vehicle-following distance is an important factor to evaluate the traffic and the driving manner. Statistics is focused on the time gap [21] calculated by the velocity of self-vehicle and the distance from the front labelled vehicle, which can be obtained from CAN data and estimated by the method in Section 5.1.2 separately.

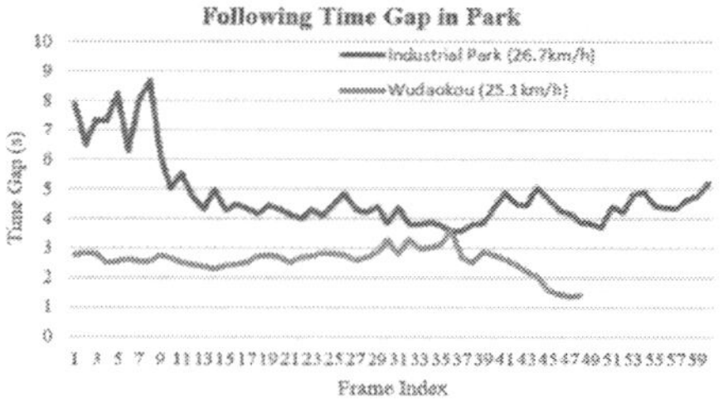

Figure 13. The time gap curves in two scenes.

Two Beijing scenes are selected to show vehicle-following characteristic. In Wudaokou, our vehicle followed the front vehicle until reaching the intersection. In the industrial park, the front vehicle is far away from our vehicle at the beginning. The driver tried to catch up with it and keep a safe following distance. The change of the time gap is shown in Figure 13.

By comparison, although these two scenes have similar average speed, the following time gaps are different. In the industrial park, the time gap is nearly two times of that in the Wudaokou. This result can reveal the strategies for vehicle following would vary in varied scenes. For vehicle-following research, we should consider not only the related distance and velocity, but also the scene types.

5.3 Road user behavior analysis in China

5.3.1 The actions of violation

Not only the objects are labelled in the dataset, but also some actions of violation are listed in Table 4. We select some scenes and count the violation in Figure 14.

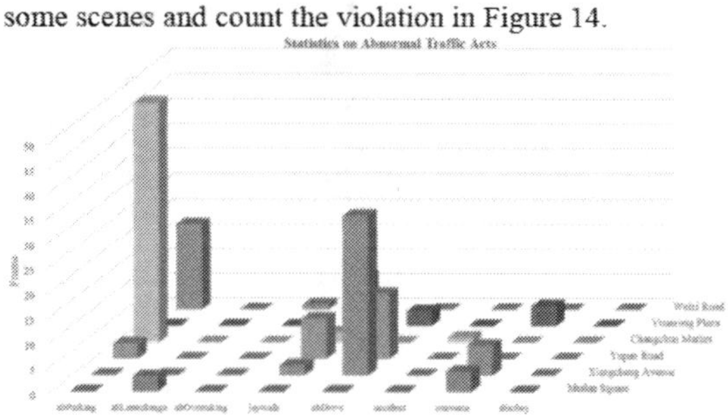

Figure 14. Statistics on abnormal traffic acts.

When conducting research on AD, some violations are worth considering. It can help researchers better understand the real situation in China. In the statistics, illegal parking is the most common violation in China followed by driving in other lanes, jaywalking and converse driving, especially on the road without digital surveillance (e.g. Changchun Market).

5.3.2 Driving manner in expressway

The dataset can also be used to analyze driving manner of different drivers. We selected Suzhou Su-Shao expressway to analyze driving manner of three drivers:

Table 5. Information of three drivers.

Driver	Gender	Miles/Year	Self-estimated driving manner
Driver A	Female	10000km	Passionate
Driver B	Male	8000km	Moderate
Driver C	Male	300km	Moderate

Driving velocity is used to demonstrate the driver manner, which can be extracted from CAN data. Figure 15 shows the change of velocity for each driver -- Driver A keeps velocity under 90km/h all the way, showing more consistence although her self-estimated driving manner is passionate; Whereas, driver B seems to be more passionate as his driving velocity keeps around 110km/h for a long time; Compared with driver B, driver C shows different driving manner, as his velocity was relatively smaller-in-amplitude but fluctuated frequently. This may be due to lack of driving experience as he only drives 300km per year on average.

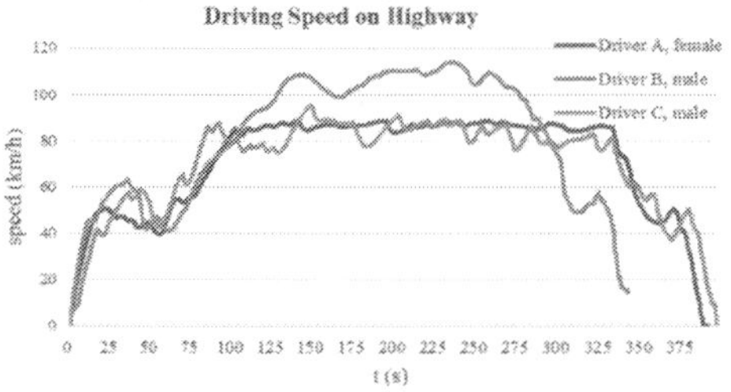

Figure 15. The velocity curves of three drivers.

Learning from collected data for decision-making is very effective, such as using deep learning or some anthropomorphic algorithms. However, driver characteristics are complex factors and how to model them is also a challenge [23]. This example to some extent reflects that not all drivers' data would be appropriate in that some may be not representative. Therefore, it is necessary to choose excellent drivers for data collection to provide typical driving data.

6 Conclusion and future work

In this paper, the whole process of the establishment of an automated driving benchmark in China is introduced, ranging from layout of sensors, scene design to labelling objects selection. Furthermore, from specific scenarios of 7,000 labelled frames, both road features and naturalistic driving manners are analyzed. Not only road parameters could be extended to other potential automated driving scene information, but also naturalistic driving manners might be utilized as a learning target for automated vehicle. With all the methodologies and the datasets yielded, researches on automated driving can acquire more supports.

Concretely, based on the work above, at least following researches can be carried on -- moving target detection, e.g. car, pedestrian and other VRUs; road marking recognition, like lane markers, traffic lights and signs; road segmentation based on boundary and lane line labelling; depth information prediction with point cloud data or multi-sensor fusion; identification and study of drivers' intention based on CAN data from different drivers' data. Moreover, the underlying research of risk assessment, automated vehicle decision and control will be in progress.

7 Acknowledgment

This research is supported by the project *NO.20173000179* from Tsinghua and NISSAN Joint Research Center for Intelligent Mobility. The authors would like to thank all partners within for their cooperation and valuable contribution.

References

1. A. Geiger, P. Lenz, C. Stiller, R. Urtasun. Vision meets robotics: The KITTI dataset. International Journal of Robotics Research, **32(11)**, 1231-1237 (2013)
2. J. Ziegler, P. Bender, M. Schreiber, et al. Making Bertha Drive — An Autonomous Journey on a Historic Route. IEEE Intell. Transport. Syst. Mag., **6(2)**, 8-20 (2014)
3. A. Rangesh, K. Yuen, R. K. Satzoda, et al. A Multimodal, Full-Surround Vehicular Testbed for Naturalistic Studies and Benchmarking: Design, Calibration and Deployment. arXiv preprint arXiv: 1709.07502. (2017)
4. B. Romera-Paredes, P. H. S. Torr. Recurrent Instance Segmentation. *European Conference on Computer Vision*, Springer, Cham, 312-329 (2016)
5. C. Stiller, F. P. León, M.Kruse. Information fusion for automotive applications – An overview. Information Fusion, **12(4)**, 244-252 (2011)
6. J. Suh, H. Chae, K. Yi. Stochastic Model Predictive Control for Lane Change Decision of Automated Driving Vehicles. IEEE Transactions on Vehicular Technology, **67(6)**, 4711 (2018)
7. D. Silver, J. A. Bagnell, A. Stentz, Learning Autonomous Driving Styles and Maneuvers from Expert Demonstration. *Experimental Robotics*, Springer, Heidelberg, 371-386 (2013)
8. M. Quigley, K. Conley, B. Gerkey, et al. ROS: An open-source Robot Operating System. *ICRA workshop on open source software*, **3(3.2)**, 5 (2009)
9. M. Veľas, M. Španěl, Z. Materna, et al. Calibration of RGB Camera with Velodyne LiDAR. In *Proc. Conference on Computer Graphics, Visualization and Computer Vision* (2014)
10. A. Geiger, P. Lenz, R. Urtasun. Are we ready for autonomous driving? The KITTI vision benchmark suite. *IEEE Conference on CVPR*, 3354-3361 (2012)
11. W. Maddern, G. Pascoe, C. Linegar, et al. 1 year, 1000 km: The Oxford RobotCar dataset. The International Journal of Robotics Research, **36(1)**, 3-15 (2017)
12. M. Cordts, M. Omran, S. Ramos, et al. The Cityscapes Dataset for Semantic Urban Scene Understanding. In *Proc. IEEE Conference on Computer Vision and Pattern Recognition*, 3213-3223 (2016)
13. E. Santana, G. Hotz. Learning a Driving Simulator. arXiv preprint arXiv: 1608.01230 (2016).
14. X. Li, F. Flohr, Y. Yang, et al. 2016. A new benchmark for vision-based cyclist detection. In *Proc. Intelligent Vehicles Symposium*, 1028-1033 (2016)
15. V. Badrinarayanan, A. Kendall, R. Cipolla. Segnet: A deep convolutional encoder-decoder architecture for image segmentation. IEEE Trans. on PAMI, **39(12)**, 2481-2495 (2017)
16. Y. Chen, X. Yang, B. Zhong, et al. CNNTracker: Online Discriminative Object Tracking via Deep Convolutional Neural Network. Applied Soft Computing, **38**, 1088-1098 (2016)
17. R. Quintero, I. Parra, D. F. Llorca, et al. Pedestrian Intention and Pose Prediction through Dynamical Models and Behaviour Classification. In *Intelligent Transportation Systems (ITSC), IEEE 18th International Conference on*, 83-88 (2015)
18. D. Dominic, S. Chhawri, R. M. Eustice, D. Ma. Risk Assessment for Cooperative Automated Driving. In Proc. *2nd ACM Workshop on Cyber-Physical Systems Security and Privacy ACM*, 47-58 (2016)
19. R. Song, H. Chen, X. Xiao, et al. Lane Detection Algorithm Based on Geometric Moment Sampling. Scientia Sinica, **47(4)** , 455-467 (2017)
20. S. Mecheri, F. Rosey, R. Lobjois. The effects of lane width, shoulder width, and road cross-sectional reallocation on drivers' behavioral adaptations. Accident Analysis & Prevention, **104**, 65-73 (2017)
21. A. Tordeux, S. Lassarre, M. Roussignol. An adaptive time gap car-following model. Transportation Research Part B Methodological, **44(8)**, 1115-1131 (2010)
22. D. Gaither, H. Do, B. R. Bryant. Toward Detection of Abnormal Behaviors in Timing and Security Requirements. In *Proc. 24th Asia-Pacific Software Engineering Conference Society (APSEC)*, IEEE, 707-712 (2017)
23. K. C. Dey, L. Yan, X. Wang, et al. 2016. A Review of Communication, Driver Characteristics, and Controls Aspects of Cooperative Adaptive Cruise Control (CACC). IEEE Intell. Transport. Syst. Mag., **17(2)**, 491-509 (2016)

Analysis on current situation of China's intelligent connected vehicle road test regulations

Bowei Zou , Wenqiang Li and Danni Wang

China Automotive Technology and Research Center Co., Ltd., Tianjin, China

Abstract. This paper first elaborated the technological development status of the intelligent connected vehicles (ICVs) and sorted out the "intelligent connected vehicle road test management regulations" jointly formulated by the Ministry of Industry and Information Technology (MIIT), the Ministry of Public Security and the Ministry of Transport, and related regulations for major cities in China. The domestic is compare with foreign ICVs road test status and policy. The differences between the national regulation and municipal regulations for the test subjects, test drivers and test vehicles are analysed and enumerated. By comparing the domestic and international road test regulations, it is concluded that the safety is the prerequisite of vehicle road test. It is compulsory to conduct the closed field testing before conducting public road tests.

1 INTRODUCTION

The Chinese government attaches great importance to the development of the intelligent connected automobile industry. "China Manufacturing 2025" has listed the smart Intelligent and Connected Vehicle (ICV) as one of the ten key development areas, and clearly promoted the development of the smart intelligent Connected Vehicle to the national strategic height as a standard of market behaviour, leading the foundation for the healthy development of automobile industry. With the maturity of domestic intelligent connected technology and the rapid expansion of market scale, the development of ICV has become the primary issue that the government needs to face. In the past two years, all circles of industry, academia and research institutes and governments have focused on this new technology and sought breakthroughs from all aspects, including research and development of new technologies, improvement of the top-level design of intelligent network- connected technologies, and development of related policies and standards

According to the road map for energy-saving and new energy vehicles issued by the Chinese Society of Automotive Engineers, intelligent and connected vehicles has been clearly defined. Intelligent and Connected Vehicles are equipped with advanced onboard sensors, controllers, actuators and other devices, and the integration of modern communications and network technology, to achieve intelligent information exchange and sharing between vehicle and X (cars, roads, people, clouds, etc.), with complex environmental awareness, intelligent decision-making, collaborative control and other functions, can achieve safety, high-efficiency, comfort, energy saving driving, and finally realize a new generation of cars to replace people to operate [1]. Intelligent Connected Vehicle can provide safer, more energy-efficient, environmentally friendly, and more convenient travel modes and integrated solutions, which is an internationally recognized future development direction and focus of attention [2]. Road testing is an indispensable link in the development and application of intelligent connected vehicle technology. The autopilot function requires a large amount of tests and verifications, and must be fully tested in a real traffic environment before it is officially introduced to the market.

2 Intelligent connected vehicle industry technology overview

According to the National Intelligent Connected Technology Roadmap for Automotive Technologies, the market share of vehicles with driving assistance (DA) and partial autopilot (PA) in China will be about 50% from 2015 to 2020; the occupancy of DA and PA vehicles will be maintained from 2020 to 2025. The percentage of stable, highly automated (HA) vehicles is about 15%; by 2025-2030, the market share of fully automated (FA) vehicles is close to10%. Through "three horizontal and two longitudinal", we realize the vision of safety, efficiency, energy saving, emission reduction, comfort, convenience and humanity. [1]

The "three horizontal and two vertical" technical architecture not only summarizes the key technical conditions of the intelligent networked vehicle itself, but also describes the external infrastructure of the vehicle and the vehicle platform, as shown in Figure 1.

https://doi.org/10.1051/mateconf/201925902003

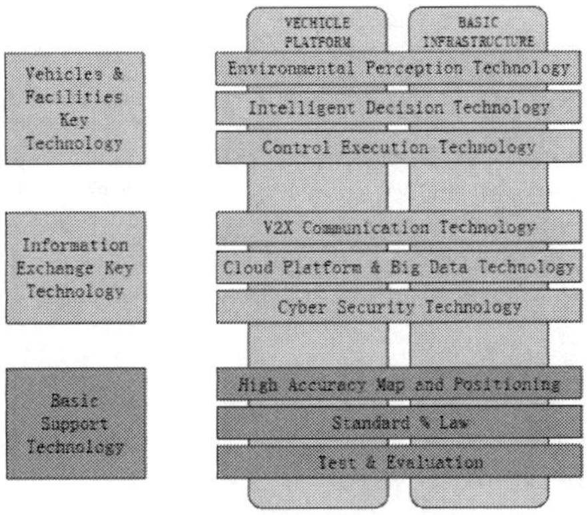

Figure 1. Three horizontal and two vertical technical architecture.

The environment-aware system uses vehicle sensors to detect and calculate the vehicle environment, which mainly consists of machine vision recognition system, radar system, ultrasonic sensors and infrared sensors. Machine vision has a wide range of detection and large information capacity, but it has a large amount of data to process and is widely used in the research field of intelligent vehicles. The millimeter wave radar system can obtain the target distance, speed, azimuth, etc., and the operation is in good performance of real-time, and the measurement range is far. However, there is no concrete and accurate description of the appearance of the detected object. The laser radar can just make up for the insufficiency of the millimeter-wave radar. The measurement accuracy is high and the speed range is high, but the measurement distance is relatively short. In addition, the cost of laser radar in the market is high, especially multi-line laser radar. In order to maximize the advantages of various types of sensors and cost reduction, many R&D institutions use sensor fusion solutions. For instance, HESAITECH and BAIDU Apollo released a "Lidar + camera" integrated sensor can provide multi-sensor fusion and perception algorithms, and finally in the form of sensor fusion suite (shown in Figure 2).

Figure 2. The effect of sensors fusion.

The intelligent decision system makes use of the sensor's traffic environment and high-precision maps and positioning systems to make decisions on the vehicle's driving path and vehicle movement posture. Currently used decision-making methods include state machines, decision trees, deep learning and enhanced learning [3-5]. The control execution system is responsible for executing the instructions of the intelligent decision system, acting on the vehicle power, braking and steering systems to follow the planned paths. At present, the main control system technologies include PID control (proportion-integral-derivative), sliding mode control, fuzzy control, model predictive control, adaptive control, and robust control.

The vehicle-mounted communication technology will adopt LTE-V2X technology in the future. Compared with the traditional V2X technology, LTE-V2X has a wider propagation distance, shorter delay, and greater communication capacity on the one hand, especially the reliability of multi-vehicle communication in high-density scenarios has been greatly improved, solving the problem of the limited capacity of multi-vehicle scenarios in the traditional way. On the other hand, LTE-V2X is able to freely connect to the Internet anywhere and anytime with the help of an optimized mobile broadband network. On the other hand, LTE-V2X is able to freely connect to the Internet anywhere and anytime with the help of an optimized mobile broadband network, and to provide cloud network collaborative data capabilities and implement applications such as traffic monitoring, remote alarms, and information push. Currently, many automotive industries have carried out extensive vehicle networking (LTE-V2X) test validation to promote the maturity of technology and rapid development of the industry.

In addition, automotive information security systems have become an important area of development for the automotive industry. Traditional vehicle information framework is generally divided into body control system, entertainment information systems, powertrain and chassis safety systems. The high-speed CAN network is used for the powertrain and chassis safety system, the medium-speed CAN network is used for the body control and entertainment information system, and MOST network with higher communication speed is used in the entertainment system of some in-vehicle networks. Some intelligent execution modules use low-speed LIN network connection, High-speed CAN and medium-speed CAN network communicate through the gateway. Traditional vehicle information system is relatively closed, which does not need to ensure the security of information technology to ensure the reliability of vehicle network systems [10].

The port of the vehicle information can be divided into three parts according to the functional classification. The first part is the basic control function of the vehicle, including the power control, steering control and energy control of the vehicle. The second part of the auxiliary driving function of the vehicle, which is the future of intelligent vehicles, the most important part of the intelligent driving class with the development of the vehicle content and constantly expand the content of the current industry generally recognized intelligent driving vehicle grading using the United States the SAE classification method is classified into six grades. The third part of the vehicle infotainment function, consumers, especially Asian consumers believe that Internet entertainment is the basic function of the car, smart phones and wearable equipment can be perfectly integrated with the car, the current specific products such as music downloads, restaurant recommendations and other functions; The car will serve as a mobile WI-FI hotspot. The fourth part is the V2X function part, which is the main function of the intelligent network vehicle. It is mainly used to realize the mutual communication between vehicles (V2V), vehicle and transportation infrastructure (V2I) communication and vehicle cloud communication (V2C) [11].

3 Typeset text the status of foreign intelligent network road testing in foreign countries

3.1 United states of america

The U.S. Federal Government has issued policy documents such as the Autopilot Car Policy Guide and Automated Driving System 2.0: Security Vision. The House of Representatives has also passed the Automated Driving Act; more than a dozen U.S. states have opened public road tests, such as California, which has begun to allow driverless tests on public road. As of the end of 2017, Washington, DC, and 21 states have passed legislation on autonomous vehicles. At present, the two largest piloted demonstration areas in the United States are located in Detroit in the east and Silicon Valley in the west. The M-City is the world's first area for intelligent vehicles and V2X vehicles test.

3.2 Europe union countries

In 2014, the European Union launched a "Adaptive" (smart vehicle automated driving application and technology) project with more than a dozen automobile manufacturers and parts suppliers in Europe, aiming to develop partial or complete driving on urban roads and highways. The European Union has issued a series of policies and a self-driving roadmap to promote the research, development, and application of intelligent connected vehicles, and to guide the development of the intelligent connected automotive industry in member countries. Germany amends current road traffic laws and regulations to allow highly autonomous or fully automated driving systems to replace autonomous driving by humans and has specifically developed test sections on highways. In May 2017, Sweden issued a decree on the testing of self-driving cars, which is reviewed and conditionally granted by the Swedish Transport Agency.

3.3 Japan

The Japanese government has actively played an inter-departmental synergy to promote the implementation of the intelligent connected vehicle project. In May 2016, Japan's IT comprehensive strategy headquarters established the road map for autonomous driving, which will allow unmanned passenger vehicles to be used in some areas in 2020. In the same period, the Japanese National Police Agency issued the Guidelines for Road Testing of Self-driving Vehicles. It also initiated the revision of the "Road Traffic Law" and "Road Transport Vehicles Law". In April 2017, the government concluded motor traffic accidents during autopilot as a target for auto insurance. In June 2017, the Japanese Police Agency issued the "Standard for Processing Permissions for Road Test of a Remote Autopilot System" to allow vehicles to perform on-the-spot tests while driving without a driver.

3.4 United nations

In March 2016, the United Nations also formally revised the "Convention on Road Traffic in Vienna" to allow automatic driving technology to be used in transportation, and it has also swiftly established technical regulations related to autonomous driving technology. Singapore set up a self-driving car motion committee to supervise the research and testing of self-driving cars. In early 2017, Singapore's automated driving shuttle bus service started operations. At the end of 2017, an automated driving test route was opened to allow self-driving cars to carry out public road tests within the permitted range.

4 Sections analysis of china's icv road test policie

This year, the Ministry of Industry and Information Technology, the Ministry of Public Security, and the Ministry of Transport jointly issued the "Intelligent Networked Vehicle Road Test Management Regulations (Trial)", in which make a series of requirements on test applications, audits and test management subjects, test drivers and test vehicles. Before this, Beijing, Shanghai, Shenzhen, Chongqing and other places have successively issued guidance documents for ICVs on-the-road testing. Subsequently, Changsha and Changchun cities have also issued ICVs road test management regulatory documents.

4.1 Test management agencies and responsibilities

In the national policy, it is defined that the provinces, autonomous regions, municipalities directly under the Central Government and separately listed cities, the Xinjiang Industry and Construction Corps' industrial and informatization administrative departments, the public security traffic management departments and the transportation and transportation departments as the management department.

In cities that have already issued road test management methods, the local municipal Economic and Credit Commission, the Municipal Public Security Bureau, and the Municipal Communications Bureau have established relevant working groups. For instance, Shanghai and Chongqing established the advancing working group, and Beijing, Changsha, Chongqing and Shenzhen established joint working group. Although the names are different, the nature and duties of the working group are basically the same.

The basic responsibilities of the working group are as follows

1. Review the company's test application
2. Issuing test notices
3. Organize relevant assessment work
4. Matters related to the implementation of the coordination method

The responsibilities of expert Committee are to regularly convene an expert group review meeting to demonstrate the road test application submitted by the test subject and finally provide an expert group opinion.

The third-party institute is responsible for the whole process supervision and closed testing of intelligent networked automotive, including application acceptance, material review, test tracking, data collection and analysis, and supervision and management of testing. Third-party agencies are the specific implementing agencies of the regulatory agencies, but not all cities have yet identified specific units.

4.2 Requirements of test subjects, test drivers and test vehicle

4.2.1 Test subjects:

In the national management specification, the seven point requirements are put forward for the test unit's unit character, business scope, accident compensation ability, test evaluation ability, remote monitoring ability, event record analysis ability and compliance with laws and regulations, as shown below:

•Independent legal entity registered in the territory of the People's Republic of China.

•Competence of intelligent ICV-related businesses, such as automobile and parts manufacturing, technology research and development, or testing and inspecting.

•Sufficient civil compensation ability for the loss of personal and property that may be caused during the test of the intelligent networked car.

•Having intelligent network connected auto driving function test evaluation procedures.

•Ability to perform real-time remote monitoring of test vehicles.

•Ability to record, analyze, and reproduce test vehicle events.

•Other provisions stipulated by laws, regulations and rules [12-16]

The requirements of test subjects in these six cities' local management methods are different from those in the national management standards. The management methods in Shanghai and Changchun directly refer to the national regulations. See the Table I below.

Table 1. Table captions should be placed above the table.

City	Requirement
Shanghai	Same as national regulations
Beijing	Only make explicit requirements for the nature of the independent legal entity of the test subject; at the same time, the test subject is required to promise to install the supervision device and accept the daily supervision of the third-party authorized institution.
Shenzhen	There are only three requirements for the test subjects. Their requirements are the same as those of the State's "Management Regulations," and their expressions are slightly different.
Changsha	On the basis of the seven articles of the "Regulations for Management" of the country, an article was added to the test subject "Need to purchase a certain amount of production safety liability insurance."
Changchun	Same as national regulations
Chongqing	There is only one requirement on drive subject that test subject is a complete vehicle company, a modified vehicle manufacturer, a parts and components company, an electronic information enterprise, a research institute/college, a transportation company, or other technology-based enterprise that has autonomous driving technology and product development or production capabilities.

4.2.2 Test driver and test vehicles:

Test driver and test vehicles: the test drivers should sign labor contracts or service contracts, pass autopilot training, no major traffic violation records, etc. The test vehicles should be registered and pass mandatory inspection, and the vehicle should have man-machine control mode conversion and data recording function.

Table 2. The difference of requirements on test driver and vehicles of 6 cities.

City	Requirement
Shanghai	1. There is no clear requirement for signing labor contracts with test subjects 2. Add the requirement that test driver's experience in autopilot system operation with more than 50 hours, of which more than 40 hours apply for the test project's driving experience
Beijing	Comparing with national management regulations, add the requirement that the test driver should take over the test vehicle under dangerous scene conditions operating time more than 50 hours
Shenzhen	The test driver should meet basic requirements" without clear rules
Changsha	Same as national regulations
Changchun	There are two major differences with respect to the national regulations for management. The following requirements are added: 1. Severe traffic violations such as non-driving passenger vehicles over the past year 2. Testing the driver's operation of more than 50 hours of smart network system certification (should include more than 40 hours specified smart network connected car test scenarios driving example)
Chongqing	There are two major differences with respect to the national regulations for management. The following requirements are added: 1. The test driver shall undergo professional training in road testing for not less than 50 hours, including professional training for

https://doi.org/10.1051/matecconf/201925902003

City	Requirement
	taking over test vehicle control in dangerous scenarios, where the driving management training time for the requested test function is not less than 40 hours. 2. Test drivers must fully understand the automatic test vehicle performance and test driving requirements

In the requirements for test vehicles, test safety is regarded as the primary requirement in both national and local test management regulations. It is mainly reflected in the fact that the test vehicle must meet mandatory inspection project requirements, and it must not reduce vehicle safety performance. Furthermore, it must have manual operation and autopilot modes, which must be ensured that the driver can immediately switch the manual mode in dangerous situations. In addition, in order to ensure the safety of public road tests, Shanghai, Beijing and Changsha also require that vehicles be equipped with devices that provide reminder functions. When an autopilot system fails to act, the device should immediately remind the test driver to take over the vehicle.

Another important requirement is that the test vehicle must have monitoring equipment installed to monitor the vehicle status and to store records. This requirement is the same as the national regulation in the specifications of the three cities of Changchun, Changsha and Shanghai. The stored data includes the vehicle control mode, vehicle position, and vehicle motion status, so that accident analysis and processing can be performed in the event of a traffic accident.

4.3 Test application and audit requirements

In order to strengthen the normative participation in the road testing enterprise's behavior, the national and local management requirements separately request the company to apply for the testing process, mainly to request the integrity of the application materials. According to the specifications of the test subjects, test vehicles and test drivers, it is required to submit a series of materials to prove the technical ability, security and responsibility of test subjects and vehicles. In terms of technical capabilities, companies need to provide certification materials for test subjects who had carried out actual vehicle tests in certain areas such as closed roads and venues, and the test reports of autopilot functions issued by third-party test agencies that are recognized by the national or provincial authorities for conducting automotive-related business. Test programs including test road sections, test times, test items, test procedures, risk analysis, and response measures are also needed.

4.4 Test Management Requirements

In order to strengthen the road traffic management and safety supervision during the testing process, the "Management Rules" also put forward requirements for the test parties to carry notifications, apparent target indications, takeover operation, vehicle transitions, and reporting test summaries during the testing process. Furthermore, the situation of provincial and municipal government should cancel the test notice and recall the temporary number plate is defined.

5 Summary

Intelligent connected vehicles (ICVs) can provide safer, more energy-efficient, environmentally friendly, and more comfortable travel modes and integrated solutions. They are an important part of the urban intelligent transportation system and a core element in building a green car society. Its significance lies not only in the upgrading of automotive products and technologies, but also in the reshaping of the entire industry and value chain system in automobiles and related industries. The implementation of intelligent network-linked public road testing will help China's intelligent network-linked automotive technology upgrade and product promotion. By comparing and analyzing domestic and foreign road test regulations, the OEMs should conduct abundant of closed field test before conducting public road test. Furthermore, the OEMs should also have the capability of assuring the safety of vulnerable group and compensation capacity when traffic accidents happen. This paper provides a brief description of current technology status and policy of intelligent connected vehicles, and analyzed and compared the public road test regulations of six major cities in China, aiming at promoting the comprehensive technology capability of ICVs and leading to the right direction of development of ICVs in China.

References

1. China-SAE. "Technology roadmap for energy saving and new energy vehicles [M]," Beijing: Mechanical Industry Press, 2016. (in Chinese)
2. "State-of-the-art and technical trends of intelligent and connected vehicles [J]." Automotive Safety and Energy, 2017, Vol. 8 No. 1.
3. Li S, Peng H. "Strategies to minimize the fuel consumption of passenger cars during car-following scenarios [J]." Proc Instit Mech Eng, Part D: J Automobile Eng, 2012, 226: 419-429.
4. Barkenbus, J. "Eco-driving: an overlooked climatechange initiative [J]." Energy Policy, 2010, 38: 762–769.
5. Zhou J, Peng H. "Range policy of adaptivecruise control vehicles for improved flow stabilityand string stability."
6. Li Shengbo, Li Keqiang, Wang Jianqiang, et al. "Nonsingular and fast terminal sliding mode control method [J]." Info & Contr, 2009, 38(1): 1-8. (in Chinese).
7. Wang Y, Bin Y, Li K. "Longitudinal acceleration tracking control of low speed heavy-duty vehicles [J]." Tsinghua Sci& Tech, 2008, 13(5): 636-643.
8. Hellstrom E, Ivarsson M. "Look-ahead control for heavy trucks to minimize trip time and fuel

consumption [J]." Contr Eng Practice, 2008, 17(2): 245-254.
9. Ferrara A, Pisu P. "Minimum sensor second-order sliding mode longitudinal control of passenger vehicles [J]." IEEE Trans Intell Transp Syst, 2004, 5(1): 20-32.
10. Luo Feng, Su Jian, Yuan Dahong. "Vehicle electronic networks and bus standards [J]." Automobile Engineering, 2003, 2594):372-376. (in Chinese).
11. Zou Bowei, Gao Mingqiu, Cui Xiaochuan. "Research on Information Security Framework of Intelligent Connected Vehicle." ICCSP 2017. Wuhan, China, pp. 91-65.
12. Shanghai Intelligent Network Linked Vehicle Road Test Management Measures (Trial). (in Chinese).
13. Chongqing Autopilot Road Test Management Regulations. (in Chinese).
14. Beijing Municipality Automated Vehicle Road Test Management Regulations (Trial) . (in Chinese).
15. Changchun Intelligent Network Linked Vehicle Road Test Management Measures. (in Chinese).
16. Intelligent Networked Vehicle Road Test Management Specification (Trial).

MATEC Web of Conferences **259**, 02004 (2019) https://doi.org/10.1051/matecconf/201925902004
ICTLE 2018

Signal Timing Optimization Model of Urban Road Intersection Based on Multi-factor

Qiuping Wang, Kai Wang, Shuaiqi Zhou, Qiongjie Shi, Qi zhang and Hao Sun
University of Architecture &Technology, China, Xi'an

Abstract: As the congestion point of urban road network, it is critical to keep the intersection traffic clear. The model was built targeting on decreasing vehicle delays and vehicle emissions, taking traffic queens,stopping time,road capacity and capacity utilization as verification indicators with two constraint conditions which are cycle time length and green ratio by means of genetic algorithm. And the dynamic vehicle emissions rate was taken into account. According to the data of Ke Ji Road—Feng Huinan Road intersection and Tong Yi Road—Tong De Road intersection, which show that the method is practical, and traffic queens,stopping time,vehicle delays,vehicle emissions decrease 34.03%,28.79%,48.73% and 28.04% at most,road capacity and capacity utilization increase 15.67% and 7.74% at most.

1 Introduction

As the congestion point of urban traffic network, traffic problems at the intersection are becoming more and more serious, how to maintain the smooth traffic at the intersection is an urgent problem to be solved in urban traffic. At present, signal control is widely used to manage traffic flow in order to reduce streamline interference and improve the operating environment of intersections.

Many domestic and foreign scholars have studied the signal timing method at the intersection, Yuliu and others[1] set up a signal timing optimization model aimed at reducing emissions and average delays, and developed a corresponding algorithm program for the model; From the economic point of view, Wang Yuquan and others[2]have established a cross-intersection signal cycle selection model with the goal of the lowest cost of travel per vehicle; Wang Qiuping[3] established a nonlinear function model for signal timing optimization in urban single-intersection intersections with the shortest average delay time and the average number of parking minimum as the goal and phase effective green light time, saturation and period lengthas the constraint; Based on the exhaust emissions of motor vehicles, Zhou Shenpei[4]have established atwo-layer multi-objective optimization model for signal control of intersections; Zhang Yingying[5] studied the characteristics of exhaust emissions under different signal timing schemes; Unal[6] found that the signal timing scheme will affect the emissions of motor vehicles at the intersection. A reasonable signal timing plan will effectively reduce vehicle exhaust emissions. For the study of signal timing methods, you can also refer to the literature [7-16]. In summary, there have been few studies that have comprehensively considered delays, emissions, queue lengths, number of stops, capacity, capacity utilization, and other factors from the perspective of comprehensive benefits.We generally use fixed values for the emission rate of motor vehicles and have neglected the relation between exhaust emission rate and idle time, It causes the signal timing plan to be unreasonable. Therefore, this article introduces dynamic vehicle exhaust emission rates and established a signal timing optimization model for urban road intersections based on comprehensive benefits under multi-factor conditions. And this article combined the different types of intersections to verify the model. The research results have a certain significance for improving the signal timing theory and coordinating the urban traffic network.

2 The implementation target of signal timing optimization at urban road intersections

(1) Reduce the queue length of the vehicle: The longer the vehicle queues is, the longer it takes for the vehicle to cross the intersection. Therefore, when the signal timing is optimized, the traffic congestion can be reduced by reducing the queue length to improve the service level of the intersection.

(2) Reduce the number of vehicle stops: When the traffic passing through the intersection is more continuous, there is less parking. On the contrary, the parking frequency is relatively high.Considering that the continuous driving environment is easily accepted by the drive, the optimization goal can be characterized asthe number of vehicles passing through the intersection is significantly reduced.

(3) Improve the capacity of intersections: Capacity is one of the important indicators that affect the service level of intersections, In general, the greater the capacity is, the greater the number of vehicles passing through the intersection per unit of time are, the smoother the traffic flow is, and the more conducive to reducing urban traffic pressure. In order to avoid the urban road intersection due to the traffic capacity too small to affect the operation of the regional traffic, We should improve traffic capacity when optimizing signal timing.

(4) Improve utilization of traffic capacity at intersections:Intersection capacity needs to be adapted to actual vehicle traffic conditions, otherwise it will cause waste of resources. We introduced the utilization of intersection capacity to assess the use of intersections here. Improving the utilization rate of intersection capacity to improve the use efficiency of intersections.

(5) Reduce the average delay of vehicles: Drivers often judge the use of intersections with average delays. Average delay is inversely proportional to service level,the greater the delay, the lower the level of service at the intersection, the less easily accepted by the drivers. When the delay is reduced, the operating efficiency of the intersection is greatly improved.

(6) Reduce vehicle exhaust emissions: Exhaust, as an important part of atmospheric pollution, can be reduced through traffic management. There are a large number of urban road traffic network intersections, unreasonable signal timing solution increases vehicle emissions at Intersections. we select CO and HC emissions as the optimization target here to minimize exhaust emissions.

After considering the above indicators, this paper uses the reduction of the queue length, the number of stoppages, the improvement of traffic capacity and the utilization of traffic capacity as verification indicators to reduce average delays and reduce emissions as an optimization goal. In order to achieve the rationalization of the signal timing scheme for Urban Road intersection,this paper uses the cycle time length and the green time ratio to contact the verification index and the optimization target.

3 Signal timing optimization model for urban road intersections

3.1 Verification indicator

(1) The length of queue

The queuing length of each lane at the beginning of the green light is the sum of the remaining number of vehicles Q1 in the previous green time and the number of vehicles Q2 arriving in the red time. [17] which is:

When saturation x_{ij}>0.5:

$$Q_1 = 0.25cap[(x_{ij}-1)+\sqrt{(x_{ij}-1)^2+\frac{8(x_{ij}-0.5)}{cap}}] \quad (1)$$

otherwiseQ_1=0.

$$Q_2 = T\frac{q_{ij}(1-\lambda_i)}{3600(1-\lambda_i x_{ij})} \quad (2)$$

$$Q_m = Q_1 + Q_2 \quad (3)$$

The average queuing length $Q_j$of the j-th inlet lane is the average of the average queuing length Q_m for each lane of the inlet lane, the average queue length Q_{ave}at the intersection is the average of the average queue length Q_j for each entrance lane. The goal of signal timing optimization is to reduce the average queue length Q_{ave}, which is:

$$MinQ_{ave} = Min(\sum_j Q_j / Maxj) = Min[\sum_j(\sum_m Q_m / Maxm)/Maxj] \quad (4)$$

In Formula(1)~Formula(4), *cap is* the lane capacity, pcu·h^{-1}; λ_i is the green time ratio of the i-th phase; x_{ij} is the saturation of the j-th phase of the i-th phase; *T is* the cycle length, s.

(2)Parking times

The number of parking times for the i-th phase can be expressed as[17]:

$$h_i = \sum_i 0.09\frac{T-g_{ei}}{1-y_{ij}} \quad (5)$$

The average number of stops at an interperiodic intersection is calculated by the weighted evaluation of the actual traffic volume of each phase, sothe objective function to reduce the number of stops is:

$$Minh = Min\frac{\sum_i h_i \cdot q_i}{\sum_i q_i} \quad (6)$$

In Formula(5)~Formula(6), $y_{ij}$is the flow ratio of the j-th inlet of the i-th phase; $g_{ei}$is the effective green time of the i-th phase, s; q_i is the sum of the actual arrival-equivalent traffic volume for each i-th phase ingress road, pcu·h^{-1}.

(3)Traffic capacity

Intersection capacity can be weighted according to each entrance, Therefore, the objective function for improving the capacity of intersections is[18]:

$$MaxCAP = Max\sum_{j=1} w^j c^j = Max\sum_{j=1} w^j s^j \frac{g_{ei}}{T} \quad (7)$$

In Formula(7), $c^j$is the capacity of the j-th entrance at the intersection, pcu·h^{-1}; $w^j$is the weight coefficient of the j-th inlet of the intersection; $s^j$is the saturated flow rate at the j-th inlet of the intersection, pcu·h^{-1}.

(4)Capacity utilization ratio

Intersection capacity utilization*ICU* indicates steady state of traffic flow, Its mathematical expression is[19]:

$$ICU = \frac{\sum_{i=1} Max(tMin(v/s)_i T) + tL_i}{T} \quad (8)$$

The goal of signal timing optimization is to improve the utilization of traffic capacity at intersections, which is:

$$MaxICU = \frac{\sum_{i=1} Max(tMin(v/s)_i T) + tL_i}{T} \quad (9)$$

In Formula(8)~Formula(9), tL_i is the green light loss time of the i-th phase, s; *t Min*is the shortest green light duration, s; The remaining parameters are the same as above.

3.2 The optimization goal

(1)Average delay

The average delay *d*at the intersection consists of the consistency delay d_{ui} and the random delay d_{ri},which is[3]:

$$d_{ui}=\sum_i \frac{T(1-\lambda_i)^2}{2(1-y_{ij})} \quad (10)$$

$$d_{ri}=\sum_i \frac{x_{ij}^2}{2q_{ij}(1-x_{ij})} \quad (11)$$

$$d_i=d_{ui}+d_{ri} \quad (12)$$

The average delay in one cycle intersection is calculated by the weighted evaluation of the actual arrival equivalent traffic of each phase, Therefore, its optimization goal is:

$$Mind = Min\frac{\sum_i d_i q_i}{\sum_i q_i} \quad (13)$$

In Formula(10)~Formula(13), *d* is the average delay in a period intersection, s; d_i is the average delay time of the i-th phase, s; $y_{ij}$is the flow ratio of the j-th inlet of the i-th phase; $x_{ij}$is the saturation of the j-th inlet of the i-th phase; The remaining parameters are the same as above.

(2)Exhaust emission

Exhaust emissions from vehicles include *CO*、*HC* and NO_X[20], Considering that NO_X emission is less during idling, and the idle time of the vehicle at the intersection is longer, Therefore, this article mainly considers *CO and HC emissions.*

Second, this paper introduces dynamic vehicle exhaust emission rates, that is the standard model exhaust emissions per unit of time, and this article comprehensively considers the relation between exhaust emission rate and idle time[21]:

$$E_{CO}=\begin{cases}151.0967/d-5.965252(0s<d\le 24s)\\ 0.330443(d>24s)\end{cases} \quad (14)$$

$$E_{HC}=\begin{cases}3.487390/d+0.285907(0s<d\le 24s)\\ 0.431215(d>24s)\end{cases} \quad (15)$$

$$E=\alpha E_{co}+(1-\alpha)E_{HC}(0<\alpha<1) \quad (16)$$

The total amount of pollutants discharged when vehicles pass through the intersection is:

$$E_{Tol}=x_a E d \quad (17)$$

Minimization of pollutant emissions required for signal timing optimization, Therefore, the optimization goal is:

$$Min\ E_{Tol}=Min\ x_a E d \quad (18)$$

In Formula(14)~Formula(18), E_{Tol} is the total pollutant emissions,mg·h^{-1}; x_a is the sum of the equivalent amount of traffic actually arrived at each entryway in a cycle, pcu·h^{-1}; *E*、E_{CO}、E_{HC} are the integrated emission rate, *CO* emission rate, and *HC* emission rate of *CO* and *HC* at the intersection per unit time for the standard model at idle speed, mg / pcu·h; α is the weight factor ($0<\alpha<1$); *d* is idle time,s. general approximation equals average delay, therefore, the multi-objective model can be converted into a single-objective model for solving.

3.3 Constraint condition

The constraints of the model include the signal duration *T* and the green ratio λ_i of each phase. Reasonable *T and* λ_i are conducive to reducing vehicle congestion and improving the operating environment of the intersection. It also helps to reduce vehicle congestion and improve the operating environment of the intersection. The expression of the constraint is:

$$s\cdot t\begin{cases}0.05\le\lambda_i\le 0.95\\ 25\le T\le 180\end{cases} \quad (19)$$

In summary, an integrated multi-factor signal timing optimization model for urban road intersections can be established. which is:

$$\begin{cases}\text{Verify indicators}\begin{cases}Min\ Q_{ave}\\ Min\ h\\ Max\ CAP\\ Max\ ICU\end{cases}\\ \text{optimization goal}\begin{cases}Min\ d\\ Min\ E_{Tol}\end{cases}\end{cases} \quad (20)$$

$$s\cdot t\begin{cases}0.05\le\lambda_i\le 0.95\\ 25\le T\le 180\end{cases} \quad (21)$$

3.4 Model solving

The above model is a complex nonlinear optimization model. This paper uses genetic algorithm to solve the model. Compared with other algorithms, the genetic algorithm has a fast convergence rate, a high calculation accuracy, a wide range of applications, and has a strong robustness. According to the characteristics of the model, the number of individuals in the algorithm *M* takes 100, terminate the evolutionary algebra *T* take 500. The crossover probability P_c is 0.9. The probability of variation P_m is 0.1. The specific solution steps are as follows:

Step 1: Parameter initialization, make evolutionary algebra *T* equal to 0;

Step 2: Coding, randomly generate 100 initial individuals and substitute the survey data at the intersection into the model to program the formulas in the model;

Step 3: Calculate individual fitness in the population, and choose to replicate individuals based on fitness;

Step 4: Determine whether the crossover probability P_c and the mutation probability P_m meet the requirements, If so, let T=T+1, Otherwise do crossover operation, until the crossover probability and the mutation probability meet the requirements;

Step 5: Repeat iterative calculations, Until the termination of evolutionary generation T is greater than 500, the cycle time and green time ratio at this time are the results of the model.

Step 6: Calculate average delays and tail gas emissions using cycle time and green time ratio, substitute solution results into verification indicators, calculate queue length, number of stops, capacity and capacity utilization, and compare the optimization results.

4 Examples demonstrate

In this paper, two different types of intersections are selected for example verification. The first kind of intersection is the intersection of technology road - Fenghui south Road (As shown in figure 1). The intersection is located in the high-tech zone of xi 'an, The peak traffic flow and traffic density are large, Long vehicle idle time, The service level is D, Traffic jams are easy; In terms of traffic organization, there is a two-way traffic from east to west and one-way traffic from north to south. Therefore, this paper optimizes the signal timing of the intersection based on the existing traffic flow.

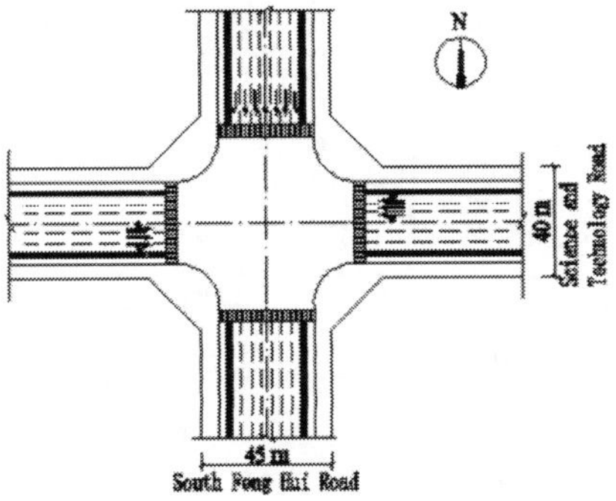

Figure 1. Figure 2 the intersection of South Fenghui Road and Keji Road

Through the investigation of the intersection, the survey of traffic volume was video recording. The instrument used was Panasonic HC-V270GK HD digital camera.The weather is all sunny.On this basis, the weekend morning peak(7:30-8:30) and the evening peak(18:00-19:00) are chosen to shoot one hour in succession; The current situation signal timing and road facilities investigation are all carried out by artificial method.

By collating the survey data, we can get the early and late rush hour traffic of each entry channel. Here, in order to facilitate the calculation of the model, the traffic volume needs to be converted into standard car equivalents, conversion factor see Table 1, And use the average value of standard car equivalents of morning and evening peak hours as the final reference data; Using formula (4) to calculate the current capacity, The weight value is calculated as the ratio of the traffic volume of each entrance road to the total traffic volume; The saturation equals the amount of traffic divided by the capacity of passage, The flow ratio is the ratio of traffic to saturated flow, The service level is referenced in table 2, The final statistical analysis results are shown in Table 3.

Table 1. Conversion coefficient

Model	Conversion factor	Model	Conversio n factor
Car、taxi	1.0	small truck	1.5
Medium bus, small bus	1.5	Medium truck	2.0
bus	2.0	big truck	2.5

Table 2. Service level of intersection

Service Level	saturation
A	≤0.25
B	0.25<S≤0.50
C	0.50<S≤0.70
D	0.70<S≤0.85
E	0.85<S≤0.95
F	0.95<S

Table 3. Parameters of Ke Ji Road—Feng Huinan Road intersection

Parameter / Inlet	Traffic volume/(pcu·h^{-1})	Traffic capacity/(pcu·h^{-1})	Weights	Saturation flow/(pcu·h^{-1})	saturated degree	Traffic ratio	Service Level
East imports	1 223	1 805	0.209	4 885	0.678	0.250	C
West imports	1 447	1 808	0.248	4 900	0.800	0.295	D
North imports	3 171	5 732	0.543	11 300	0.553	0.281	C

https://doi.org/10.1051/matecconf/201925902004

The data of Table 3 is used in the optimization model, and the model is solved with Matlab software and genetic algorithm. After 100 generations of iteration, it tends to the optimal solution, and the solution results are substituted into the verification index. Considering that the value of the weight coefficient α is related to the composition of the traffic flow, and there are differences in the values of different types of vehicles. This article combines the traffic conditions and spatial composition of intersections, refer to related documents [1，2]. Determine the weight coefficient α equal to 0.5, And compare the optimization results. The specific results are shown in Figure 2 and Table 4.

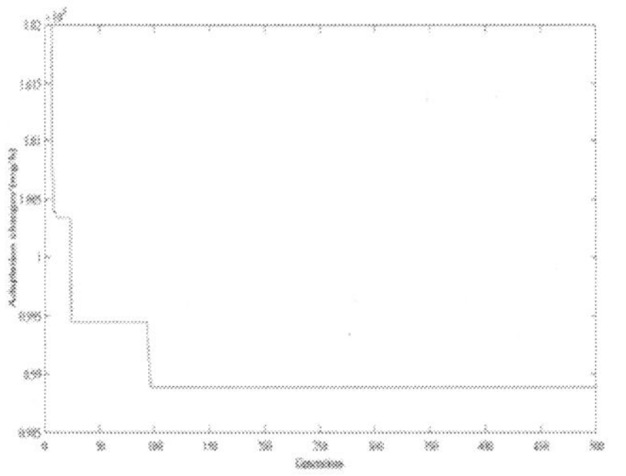

Figure 2. Optimization result diagram of model

Table 4 Comparison of optimization results of Ke Ji Road—Feng Huinan Road intersection

Compared index		**status quo**	**Optimi zed**	**Change rate**
Signal Timing Scheme	Cycle time/(s)	136	101	---
	First phase effective green light time/(s)	60	49	---
	Second phase effective green time/(s)	73	52	---
Verification indicator	Queue length/(vehicle)	13.31	8.78	−34.03%
	Number of parking / (times)	0.86	0.63	−26.74%
	(weighted) capacity/(pcu·h^{-1})	3 770	3 850	2.12%
	Capacity utilization	59.71%	64.33%	7.74%
optimization goal	Average delay/(s)	50.50	25.89	−48.73%
	Emissions/(mg·h^{-1})	112 314	98 895	−11.95%

It Can be seen from Table 4: Compared with the signal timing at the intersection status, the queue length of the intersection optimized by the signal timing is reduced by 34.03%, the number of parking stops is less than 26.74%, the average delay is less 48.73%, and the tail gas emission is reduced by 11.95%. Increased traffic capacity by 2.12%, increase utilization capacityby 7.74%, Having achieved better optimization results. It shows that when the current signal timing scheme is not reasonable, the optimization model can effectively improve the intersection operation status and improve the traffic control efficiency at the intersection. Therefore, it is suggested that the signal cycle time at this intersection should be 101 s, and the first phase effective green time and the second phase effective green time should be 49 s and 52 s, respectively.

The second type of intersection is a typical four-inlet cross-shaped intersection. This article selects the intersection of TongyiRoad and Tongde Road(See Figure 3). The intersection is located in Xifeng New City, Xixian New District. The service level of the east and west imports during the peak hours is C, and vehicle congestion may occur. Therefore, this paper optimizes signal timing and further tests the utility of the model.

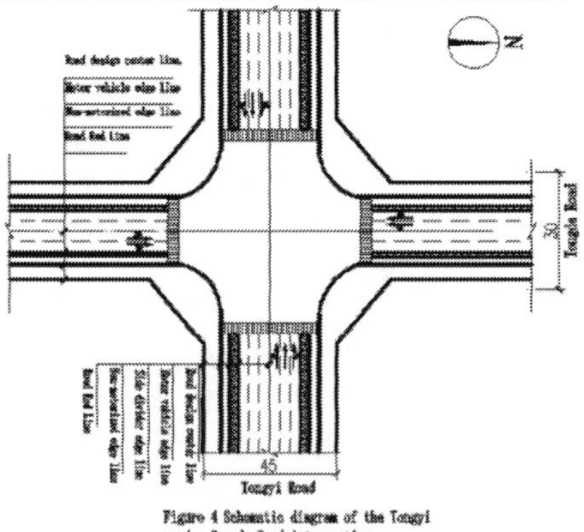

Figure 3. Tong Yi Road—Tong De Road intersection diagram

Through field investigation and statistics, the survey conditions, equipment, and sampling methods are the same as those of the first intersection. The available traffic data analysis parameters of the intersection are shown in Table 5.

Table 5. Parameters of Tong Yi Road—Tong De Road intersection

Parameter	**Traffic volume/(pcu·h^{-1})**	**Capacity/(pcu·h^{-1})**	**Weights**	**Saturation flow/(pcu·h^{-1})**	**saturated degree**	**Traffic ratio**	**Service Level**

Inlet							
East imports	1 105	1 857	0.327	4 802	0.595	0.230	C
West imports	1 041	1 853	0.308	4 796	0.562	0.217	C
South imports	587	1 253	0.173	3 152	0.468	0.186	B
North imports	651	1 304	0.192	3 171	0.499	0.205	B

Table 6. Comparison of optimization results of Tong Yi Road—Tong De Road intersection

Compared index		status quo	Optimized	Change rate
Signal Timing Scheme	Cycle time/(s)	111	91	---
	First phase effective green light time/(s)	53	57	---
	Second phase effective green time/(s)	55	34	---
Verification indicator	Queue lengt/(vehicle	7.48	5.56	-25.67%
	Number of parking / (times)	0.66	0.47	-28.79%
	(weighted) capacity/(pcu·h^{-1})	1 646	1 904	15.67%
	Capacity utilization	45.71%	48.26%	5.58%
optimization goal	Average delay/(s)	38.46	27.73	-27.90%
	Emissions/(mg·h^{-1})	49 555	35 658	-28.04%

Substituting the data in Table 5 into the optimization model, using Matlab software and genetic algorithm to solve the model, The optimal solution was obtained after 300 iterations, and the results before and after the optimization were compared. See Figure 4 and Table 6 for details.

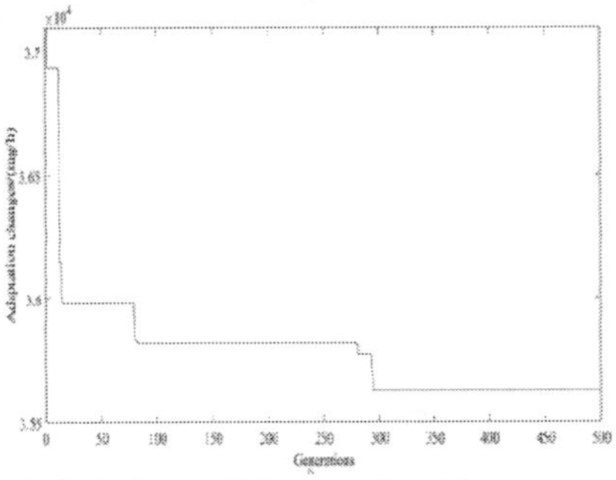

Figure 4. Optimizationresult diagram of model

It can be seen from Table 6: Compared to the signal timing in the status quo at the intersection, The optimized queue length of the intersection was reduced by 25.67%, the number of stoppages was reduced by 28.79%, the average delay was reduced by 27.90%, the tail gas emissions decreased by 28.04%, the traffic capacity increased by 15.67%, and the capacity utilization increased by 5.58%.The optimization result is better. This shows that the current signal timing plan is not reasonable and needs further improvement. Therefore, it is suggested that the signal cycle time at this intersection should take 91 s, and the first phase effective green time and the second phase effective green time should take 57 s and 34 s, respectively.

Comparing Table 4 and Table 6, the magnitude of change of each indicator after two types of intersection optimization is different. The calculation value of each indicator in the intersection of Technology Road and Fenghui South Road is greater than that at the intersection of Tongyi Road and Tongde Road. North entrance road has left and right turn dedicated lanes, reducing the interference of vehicle flow lines. Through the signal timing optimization, the green light duration is reasonably allocated, so the queue length and average delay decrease more, while the utilization capacity and capacity utilization increase less, Explaining that signal timing optimization is not easy to significantly improve the traffic capacity and capacity utilization of the intersection. In the future, the intersection can be improved by adding lanes or optimizing the canalization scheme; At the intersection of Tongyi Road and Tongde Road, the level of service is relatively high. The traffic volume of each road is relatively balanced, and the change in queue length and average delay is small. However, the change in capacity and tail gas emissions is greater than the intersection of Technology Road and Fenghui South Road. Explaining that the current conditions of use of the intersection are poor. The optimization of signal timing will benefit the traffic organization and management of the intersection.

In summary, through the verification of two different types of intersections, the results show that the model has a certain degree of practicality and can obtain a more comprehensive value of comprehensive benefits, and It is feasible to solve the model by genetic algorithm.

5 Conclusions

(1) This article takes the urban road intersection as the research object. Analysis six factors including queue length, number of stops, capacity, capacity utilization, average delays, and tailpipe emissions. With the introduction of dynamic exhaust emission rate, and with the constraint of cycle duration and green letter ratio, a signal timing optimization model for urban road intersections based on comprehensive benefits under multi-factor conditions was established.

(2)Selecting the intersection of Science and Technology Road-Fenghui South Road and the intersection of Tongyi Road-Tongde Road in Xi'an to optimize the signal timing, and achieved good value of benefits. It shows that the practicality of the model is verified, which can provide a theoretical reference for selecting the signal timing plan for urban road intersections.

(3)Weather conditions will also affect the vehicle's operation at the intersection. How to fully consider various influencing factors and reasonably determine the weight value of each optimization goal is the direction of further research.

6 Acknowledgment

This work is financially supported by the National Natural Science Foundation of China (Grant No. 51278396).

References

1. Yu Liu, Yu Lei, Wan Tao, et al. Simulation study on traffic signal timing optimization integrating environmental factor and delay. Journal of System Simulation,2008; 20(11): 3016—3019
2. Wang Yuquan, Jiang Xinyue, Guo Weiwei. Traffic signal cycle optimization method in urban intersections based on travel cost. Journal of Chang'an University(Natural Science Edition), 2015; 35(sup): 104—107
3. Wang Qiuping, Tan Xuelong, Zhang Shengrui. Signal timing optimization of urban single point intersections. Journal of Traffic and Transportation Engineering, 2006; 6(2):60—64
4. Zhou Shenpei, Wu Chaozhong, Yan Xinping. Bi-level optimization model for signal timing of intersections under environmental objective. Journal of Wuhan University of Technology(Transportation Science & Engineering), 2009; 33(04): 715—717
5. Zhang Yingying, Chen Xumei, Zhang Xiao, *et al.* Assessing effect of traffic signal control strategies on vehicle emissions. Journal of Transportation System Engineering and Information Technology, 2009; 9(1): 150—154
6. Unal A, Rouphail N, Frey H C. Effect of arterial signalization and level of service on measured emission. Transportation Research Record. Journal of the Transportation Research Board,2003; 2(1842): 47—56
7. Shen Feng, Yang Xiaoguang. Research on multi-object optimization algorithm of urban road tafficlights. Journal of Tong Ji University(Natural Science), 2009; 37(7): 898—901
8. Cheng Wei, Yuan Manrong, Ji Lina. Signal timing model based on particle swarm optimization: a case study of kunming. Urban transport of China, 2013; 11(3): 27—33
9. Li Shiwu, Wang Yunpeng, Fu Jianping, *et al.* Signal timing optimization simulation on urban road intersection based on vehicle emissions. Journal of Ji Lin University(Engineering and Technology Edition),2007; 37(6):1268—1272
10. Zhou Shenpei. Research on traffic signal control strategies in urban intersection based on emission factors. Wuhan: Wuhan University of Technology, 2009
11. Liu Quanfu, Lu Baichuan, Ma Qinglu, *et al.* Traffic signal control multi-objective optimization at intersection. Technology & Economy in Areas of Communications, 2014; 16(1): 47—50
12. Gao Yunfeng, Xu Lihong, Hu Hua, *et al.* Multi-objective optimization method for fixed-time signal control at intersection. China Journal of Highwav and Transport, 2011; 24(05): 82—88
13. Li Zhenlong, Dong Wenhui, Han Jianlong, *et al.* Multi-object optimization for intersection signal control based on genetic algorithm. Journal of Computer Applications, 2016; 36(02): 82—84
14. Shou Yanfang, Xu Jianmin. Multi-objective dynamic decision-making model of signalized intersection and its optimization method. Journal of Highway and Transportation Research and Development, 2012; 29(11): 92—97
15. Tang Dehua, Xu Lunhui, Lin Quan. Multi-objective control model of over-satutated signalized intersection. Science Technology and Engineering, 2009; 9(19): 5726—5729
16. Chen Xiaohong, Qian Dalin, Shi Donghua. Multi-objective optimization method of signal timing for the non-motorozed transport at intersection. Journal of Transportation Systems Engineering and Information Technology, 2011; 11(2): 106—111
17. Yang Peikun, Zhang Shusheng. Traffic management and control. Beijing: China Communications Press, 1995: 92—94
18. Yang Xiaoguang. Manual of urban traffic design. Beijing: China Communications Press, 2003: 80—84
19. Yang Jie, Guo Xiucheng. Dynamic traffic coordination control method for intersection group. Nanjing: Publishing House of Southeast University, 2013: 71—73
20. Yang Dongbo, Wang Min, Qiu Zhaowen. Study on vehicle fuel consumption and emissions estimation under the different road grade. Science Technology and Engineering, 2017; 17(12): 108—114
21. Sha Xuefeng. Research on vehicle dynamic emission model on urban road. Changchun: Jilin University, 2007

ATM change management: An evolutionary and agent-based approach

Gabriella Gigante[1], Domenico Pascarella[1], Marta Sánchez Cidoncha [2], Miquel Angel Piera[3], Gabriella Duca[4], Nicola Lettieri[4], Luiz Manuel Braga[5] and Jose Luis Muñoz Gamarra[6]

[1]*Italian Aerospace Research Centre, 81043 Vai Maiorise Capua, Italy*
[2]*Centro De Referencia Investigacion Desarrollo e Innovacion ATM A.I.E. Avda. de Aragón 402, Madrid, Spain*
[3]*Universitat Autònoma De Barcelona C/ dels Emprius nº 2, Barcelona, Spain*
[4]*Institute For Sustainable Society And Innovation, Via F.Napolitano 219 Nola, Italy*
[5]*Projectos, Empreendimentos, Desenvolvimento E Equipamentos Cientificos E De Engenharia*
[6]*ASLOGIC, Avda. Electricitat 1-21, Barcelona, Spain*

Abstract. Air Traffic Management (ATM) is a complex socio-technical system, whose behaviour depends on a combination of various subsystems of different nature: societal, technical, and human. Due to such aspects, it is difficult to understand which could be the part to be changed in order to improve performances, or which is the impact of a change on the overall performances. Such tasks are though issues, and cannot be easily performed. In this work, a new approach for the ATM change management process is proposed. It aims to introduce an innovative multidisciplinary process by combining the following different paradigms: the agent-based paradigm for the modelling of a change solution and the assessment of the achieved performances; the evolutionary computing paradigm for the tuning of the change: the sensitivity analysis to understand which part of the ATM system should be changed in order to match the targeted performance levels.

1 Introduction

Despite the excellent services supported by present Air Traffic Management (ATM) system, present architecture does not provide an efficient mechanism to adapt the capacity to the demand by exploiting the flexibility of the air transport system. To enhance ATM and meet the future traffic demand and environmental requirements, present ATM system is under evolution. The Single European Sky Air Traffic Management Research Program (SESAR) aims to transform European ATM into a more modular, automated interoperable, flight and flow-centric system.

The key objectives are to restructure European airspace as a function of air traffic flows; to create additional capacity; to increase the overall efficiency of the ATM system. In order to fulfil these objectives, the European Commission set some performance driven high-level goals on capacity, safety, environment, costs for airspace users. Hence, the change management process becomes crucial in order to design the new solutions to achieve the expected performances. In addition, the design and the validation of the changes are tough issues due to the fact that the evaluation of the impact of a possible change on the overall ATM system cannot be easily performed. Indeed, the global behaviour of the system is not easily predictable, since ATM is a socio-technical complex system joining three major interacting subsystems: the societal subsystem that contains the society constraints such as airlines; the technical subsystem that contains all technical infrastructures supporting the functioning of the ATM system; the human subsystem that includes all human components of the whole system, from flow managers to supervisors, pilots and controllers. This work proposes an innovative process to perform the change management by combining the agent-based paradigms with evolutionary computing and sensitivity analysis. It discusses preliminary considerations on the approach itself, the Agent-Based modelling paradigms to adopt, the sensitivity analysis, and the possible Evolutionary algorithms to use; the referring case study to validate the approach; the performances to apply.

2 Related work

The ATM change impact assessment is addressed by other SESAR exploratory research projects in several ways. For example, PACAS project is structured around a collaborative platform and expert assessments on different performance areas by using different formalisms, like fault trees and security networks. Other research projects have been focused on the performance evaluation of an ATM solution. APACHE project proposes a new framework to assess European ATM performance based on simulation, optimization and performance assessment tools that will be able to capture complex interdependencies between Key Performance Areas (KPA) at different modelling scales (micro, meso and macro). AURORA project proposes advanced metrics to assess the operational efficiency of the ATM system. These new metrics will be developed with the aim of encapsulating the airspace users' operational objectives, considering fuel consumption, schedule adherence and cost efficiency of the flights. User-preferred trajectories are defined as references for performance analysis purposes.

Other research projects investigate the performance evaluation by using the agent-based modelling and simulation technique, in order to take advantage of its high flexibility in terms of level of abstraction. In these projects, the agent-based simulation is used (usually in combination with other techniques) either as an analysis tool to understand emergent behaviours of the ATM system, or to study resilience and disturbance propagation (see projects CASSIOPEIA, ELSA and MAREA). For what concerning the Agent-based Evolutionary Search (AES), it is an emerging research paradigm. Depending on the research area taken into account, the combination of EC and ABM is still relatively unexplored, especially if it is considered the study of socio-technical systems in which the interaction between individual/cognitive, social, technical and environmental factors, generates complexities that are difficult to predict and dealt with [1]. A recent edited volume by Rennard [2] describes numerous applications in economics, but applications of EC to other social sciences (anthropology, political science, sociology, psychology) have been relatively fewer. Even if the first attempts to exploit evolutionary approaches in the analysis of air traffic assignment date back to the early 1990's, most interesting works about ATM and ABM have been published only very recently in more limited contexts and with exploratory goals limited in scope [3].

3 Change management approach

The change assessment process is made up of different steps: understand which are the targeted performances, which element has to be changed to obtain such performances; design at high level the change and choose on some decision variables which could be the best tuning to match the targeted performances; implement the change and verify and assess the impact, that is, the obtained performances. The proposed work starts from the current state-of-the-art in agent-based modelling for ATM. It aims to extend the state-of-the-art on ABM in ATM and the respective simulation methodology, by introducing: 1. a full agent-based model of the direct routing operations; 2. possible different paradigms according to the nature of the specific agent; 3. the evolutionary computing approach for the optimal tuning of the modifications; 4. a sensitivity analysis strategy for the identification of the most critical parameters from a performance-oriented point of view (see Figure1). The simulation may "discover" the emergent (i.e., unpredictable) aspects of the change, also in terms of failures and performance degradation of the global system.

As a consequence, a set of KPIs for the ATM domain shall have to be rigorously established in order to include all the relevant performance dimensions. They may be directly measured or they may be calculated from supporting metrics. In addition, the agent-based simulation may be used as a feedback to the design process of the new ATM solutions for their optimal tuning and calibration, in order to comply with the requirements for integration. The calibration of the modification implies the fine-tuning of the model that integrates the modification. The parameters that are related to the modification (a new or modified component, a new procedure, etc.) are chosen so that the performance of the designed system is optimal in some way, e.g., with respect of some reference KPIs. In detail, the introduction of evolutionary computing approaches for the optimization allows for a proactive approach to the design of a modification enabling the emergence of the most suitable solution. The optimization shall be performed by means of evolutionary agents, i.e., agents whose behaviour rules are processed by means of evolutionary computing paradigms. The provision of feedback to design relies in exercising the system after the insertion of the modification by means of evolutionary agents and running the optimization algorithms in order to find the proper constraints to the inserted modification in order to fulfil the expected KPIs of the overall modelled system and other possible constraints.

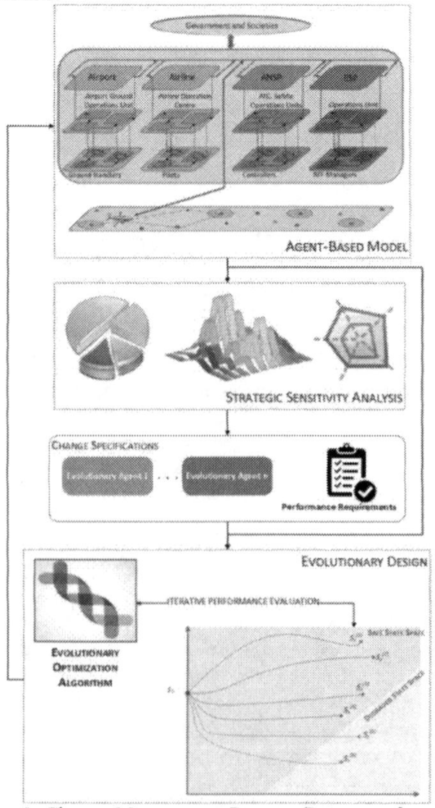

Figure 1. Change Management Process (Image partly adapted from [4]).

3.1 Agent-Based modelling

One of the main difficulties in the modelling of the ATM system architecture is the coexistence of human, technical and information managing components that must be described considering a holistic approach in which the relationship between components and actors can lead to different behavioural states. Despite most of these relationships rely on information flows that can be traced, there are several intangible flows difficult to identify (trust, knowledge sharing, experience, training, competences) that influence drastically in the overall system behaviour. Thus, the modelling framework to be used in the proposed work, to get a deep understanding of the behavioural dynamics of component/actors relationships must be enhanced with mechanisms to support complex relationship dynamics described by multiple loops and chains, loops within loops, mutual cross feed relationships connecting them, inhibitory connections and preferential reactions in front of different events.

Despite there are relatively few scientific publications of ATM modelling [5] considering ATM as a socio-technological system that could support the holistic approach mentioned, there are well accepted formalisms (see Figure 2) that could provide the right framework considering different abstraction levels.

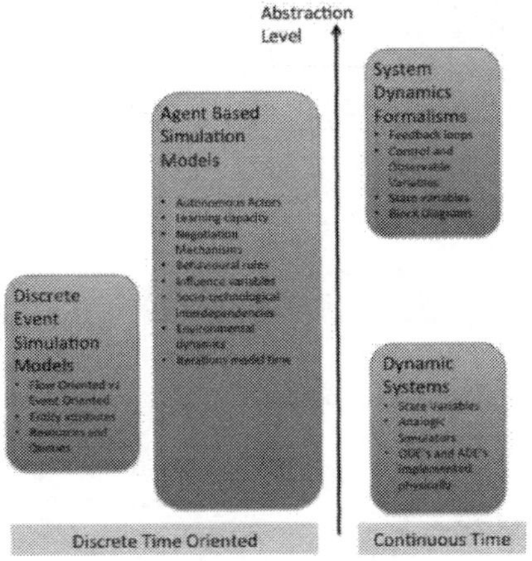

Figure 2. Modelling formalisms.

Among these dynamic modelling approaches, in the proposed work it is chosen to combine the use of Agent Based Modelling (ABM) and discrete event formulation. The first one will model the tactical layer (i.e., Aircraft and ATC) because it supports different types of tangible and intangible relationships connecting the actors of the ATM architecture and in consequence provides an excellent quantitative analysis tool. The second one it is proposed for the strategic layer (Network Manager and Air Operator) to be able to support a better understanding of the traffic behaviour at the tactical layer by suggesting new roles or changes in the architecture component and services.

3.2 Evolutionary computing

Agent-based Evolutionary Search (AES) is an emerging research paradigm lying at the intersection between Evolutionary Computation (EC) and Agent-Based modelling (ABM), two methodologies that pledge to contribute together to the analysis and the solution of challenging and complex problems in in areas that span from online trading to disaster response [6]. In general terms, the idea behind AES is to exploit a bio-inspired framework to explore the dynamic adaptation of complex systems as it evolves over time, so as to discover the significance of critical factors of stability and change. A major challenge in this field is the exploration of the operating conditions that separate sustainable and stable from fragile or unstable configurations of the target system.

The main objective of the application of EAS is to improve the AB model performance by varying it. The integration of ABM and EC usually works according to a schema that may be summarized as follows [7]: an evolutionary algorithm maintains a population of "individuals" that represent the potential ABM variations.

Each model variation is run in order to assess the quality of that particular model with respect to a given goal. Once a "fitness", based on the quality of the model it represents, has been assigned to each variation, a selection mechanism chooses the more fitting individuals to produce "offspring", representing the variations on the parents to be evaluated as part of the next generation. Less fitting individuals (model variations) are eliminated, so the average fitness of the population increases over time. In order to apply these notions, and to explore how ATM architecture, rules and procedures could be better designed, reformed or re-engineered it is obviously necessary to specify which parts of the ATM model are changing and what the selection pressure is, understanding at which level intends to apply the EAS: model variation can be represented at parametric, structural, and behavioural variations. Over the years, evolutionary algorithms have been shown to be particularly good at solving parameter optimization problems in which the parameters interact in highly non-linear ways [8]. In this sense the model variation will be specified by means of simulation parameters whose values are left unspecified within a range of pre-defined values and thus will be used for tuning the parameters. It could be chosen also to automate model refinement at a more fundamental level by allowing variations of structural elements of models like, just for instance, the network layout, or the communication patterns. The proposed for improving model performance based on this kind of structural variation. The most fundamental level of model variation will be the characterization of the behaviour of the agents that make up a system as even simple changes in these behaviours can have a huge impact on the emergent behaviour of a model over time.

3.3 Sensitivity Analysis

One of the main problems of ABM is the difficulties to trace the interactions between the agents that generates a global system behaviour from isolated component/actor

https://doi.org/10.1051/matecconf/201925902005

behaviours. Thus, the sensitivity analysis in ABM should allow to identify the most critical parameters that affect some reference performance metric as a results of complex agent relationships formalized as a complex adaptive system.

Most of the time, in a stable condition, the variation or oscillation of metrics and agent attributes around a reference value remains within a variation range with very well defined boundaries, characterized by limit values. The variation of each metric depends on the interaction with other metrics and agent attributes that exist in the system. For a specific metric, the other agent attributes constitute the environment responsible for the noise, which represents the variability of this environment. Consequently, the functional resonance method supported by FRAM (Functional Resonance Accident Model) [9] can be considered as the detectable event that appears out of the non deliberate interaction of the weak variability of many metrics and agent attributes interacting with each other. In Figure 3, it is represented the agent relationships to be implemented in the ABM framework considering the FRAM approach for sensitivity analysis.

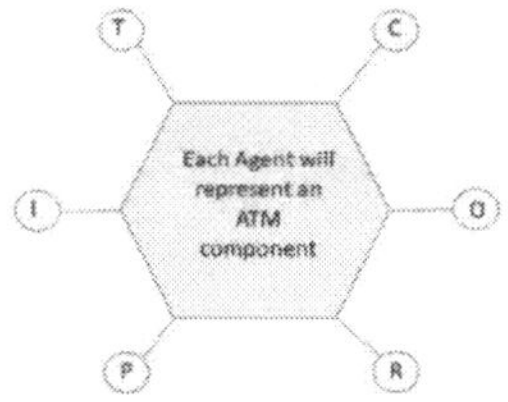

Figure 1. FRAM Agent relationships.

The agent interfaces correspond to:

- Input: Triggers an action to be implemented by a computer service, a machine or by a human;
- Time: Available time horizon to perform an action. It can be immediate or with a latency in the case of computer service, or can be a stochastic time parameterized by values of influence variables in case of a human action;
- Control: An action usually requires the adjustment of a function that can be a plan, a procedure or a human task.
- Output: The results produced by an action;
- Preconditions: State variables that must be fulfilled to proceed with the action;
- Resources: Provides an estimation of resource availability at a particular time instant, required to perform the action.

4 The case study

The validation of the approach finds its effectiveness when the a specific part of the ATM system is chosen to apply the proposed approach. The soundness of the validation relies on the representativeness of the modelled part of the ATM system in terms of complexity, involvement of human aspects, criticality and interdependencies among the subsystems. To these purposes, the reference scenario takes place in the execution phase, for en-route flights both in cruise and vertically evolving for cross Area Control Centers (ACC) borders and in medium and high complexity environments. The overall approach will be applied to a specific case study starting from the reference scenario and proposing an evolution in direct routing that introduces the revision of trajectory proposed by SESAR. Such a revision will be modelled involving a new (evolutionary) agent, airspace user, to negotiate the conflict resolution with Air Traffic Control (ATC).

4.1 The direct routing

Free Routing is intended to overcome the excess of distance flown by aircraft due to the constraints imposed by the current airspace network structure. Within Free Routing, the Direct Routing paves the way forward an environment where the user preferred routing is working, by allowing more direct route options so that the users can be closer to their preferred flight path. Direct Routing allows users to select a direct route between two published waypoints without the need of passing by the intermediate points of the fixed-route network. This operational solution is particularly relevant for cross-border operations in high and very high complexity environments.

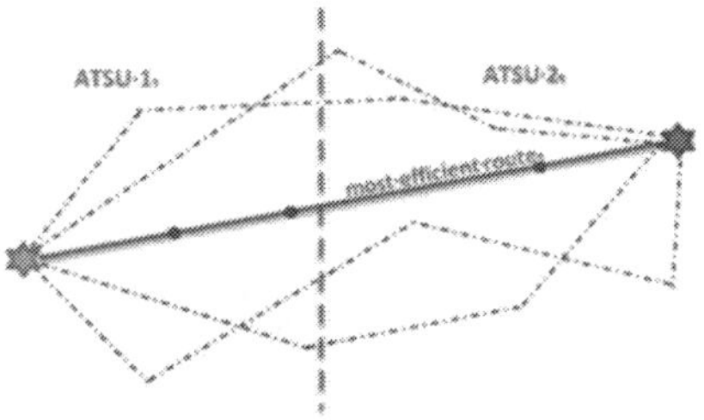

Figure 3. FRAM Agent relationships.

The implementation and deployment of Direct Routing requires operational and airspace changes as well as the evolutions of the air traffic systems, and in particular of the Flight Data Processing ground system. The selection of Free Routing as Case Study responds to a number of considerations.

- It is part of the SESAR Pilot Common Project. Free Routing will be mandatory in Europe above FL305 in 2022;
- Direct Routing has been partially implemented in some parts of Europe and is still pending in many others. This uneven implementation level makes the scenario interesting because of the data availability, on one hand, and of the remaining open issues and unclear implementation aspects, on the other hand;
- The impact for controllers is substantial, since Free Route increases the unpredictability of the traffic making human actors a critical source of its performance;

- Direct Routing allows the comparison with the evolution to free route, since it is open to several architectural evolutions, such as the introduction of a new agent, airspace user, who will be involved in the negotiation of changes to the agreed trajectory; the inclusion of the short term planning timeframe with Demand and Capacity measures.

4.2 Human aspects

In order to support the design and development of evolutionary solutions to Free Routing, it is relevant to assume a wide perspective reviewing the behavioural and cognitive aspects of the controllers working in the Free-route environment. In the presented work human aspects (pilots and Air Traffic Control Operators – ATCO) are investigated under three perspectives: learning from direct routing scenarios elements on how human agents behave, to model the transition to the free route simulation; describing at operational level the cognitive processes in ATM (desk analysis on literature review); depth analysis of the socio-cultural aspects through literature empirical data gathering and their subsequent heuristic validation form expert pilots and ATCOs. The human behaviour works as an elective adaptive system, by which we reply to needs, keep our balance, live in our environment. The chosen approach for the proposes work combines the human activity theory addressed by Leont'ev [11], the Hierarchical Task Analysis (HTA) [12] and the recent studies on the complexity in work environment [13]. The human activity is a hierarchical system where each activity comprises a set of actions including a set of operations. Activities always reply to a motivation, that at the level of the actions refer to a goal and at levels of operations are conditions. This theory includes the tools, the equipment and systems that support the action execution; the sets of conditions determining how and why individuals may act; these conditions are expressed in rules that are a result of the social conditioning [14]. The HTA enables to model the complexity cognitive levels that interplay in ATC implying at the same time knowledge-based skill-base and rule-based processes [15]. The expected description of the HTA focuses on the structure of the main goal-oriented tasks and subtasks in the free route control management, specifying the temporal and causal hierarchy and triggers such as operation procedures. Moreover, quantified variables on behavioural, cognitive (Visual search, Trust, Problem solving, Decision making, Judgement on take actions, Prioritization Situation awareness) and socio-cultural aspects (Group behaviour, verbal coordination, Shared situation awareness, Negotiation) are provided. The complexity aspects in the Free Routing environment, such as Distributed systems, Hazardous system, Automation, etc. will be considered too.

4.3 Applicable performance metrics

To the evaluation of the performances achieved by the proposed scenarios, the applicable metrics to assess will be: punctuality, fuel consumption and safety. Punctuality or timeliness can be understood as the difference between the planned and the actual departure and arrival times. It can be: (i) zero for timely arrival; (ii) positive for delayed arrival; (ii) negative for early arrival. Taking as a limit of acceptability a 6 minute delay. Fuel consumption can be very different say between full thrust at take-off and initial climb, on climb after cut-back, on approach at idle, or on cruise at different altitudes and speeds and on acceleration or deceleration. Thus fuel consumption needs to be considered by sector of flight and added for all sectors, unless all sectors refer to cruise at the same speed and altitude. The evaluation of safety can be performed as follows: (i) consider each pair of aircraft trajectories and determine the point of closest approach: (ii) if either the altitude difference meets or the horizontal separation meets the situation is safe; (iii) if both the assigned separation thresholds are violated then a potentially unsafe situation arises; (iv) the speed, heading or vertical rate of one or both aircraft must be changed until at least one of the thresholds is met. The key to this assessment is the calculation of the altitude and horizontal separation at the point of closest approach between two trajectories. This is not necessary in many cases when there is no doubt; e.g. if the aircraft fly horizontal at different flight levels separated more than the threshold or if their horizontal flight paths never come closer than the threshold.

5 Conclusions and future works

The proposed work defines a new approach in order to perform the change assessment in ATM system by supporting all foreseen steps, from the strategic thinking to the change impact assessment. The innovation potential of this proposal lies both in the innovativeness of AES application in the ATM field and in the strongly interdisciplinary vision inspiring our work. It leverages on the ABM formalism, enriching it with the EAS computing and the Sensitivity analysis techniques. The in-depth and interdisciplinary investigation of the potentialities of hybrid modelling frameworks, together with the validation results obtained by applying it to a real ATM subsystem and changed scenario will enable to get methodological and applicative results suitable of generalization. It will support the harmonization process by enabling the modelling of the ATM system according to different levels of scale, at least regional, national and trans-national. The approach has to be further specialized in the definition of the ABM paradigm and the kind of AES algorithms to be chosen, in applying the approach to the addressed case study and analysing the obtained results.

6 Acknowledgment

This work has received funding from the SESAR Joint Undertaking with grant agreement No 783189 (EVOAtm project) under European Union's Horizon 2020 research and innovation program.

References

1. Gilbert, N., and Troitzsch, K. 2005. *Simulation for the Social Scientist*, Second Edition. Open University Press.
2. Rennard, J. P. (Ed.) 2006. *Handbook of research on nature inspired computing for economics and management*. IGI Global, Hershey, PA, USA.
3. Yliniemi, L., Agogino, A. K., and Tumer, K. 2014. Evolutionary agent-based simulation of the introduction of new technologies in air traffic management. In *Proceedings of the 2014 Annual Conference on Genetic and Evolutionary Computation* (Vancouver, BC, Canada, July 12 - 16, 2014). ACM New York, NY, USA, 1215-1222. DOI= http://dx.doi.org/10.1145/2576768.2598388.
4. Gluchshenko, O., and Foerster, P. 2013. Performance Based Approach to Investigate Resilience and Robustness of an ATM System. In: *Proceedings of Tenth USA/Europe Air Traffic Management Research and Development Seminar (ATM2013)* (Chicago, USA, June 10-13, 2013).
5. Sándor, Z. 2017. Functional Modelling of the Air Traffic Control and the Integration Perspectives of the Integrated Services. *Periodica Polytechnica. Transportation Engineering*, 45, 3, 107-118.
6. Sarker, R. A., and Ray, T. (Eds.) 2010. *Agent-Based* Evolutionary *Search* (Vol. 5). Springer-Verlag Berlin Heidelberg.
7. Cioffi-Revilla, C., De Jong, K., and Bassett, J. K. 2012. Evolutionary computation and agent-based modeling: biologically-inspired approaches for understanding complex social systems. *J. Computational and Mathematical Organization Theory*, 18, 3 (Sep. 2012), 356-373. DOI= http://dx.doi.org/10.1007/s10588-012-9129-7.
8. Hansen, N., Müller, S. D., and Koumoutsakos, P. 2003. Reducing the time complexity of the derandomized evolution strategy with covariance matrix adaptation (CMA-ES). *J. Evolutionary Computation*, 11, 1 (Spring 2003), 1-18. DOI= http://dx.doi.org/10.1162/106365603321828970.
9. De Carvalho, P. V. R. 2011. The use of Functional Resonance Analysis Method (FRAM) in a mid-air collision to understand some characteristics of the air traffic management system resilience. *Reliability Engineering & System Safety*, 96, 11, 1482-1498.
10. SESAR 2017. *SESAR Solutions Catalogue*, Second Edition.

Leont'ev, A.N. 1978. *Activity, Consciousness, and Personality*. Prentice-Hall.

11. Eurocontrol 1998. *Integrated Task and Job Analysis of Air Traffic Controllers, Phase 1. Development of Methods*.
12. Carayon, P. 2006. Human factors of complex sociotechnical systems. *Applied Ergonomics*, 37, 4, (Jul. 2006), 525-35. DOI= http://dx.doi.org/10.1016/j.apergo.2006.04.011
13. Engeström, Y. 2010. Expansive Learning at Work: toward an activity theoretical reconceptualization. *J. of Education and Work,* 14, 1 (Aug. 2010), 133-156. DOI= http://dx.doi.org/10.1080/13639080020028747.
14. Rasmussen, J. 1986. *Information Processing and Human-Machine Interaction: An Approach to Cognitive Engineering*. North-Holland, New York, NY, USA.

Flight Delay Prediction Based on Characteristics of Aviation Network

Tan Zhou[1], Qiang Gao[1], Xin Chen[2] and Zongwei Xun[1]
[1] *Civil Aviation College, Nanjing University of Aeronautics and Astronautics, 211106, China*
[2] *Management Science and Engineering College, Nanjing University of Finance and Economics, 210046, China*

Abstract. In recent years, the increasingly serious flight delay affects the development of the civil aviation. It is meaningful to establish an effective model for predicating delay to help airlines take responsive measures. In this study, we collect three years' operation data of a domestic airline company. To analyse the temporal pattern of the Aviation Network (AN), we obtain a time series of topological statistics through sliding the temporal AN with an hourly time window. In addition, we use K-means clustering algorithm to analyse the busy level of airports, which makes the airport property value more precise. Finally, we add delay property and use CHAID decision tree algorithm to train the data of an airline for nearly 3 years and use the train、ing model to predicate recent half a year delay. The experimental results show that the accuracy of the model is close to 80%.

1 Introduction

With the developing of civil aviation, the number of airports and flights are increasing sharply. As a result, flight delay gets so serious that limits the development of airports and airlines.

In the perspective of aviation network, Nikolas Pyrgiotis[1] proposed a phenomena of propagated delay in large-scale aviation network (AN). In addition, Guimera[2], Jianhong Mou[3] , Liu[4], Xiaozhou Zeng[5] et al. conducted an analysis of modeling and topology of aviation network based on complex network theory. In terms of delay prediction, decision tree[6] is a kind of data mining model which builds a classification model through information entropy in an unordered and irregular data set. At present, decision tree models are applied in many areas, such as C4.5 for network traffic prediction[7], financial warning [8], remote sensing image classification [9], and Chinese character recognition [10], etc. And CHAID is used for credit card risk management [11] and personal income forecast [12]. In this paper, combined with the dynamic topology characteristics of aviation network, a prediction model based on CHAID decision tree is proposed. The study is conducted from the perspective of the airline and is tested with three years' operation data of a major domestic airline.

2 Temporal Characteristics of Topology

2.1Sources of data

We collect actual flight operation data of a major domestic airline from July 1, 2013 to July 1, 2016, a total of 2,114,122 pieces of data. Among them, there are 1,421,359 delayed flights, accounting for 67%, and total arrival delay is 125,476,116 minutes.

Each flight information includes flight date, flight number, aircraft type, agent, nature of flight, actual takeoff station, actual landing station, planned takeoff station, planned landing station, actual/planned departure time, actual/planned arrival time, flight time, takeoff delay etc.

Extract four important segments as training set.

Table 1. Training set

Training set	Number
Training set 1 PEK-SHA	18321
Training set 2 SHA-CAN	12061
Training set 3 PVG-XIY	8103
Training set 4 KMG-PEK	9801

2.2 Static characteristics of topology

(1) The topology of the network is shown in Figure 1. The average shortest path length <l>=2.412, the average clustering coefficient <C>=0.663, and the network diameter <d> is 6, indicating that the separation between the two airports is very small, and the average can be reached in as few as 2 transfers. It needs no more than four connections for any two farthest airport to reach each other. The network has a shorter path length and a higher clustering coefficient, which shows typical small-world network characteristics. Therefore, the flight delay of any node will be quickly propagated to other nodes.

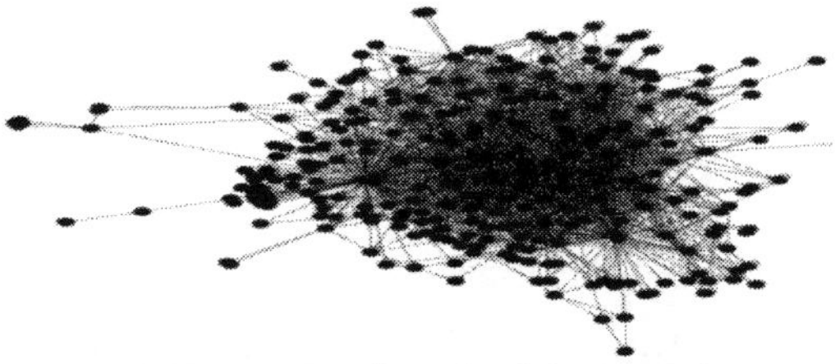

Figure 1. Topology of Aviation Network

(2) In comparison to the whole aviation network in 2005 and 2010[3][4], we can see that the network has a higher average degree <k>=22.847 and smaller clustering coefficients <c>=0.663. The increase of the average degree indicates that the accessibility of the network is improved, while the decrease of the clustering coefficient indicates that the nodes directly connected increase. In such condition, the delay in hub airports is more likely to be propagated to broad range of airport nodes.

(3) The degree distribution of nodes in the network follows the power law distribution. As shown in Figure 2. The ordinate is degree value, and the abscissa is the number of airports, indicating that it is a scale-free network. That means few airports have high degree.

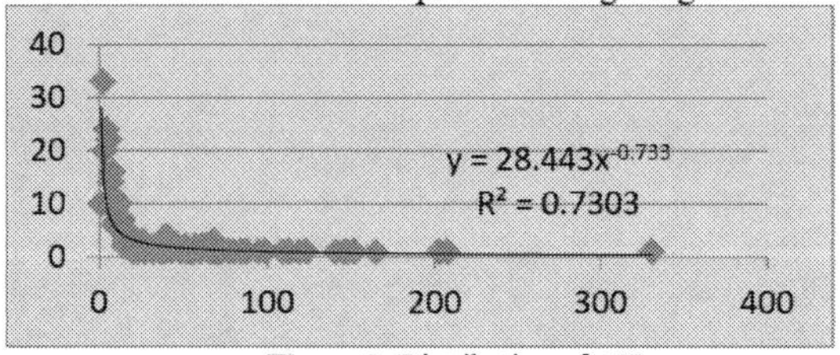

Figure 2. Distribution of AN

2.3 Dynamic Characteristic of Aviation Network

Time Window. We analyze the temporal characteristics of AN by sliding time window by one hour.

In this method, we use a contact sequence of quadruplets $(i, j, t_{ij}^s, t_{ij}^e)$ to represent aviation network, where i, j denote airports and (t_{ij}^s, t_{ij}^e) are the takeoff and landing times of the flight. The edges between nodes appear at time $t_{ij}^s = \{t_{ij}^s(1), t_{ij}^s(2), \ldots t_{ij}^s(n)\}$,which is ordered as $t_{ij}^s(a) < t_{ij}^s(b)$. Since we assume that edges are established when the flights start to end, edges maybe overlapped because of the duration. We divide the daily AN into consecutive sub-networks with each network being constructed of flights with $e_{ij}(t) = 1$.

$$e_{ij}(t) = \begin{cases} 1, t_{ij}^s \le t \le t_{ij}^s \\ 0, otherwise \end{cases} \quad (1)$$

The connections contain departure, arrival and ongoing flights. The time window is shown as:

$$\begin{cases} T_w(m) < t_{ij}^s \le T_w(m) + \Delta t \\ T_w(m) < t_{ij}^e \le T_w(m) + \Delta t \\ t_{ij}^s \le T_w(m) and t_{ij}^e > T_w(m) + \Delta t \end{cases} \quad (2)$$

Where $T_w(m)$ represents the initial time of the mth time window; Δt denotes the length of time window, and is set as $\Delta t = 1h$ in our study.

Change in the number of flights. We start analyzing the temporal traveling pattern by calculating the change in the number of flights (N_j). As shown in Figure 3. The horizontal axis is time period, and the vertical axis is the number of flights. When the number is at a high level, it is easy to make airports out of order and the delay happens. The number starts from zero at 6:00 and increases dramatically during the following 3 hours. It maintains high level from 10:00 to 17:00, then there is a relatively slight decrease and the number returns to zero from 0:00 to 6:00.

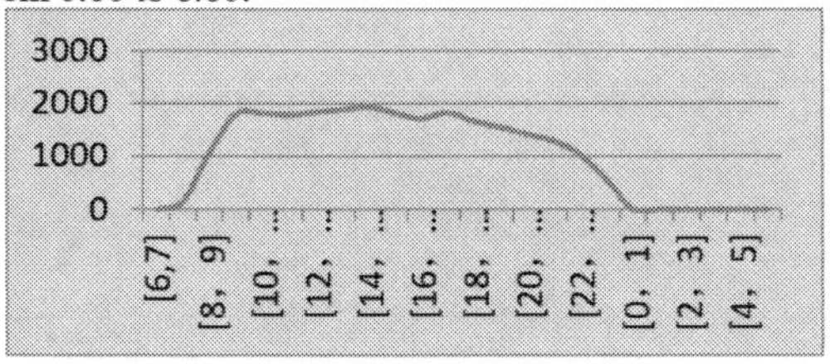

Figure 3. Change in the number of flights

3 Airport clustering analysis based on topological features

Observing the above training sets, we found delay of different airports also has a large gap. It can be speculated that the busyness of the airport has some effect on delay. The statistics of the original data were collected to obtain an average delay of 303 airports. In addition to the airports without delay, the effective data is 198 groups.

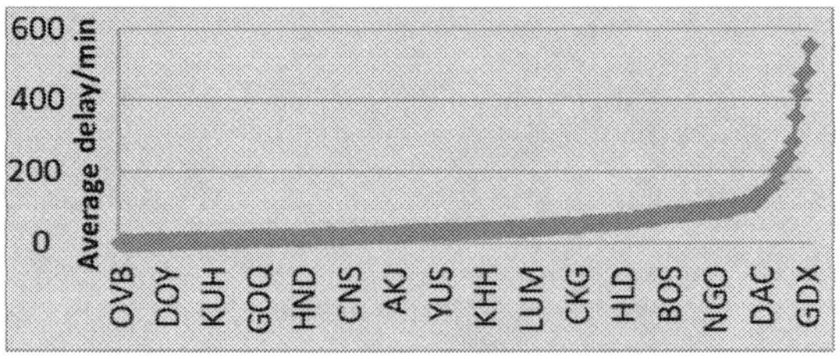

Figure 4. Average delay of different airports

Table 2. Delay time description analysis of different airports

Number of delay Airports	Average delay/min	Maximum/min	Minimum/min	Mean square error
198	56	551	1	79.84

Conclusion: By observing Figure 4, it is found that there is a large gap between the delays at different airports. Although the busy airports have a high service rate, the traffic exceeds its capacity. Thus the operating status is in a congested state and is prone to delays. Therefore, it is necessary to consider the extent of the https://doi.org/10.1051/mateccconf/201925902006

airport's busyness as an attribute that influences flight delay.

(1) Study the factors that influence airport delay and create the cluster index system

Based on the topological characteristics analyzed in Chapter 2:

1. Shorter average path length and higher clustering coefficient can both lead to higher delays, because flight delay of any node will be quickly propagated to other nodes.
2. The higher the average degree is, the higher the delay would be, because delay at the hub airports can be more likely to transmit to a wide range of airport nodes. We add in some social characteristics of the airports and build a clustering index system.

Table 3: Airport clustering index system

Num	Influencing factors	Num	Influencing factors
1	Path length	5	GDP
2	Clustering coefficient	6	Total population
3	Degree	7	Takeoffs and landings
4	Airport throughput capacity		

(2)Select airports with severe delay and establish airport samples

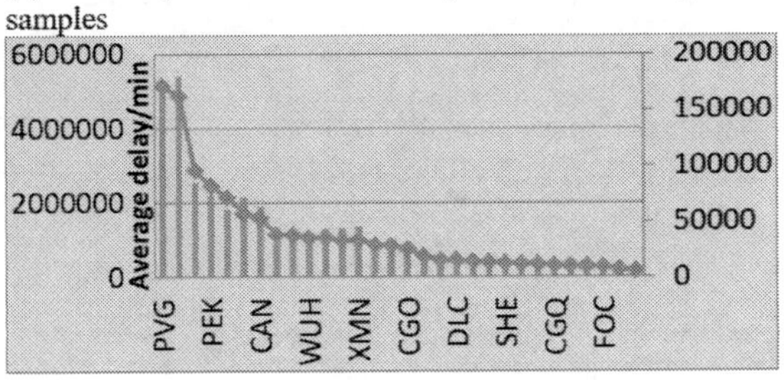

Figure 5. The number and sum of delay

Since the training set in this paper is selected from first-line airports, if all airports are used for classification, these airports will be difficult to divide. Taking the difficulty of data acquisition into account, the top ten airports with high delay are selected for clustering. The result is shown in table 4.

Table4. Airport clustering result

Airport	Clustering	Airport	Clustering
PVG	1	NKG	3
SHA	2	CAN	2
KMG	2	TAO	3
PEK	1	CTU	2
XIY	3	WUH	3

After calculation, Sig. of the cluster analysis is not higher than the significance level of 0.05, so it can be considered that there are significant differences among the above 7 variables. The clustering is ideal, and then we can use the airport cluster for delay forecasts at different airports.

4 Flight delay prediction based on decision tree

4.1 Attribute selection and division

In the flight forecasting problem, we need to filter attributes that are related to flight delay to ensure good forecasting results. We filter the attributes of the original data to obtain 9 available attributes. The specific attributes are shown in Table 5. The following describes the classification basis and classification results.

Table 5. Data set property description

Number	Attribute	Description
1	Solarterm	The month of the flight date, an integer in the range of 1-12
2	PlaneType	Aircraft types, classified into 3 categories according to the number of seats, and valued as 1, 2, and 3.
3	DepTimeofPlan	Planned departure time, valued as 1, 2, 3, 4
4	ArrTimeofPlan	Planned arrival time, valued as 1, 2, 3, 4
5	Origin	Original airport, according to the result of clustering, the values are 1, 2, 3
6	Destination	Destination airport, according to the result of clustering, the values are 1, 2, 3
7	FlightTask	Flight mission attributes, such as regular shift, overtime, charter, etc.
8	FlightType	Flight attribute, the values are 5 kinds of attribute codes, such as international f flight J, domestic flight N
9	Delay	Delay level, valued as 1, 2, 3, 4

(1) Classification of flight dates

It is divided by month, divided into 1-12 months.

(2) Classification of aircraft type.

Aircraft types are divided according to the number of seats, as shown in Table 6.

Table 6. Classification of aircraft type

Aircraft types	Identification	Number	Proportion
A320 B737	1	3064	15.8%
B757 B767 B777	2	2833	14.8%
A330 33E	3	13542	69.4%

(3)&(4)Classification of planned departure/arrival time

The number of flights at the airport during the time slot determines the airport's peak hours. Referring to the analysis in section 2.3, we divide the time period as Table 7.

Table 7. Classification of scheduled departure/arrival time

Time period	Identification	Number	Proportion(%)
0:00—6:00	1	4450	22.9
6:00-10:00	2	5276	27.1
10:00-17:00	3	6556	33.7

17:00-24:00	4	3167	16.3

(5) & (6) Classification of departure/arrival airports

According to Chapter 4 clustering analysis, the airports are divided into 3 categories.

(7) Classification of flight properties

There are 5 kinds of attribute codes, such as international flight J, domestic flight N etc.

(8) Classification of flight mission

There are 19 kinds of flight mission codes, such as regular flight, overtime, charter flight, etc.

(9) Classification of delay

Delay attribute is the attribute of the independent variable. Delay is the difference between "actual arrival time" and "planned arrival time". Delay is divided into 4 levels, which are respectively represented by 1-4. 1 means slight delay or no delay. In practice, it can be almost ignored. 2-4 indicates the level of delay. The greater the number is, the more severe the delay, as shown in Table 8.

Table 8. Classification of delay

delay(min)	identification
<25	1
25-60	2
60-120	3
>120	4

4.2 Experimental results

Experiment 1 Analysis of a single segment. The training set 1PEK-SHA is trained using CHAID decision tree, and the decision tree is obtained as shown in Figure 6 and used to test the data from July 1, 2016 to January 2017 to obtain the confusion matrix as Table 9.

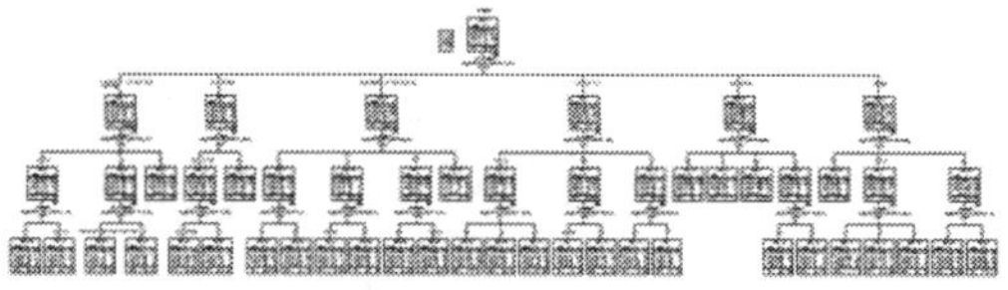

Figure 6. Decision tree of training set 1

Table 9. Confusion matrix of training set 1

Observed	Classification Predicted 1	2	3	4	Correct percentage
1	1100	6	0	0	99.40%
2	95	346	0	0	40.69%
3	16	6	84	0	47.71%
4	0	0	1	1	20.00%
Total	73.23%	21.66%	5.10%	0.01%	80.32%

In the study of delay prediction, the confusion matrix is the model evaluation result, that is, the ability to handle unknown data. This paper is a delay forecasting problem with a total of four delay degrees. The correct percentage of the rightmost column is the accuracy of degree i, that is, the percentage of data in degree i in the observed sample is correctly predicted. Ex: The accuracy calculation process of degree 1 is 1100/(1100+6+0+0)=99.4%. The correct percentage in the bottom line is also called the overall accuracy rate, which is an important evaluation parameter for the overall prediction effect of the reaction model. The calculation process is the sum of the amount of degree i data which is correctly predicted divided by the total amount of data.

Based on the results of the previous airport cluster analysis, six training sets with different levels of airport connections are selected. Take the same method as the training set 1, use the Data of the first three years to train the decision tree and use the data for the last half year for testing. The prediction results of the six training sets are shown in Table 10.

Table 10. Comparison of predictive accuracy

Inspection set	Connection airports	Connection degree	Predictive accuracy
1	PEK-SHA	1-2	80.32%
2	SHA-CAN	2-2	77.60%
3	PVG-XIY	1-3	84.20%
4	KMG-PEK	2-1	74.60%
5	PVG-PEK	1-1	79.30%
6	WUH-XIY	3-3	81.20%

Experiment 2: Mix six training sets into one training set and add attributes: "Departure airport attributes" and “arrival airport attributes” Use CHAID decision tree to train the mixed training set.The confusion matrix obtained with the data from July 1, 2016 to January 2017 is shown in Table 11.

Table 11. Confusion matrix of mixed training set

Observed	Classification Predicted 1.00	2.00	3.00	4.00	Correct percentage
1.00	10883	1323	235	871	81.8%
2.00	2095	4415	275	367	61.7%
3.00	1077	943	1324	152	37.9%
4.00	56	236	16	6006	95.1%
Total	59.8%	12.9%	2.8%	24.4%	84.7%

4.3 Comprehensive analysis

The six sets in Experiment 1 were to predict a single segment. "Flight Date", "Aircraft type", "Planned Arrival Time", "Planned Departure Time", the four attributes are taken into account, while the attributes "Flight Mission Attribute" and "Flight Attributes" are abandoned. Both properties are invalid. The average accuracy of forecast results reached 79.54%, close to 80%.

Experiment 2 is to predict multiple segments of a mixed training set. The six attributes of "flight date", "model", "planned arrival time", "planned departure time", "mission attribute", "departure airport level", and "arrival airport level" are taken into account. In the result, six properties are reserved. It is effective to analyze the network topology and add its characteristics into clustering aircrafts before prediction.

Concluded from the above seven groups of experiments, we can find that when the delay level is 1 and 4, the correct rate of the decision tree is relatively high, but the correct rate is low in the 2 and 3 categories. The reason may be related to the tilt of the training set. Decision-making path is mutually exclusive, and most of the data in the training set are at 1, 4 and it is speculated that over-fitting may occur. For this problem, pruning can be used to reduce the influence of over-fitting.

In the previous study, we divided the flight delays into four levels, which are represented by 1-4. The greater the number is, the more severe the delay. If we only consider two conditions: delay and not delay, although the accuracy rate may decrease, the correct rate of prediction should increase According to the 2016 civil aviation flight normal statistics method (the definition of flight delay refers to the condition where actual arrival time exceeds scheduled arrival time more than 15 minutes). We change the flight attributes to 0 and 1, 0 means no delay, 1 denotes delay and then obtain the correct rate of four training sets, as shown in Figure 7.

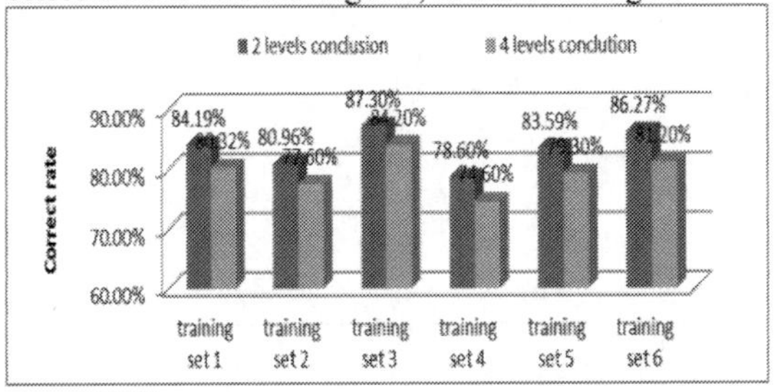

Figure 7. Comparison of the two and four types of delay attributes

In the experiment, it is found that more detailed delay attribute partitioning will lead to a decrease in the precision. The more classification values are, the more difficult to classify. For the value of delay level, the degree of subdivision should be increased as much as possible while ensuring a certain accuracy rate. If the data is skewed to a large extent, for example, 2/3 data is concentrated in <25 minutes (level 1), the level 1 should be more subdivided, and the delay of >25 should be more roughly classified.

5 Conclusion

From the perspective of the airline, this paper constructs a delay tree prediction model for forecasting delay. Firstly, we analyze the topology characteristics of aviation network, and then combine node attributes with K-means clustering algorithm to classify the busy level of airports, which improves the timeliness of the classification and the accuracy of the prediction. Finally, we use three years' operational data of a major domestic airline network to conduct experiments. The results show that the accuracy rate of the established model is close to 80%, which is of high prediction accuracy. The insufficiency of this paper lies in the tilt characteristic of the data, which is too concentrated in the delay level 1 and 4, which leads to low accuracy of the 2 and 3 levels of prediction. Our future research will focus on resolving this phenomenon of over-fitting. There are two methods to be adopted. One is to prune decision trees, the other is to classify the data sets more precisely in dense data level, and roughly divide in loose areas to solve the problem of over-fitting.

References

[1] Pyrgiotis N, Malone K M, Odoni A. Modeling Delay Propagation within an Airport Network. Transportation Research Part C: Emerging Technologies, 27: 60-75(2013).

[2] Uimera R, Amaral LAN. Modeling the World-wide Airport Network. The European Physical Journal B-Condensed Matter and Complex Systems, **38** (2): 381-385(2014).

[3] Xiaozhou Zeng. Analysis of China's Aviation Network Structure Based on Complex Network Theory. Nanjing: Nanjing University of Aeronautics and Astronautics (2012).

[4]Liu Hon KUN. Empirical study of Chinese city airline network.Acta Physica Sinica-Chinese Editin-56,106-112(2007).

[5] Quan Shao, Yan Zhu. Analysis of Flight Delay Propagation Based on Complex Network Theory Aeronautical Computing Technique (2015).

[6]Han J, Kamber M, Pei J. Data mining: Concepts and techniques (2006).

[7]XU Peng, LIN Sen. Internet traffic classification using C4.5 decision tree. Journal of Software, 20(10):2692-2704(2009).

[8]ZHAO Jing Xian. A model of financial distress early-warning based on decision tree. Journal of Harbin University of Commerce: Social Science Edition, 101(4):97-99(2008).

[9]YUAN Lin Shan, Du Pei Jun, Zhang Hua Peng, et.al. CBERS imagery classification based on decision tree and performance analysis[J]. Remote Sensing for Land & Resources, **2**:92-98(2008).

[10]QIN Wen,YUAN Chun-fa. Identification of Chinese Unknown Word Based on Decision Tree. Journal Of Chinese Information Processing..**18**, 2 (2008).

[11]PENG Shi-Feng. Design of Credit Card Risk Management System on CHAID Decision Tree. Shanghai: FU Dan University (2013).

[12]HUANG Qi. Analysis of Personal Income Based on CHAID Decision Tree. Mathematical Theory and Applications, **29**, 4(2009).

[13]Zhang Wen-Tong, Zhong Yun-Fei. IBM SPSS data analysis and mining real case. Tsinghua University Press, (2013).

Integration of Genetic Algorithm and Support Vector Machine to Predict Rail Track Degradation

Amir Falamarzi[1],Sara Moridpour[1], Majidreza Nazem[1], Reyhaneh Hesami[2]

[1]*Civil and Infrastructure Engineering Discipline, School of Engineering, RMIT University, Australia*
[2]*Asset Planning and Visualisation, Yarra trams, Australia*

Abstract. Gradual deviation in track gauge of tram systems resulted from tram traffic is unavoidable. Tram gauge deviation is considered as an important parameter in poor ride quality and the risk of train derailment. In order to decrease the potential problems associated with excessive gauge deviation, implementation of preventive maintenance activities is inevitable. Preventive maintenance operation is a key factor in development of sustainable rail transport infrastructure. Track degradation prediction modelling is the basic prerequisite for developing efficient preventive maintenance strategies of a tram system. In this study, the data sets of Melbourne tram network is used and straight rail tracks sections are examined. Two model types including plain Support Vector Machine (SVM) and SVM optimised by Genetic Algorithm (GA-SVM) have been applied to the case study data. Two assessment indexes including Mean Squared Error (MSE) and the coefficient of determination (R^2) are employed to evaluate the performance of the proposed models. Based on the results, GA-SVM model produces more accurate outcomes than plain SVM model.

1 Introduction

Nowadays public transport is regarded as an effective alternative to private transport. This mode of transportation has several advantages such as reducing road casualties, mitigating air and environmental pollution, easing traffic congestion and promoting healthy lifestyle [1, 2]. Tram as a light rail transit is implemented in different places to facilitate the movement of people. As an accessible approach of public transport option, tram has different benefits compared to conventional transit systems. For example, they are more economically efficient due to lower cost of repair and purchase than heavy rail trains. Because of their design and lower weight, trams are able to travel on steeper gradients and tighter curves compared to conventional trains. In addition, compared to buses, they are more durable and able to carry more passengers. Trams also provide more reliable services through the central business districts (CBD) than conventional buses. On the other hand, degradation of tram tracks can be increased because of population growth in big cities which leads to higher demand for the rail transport. In addition, because of sharing the routes with other private vehicles, trucks and buses, tram tracks undergo further pressure and stress [3, 4].

Based on the previous works, the degradation in tram tracks is usually measured in terms of track geometry defects including longitudinal level defects, the horizontal alignment defects, excessive gauge deviations, cross-level and track twist [5, 6]. In order to design a sustainable rail planning and management system to mitigate the degradation of rail tracks, effective preventive maintenance operations and rehabilitation actions are required. Comparative analysis conducted for light rail transport demonstrated that designing strategies and programs for tram infrastructure maintenance is a multifaceted problem in which the behaviour modelling of track degradation is an essential part. Without developing rail track degradation prediction models, designing preventive maintenance strategies is not possible [7, 8].

This research intends to create rail track degradation models by using track gauge parameter data. Investigating several studies has shown that track gauge parameter can be employed as an effective representative of degradation of whole track infrastructure. This paper is organised as follows. Section 2 reviews the existing works on modelling of rail track degradation prediction. Section 3 presents the case study which consists of the data of Melbourne tram network over multiple years. Section 4 describes the development process of the proposed models including plain SVM and GA-SVM. Section 5 yields the results of the models and discussions. Finally, Section 6 presents the conclusion of this study.

2 Literature review

Several studies based on different methods have been carried out in the field of modelling of rail track degradation. In this study, as the establishment of SVM models are decided, the literature review presented below mainly considers this type of model.

Asada et al. [9] elaborated an algorithm for developing fault detection of point machines and railway condition monitoring systems based on the SVM model. A point machine is a mechanical device used to turn the switch blades (point blades) from one point to another point in order to provide different routes for trains. In this research, two faults were addressed including over-driving and under-driving of the drive rod. Data collected from a Japanese AC point machine were applied. The results of system validation represented that the developed SVM model had accurate and valid performance in terms of classification of fault detection with different severity levels by using it in the electrical active power data of the point machine.

Ebrahimi et al. [10] elaborated SVM and a mechanistic model for prediction of ballast deformation of railway substructure in order to schedule the maintenance of ballast. Both models used the data collected from a large-scale cyclic tri-axial apparatus. Two distinguished phases were carried out including an initial compaction phase and a fouling impact phase. A gradual integration of plastic deformation of railway ballast in different fouling, traffic loading and moisture is used in conjunction with a subgrade deformation model to estimate the surface deviation of the railway track. Five-fold cross-validation was applied in this study. Based on the results of this study, the SVM method predicted the rate of plastic strain of ballast more accurately than the mechanistic model.

Cárdenas-Gallo et al. [11] conducted a research to classify different types of rail track defects such as alignment, profile and cross-level into yellow tag defects and red tag defects by applying SVM method. Red tag defects are defects in which the value of track geometry parameters exceed the track safety thresholds provided by Federal Railroad Administration (FRA) and must be fixed as soon as possible. Yellow tag defects satisfy FRA thresholds but they will ultimately turn into red tag defects if they are not treated. Capability to predict when yellow tag defects would progress toward red tags, assists railroads administrators in delivering more efficiently rail maintenance strategies. With respect to the result of the study, the performance of SVM model in classification of profile defects was successful.

3 Case study

In this research, Melbourne tram as the one of the largest tram networks in the world has been used as a case study. This network consists of 250 km of double tracks and 25 routes. Based on Yarra Trams (the operator of Melbourne tram network) reports [12], in 2017 Melbourne tram network provided service to more than 204 million passengers, which represented an annual increase of 0.2 percent. This data set includes a wide range of parameters such as track geometry parameters (e.g. profile, alignment, gauge and twist), annual traffic in MGT, rail profile, track surface, rail type, rail support, location and the year of instalment. This data set covers the years from 2009 to 2015. For measuring track geometry parameters, a chord length of 10 m is used. As the presence of noisy and outliers in maintenance data sets is usual, data filtration technique has been applied to prepare the data set for development of the models. Data filtration is the procedure of removing noisy (out of range) and incomplete data from the main data set in order to improve the accuracy of the models. If the distribution of data set behaves like the normal distribution it can be considered that 99.7% of data are in the range of 3×SD distance from the mean value [13, 14], where SD is the Standard Deviation. In this study, statistical tests such as Shapiro-Wilk test and Kolmogorov-Smirnov test have been employed. These tests are helpful measures to examine the possibility of the normal distribution. After analysing different track sections, it has been revealed that the changes in track gauge parameter follow a normal distribution. In this context, track sections in which gauge deviation parameters deviate from $\mu \pm 3 \times SD$ values are labelled as outlier data and are eliminated from the data set. In the final stage, the remaining data is controlled again and data with incomplete specifications are removed.

4 Model development

In this research, predicting track gauge deviation as a representative of tram rail degradation based on historical data has been studied. By reviewing the relevant works, input parameters including track gauge deviation in previous year (Gt-1), MGT, curve radii, rail support (RS), rail profile (RP), rail type (RT) and track surface (TS) are selected. The output parameter is track gauge in current year (Gt).

In this research in attempt to represent a broader perspective on track degradation modelling, two models including plain SVM and GA-SVM have been applied.

In this research, as the target parameter is a continuous variable, R^2 and MSE have been used to assess the performance of the proposed models as follows:

$$R^2 = 1 - \frac{\sum_{i=1}^{N}(y_i - f_i)^2}{\sum_{i=1}^{N}(y_i - \bar{y})^2} \qquad (1)$$

where R^2 denotes the coefficient of determination and provides information about goodness of fit of the model, N denotes the number of samples, f_i denotes the value estimated by the model, y_i represents the observed data and $\bar{y}$ is mean value of y_i.

$$MSE = \frac{1}{N}\sum_{i=1}^{N}(y_i - f_i)^2 \qquad (2)$$

In this study, R Statistical Software package is used to build the SVM and GA-SVM models. In the following sections, development process of SVM and GA-SVM are represented.

4.1. SVM model

SVM model can be categorised into two major classes including Support Vector Classification (SVC) and

Support Vector Regression (SVR). SVR employs the adaptive margin-based loss functions and delivers the learning data (non-linear or linear) to higher dimensional feature space. This method explores the best possible decision rule with regards to generalisation ability [15, 16]. Regression estimation can be established by finding a function of y=f(x) based on a set of training sample $\{X = (x_i, d_i), i = 1, \dots, N\}$, where $x_i \in R^n$ is the input vector (training data set), d_i is the objective value of training sample and N represents the number of training samples. To find the variables of the SVM model, the following optimisation problem needs to be solved:

$$\min \frac{1}{2}\sum_{i=1}^{N}\sum_{j=1}^{N}(\alpha_i-\alpha_i^*)(\alpha_j-\alpha_j^*)k(x_i,x_j) + \varepsilon\sum_{i=1}^{N}(\alpha_i+\alpha_i^*) - d\sum_{i=1}^{N}(\alpha_i-\alpha_i^*) \quad (3)$$

$$s.t. \sum_{i=1}^{N}(\alpha_i-\alpha_i^*) = 0 \text{ and } \alpha_i, \alpha_i^* \in [0, C] \quad (4)$$

where α_i and α_i^* are Lagrange multipliers, ε and C are the hyper parameters and $k(x_i, x)$ denotes a definite kernel function.

Kernel function can be defined in different forms including Linear kernel function, Polynomial kernel function, Gaussian kernel function and Multilayer perception kernel [17]. In this study, Gaussian kernel function is used.

4.2 GA-SVM

GA is used in engineering to generate effective solutions for optimisation problems by simulating genetic evaluation process and applying bio-inspired operators such as crossover, mutation and selection. This mechanism commences from an initial population and by randomly continues evaluation (selecting, crossing and variation) from one generation to another generation, a group of highly-adaptive individual can be produced at last. Instead of finding local optimality, this algorithm has the ability to address global optimality [18, 19]. In this study, by employing GA, the main parameters of SVM including C and γ are searched and optimised. Parameter C governs the trade-off cost between minimising the complexity of the SVM model and the training error, whereas parameter γ is not technically considered as a SVM parameter but, it is a parameter of the kernel function. In the RBF kernel, it determines how similar two points should be adjusted to be regarded similar. These two parameters have a strong influence on the generalisation performance and efficiency of the SVM model.

5 Results and Discussion

In this section, the results of the models with respect to their type are provided and discussed. Also, a comparison between the two models is presented.

First, the results of development of a plain SVM model for predicting the current track gauge deviation are presented in Table1. In this regard, the values of R^2 and MSE of the SVM models for straight data sets are calculated and provided in Table 1.

Table 1. Performance of SVM model

Parameters	Adjusted R^2	MSE
G_t		
R_T	0.90	1.95
G_{t-1}		

Figure 1 demonstrates the relationship between the data predicted by SVM model versus the observed data.

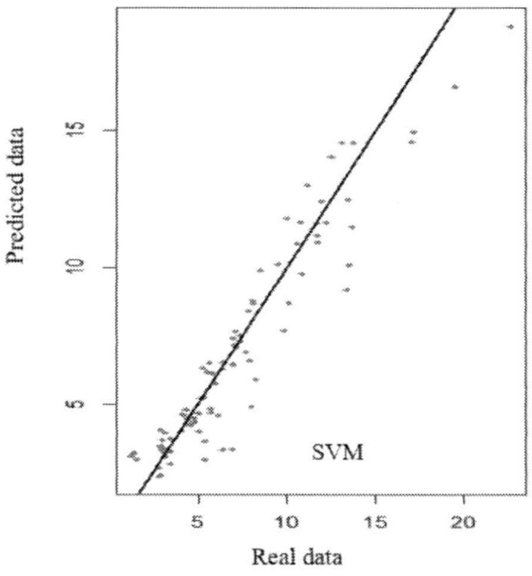

Figure 1. Observed data against predicted data (SVM)

Based on the results of the plain SVM models adjusted R^2 is 0.90 and the values of MSE is 1.95. In Table 2, the results of development of GA-SVM model for track gauge deviation prediction is presented.

Table 2. Performance of GA-SVM model

Parameters	Adjusted R^2	MSE
G_t		
R_T	0.92	1.60
G_{t-1}		

Based on the results of the GA-SVM model, adjusted R^2 is 0.92 and the values of MSE is 1.60. In Figure 2, the relationship between GA-SVM predicted data against observed data is shown.

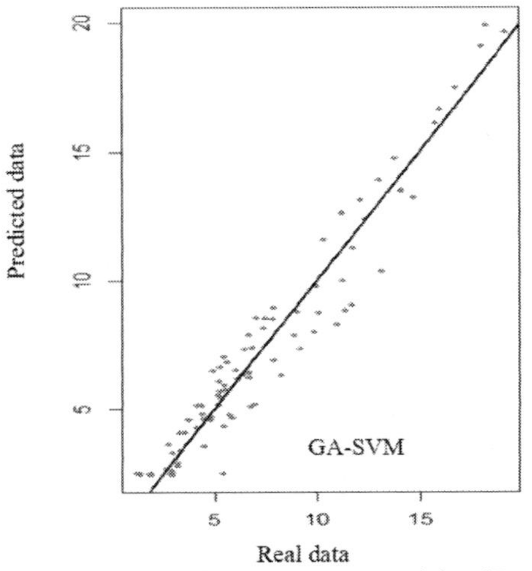

Figure 2. Observed data against predicted data (GA-SVM)

The results, which agree with the previous research works [20, 21, 22], show the strong correspondence between initial value of the track geometry parameters and current value of the parameters as well as rail type. Based on the results derived from the proposed models and in terms of adjusted R^2 and MSE, GA-SVM model represents a better performance than the plain SVM model in prediction of gauge deviation. These results are shown in Figures 1 & 2 where the predicted data of GA-SVM model are closer to the observed data.

6 Conclusion

Track gauge deviation is represented as a valuable measure of ride quality and the potential of train derailment. Practices for predication of gauge deviation is a principal prerequisite in establishment of the preventive maintenance operations. Performing preventive maintenance operations which is considered as an essential part of tram infrastructure maintenance and management systems, will result in reliable tram services along with decrease in rail maintenance cost. In this research, track data collected from Melbourne tram network was applied as a case study. Data used in this study composed of different parameters but the key parameters included track gauge deviation along with traffic data and structural parameters.

In this research for track degradation prediction, two AI approaches including SVM and GA-SVM models were used and developed. Along with model development, models were assessed based on MSE and the coefficient of determination. According to the results of this study, the performance of both SVM and GA-SVM models in predicting gauge deviation values of straight track sections is satisfying. In this regards, the values of R^2 for both models are above 0.90 and MSE values are less than 2.00. The results also showed that due to optimisation of the parameters of SVM model, the GA-SVM model is more accurate in prediction of track gauge than the plain SVM model.

References

1. Redman, L., Friman, M., Gärling, T., & Hartig, T. (2013). Quality attributes of public transport that attract car users: A research review. Transport Policy, 25, 119–127.

2. D. Knowles, R., & Ferbrache, F. (2016). Evaluation of wider economic impacts of light rail investment on cities. Journal of Transport Geography, 54, 430–439.

3. Gadziński, J., & Radzimski, A. (2016). The first rapid tram line in Poland: How has it affected travel behaviours, housing choices and satisfaction, and apartment prices? Journal of Transport Geography, 54, 451–463.

4. Ahac, M., & Lakusic, S. (2017). Track Gauge Degradation Modelling on Small Urban Rail Networks: Zagreb Tram System Case Study. Urban Transport Systems, (January).

5. He, Q., Li, H., Bhattacharjya, D., Parikh, D. P., & Hampapur, A. (2015). Track geometry defect rectification based on track deterioration modelling and derailment risk assessment. Journal of the Operational Research Society, 66(3), 392–404.

6. Andrade, A. R., & Teixeira, P. F. (2018). Assessing Temporary Speed Restrictions and Associated Unavailability Costs in Railway Infrastructure. International Journal of Civil Engineering, 16(2), 219–228.

7. Andrews, J. (2012). A modelling approach to railway track asset management. Proceedings of the Institution of Mechanical Engineers, Part F: Journal of Rail and Rapid Transit, 227(1), 56–73.

8. An, R., Sun, Q., Wang, F., Bai, W., Zhu, X., & Liu, R. (2018). Improved Railway Track Geometry Degradation Modeling for Tamping Cycle Prediction, 144(2015), 1–11.

9. Asada, T., Roberts, C., & Koseki, T. (2013). An algorithm for improved performance of railway condition monitoring equipment: Alternating-current point machine case study. Transportation Research Part C: Emerging Technologies, 30, 81–92.

10. Ebrahimi, A., Tinjum, J. M., & Edil, T. B. (2014). Maintenance model for railway substructure. Geotechnical Engineering, 45(1), 48–55.

11. Cárdenas-Gallo, I., Sarmiento, C. A., Morales, G. A., Bolivar, M. A., & Akhavan-Tabatabaei, R. (2017). An ensemble classifier to predict track geometry degradation. Reliability Engineering and System Safety, 161(April 2016), 53–60.

12. PTV. (2017). Public Transport Victoria Annual Report. Melbourne.

13. Grafarend, E. W. (2006). Linear and Nonlinear Models : Fixed Effects, Random Effects, and Mixed Models. Berlin: De Gruyter.

14. Gordon, S. (2006). The Normal Distribution. University of Sydney. Sydney.

15. Jiang, X., Zhang, L., & Chen, M. X. (2014). Short-term forecasting of high-speed rail demand: A hybrid approach combining ensemble empirical mode decomposition and gray support vector machine with real-world applications in China. Transportation Research Part C: Emerging Technologies, 44, 110–127.

16. Li, Q., Shi, Z., Zhang, H., Tan, Y., Ren, S., Dai, P., & Li, W. (2018). A cyber-enabled visual inspection system for rail corrugation. Future Generation Computer Systems, 79, 374–382.

17. Ben-hur, A., & Weston, J. (2010). Data Mining Techniques for the Life Sciences, 609, 223–239.

18. Li, X. Z., & Kong, J. M. (2014). Application of GA-SVM method with parameter optimization for landslide development prediction. Natural Hazards and Earth System Sciences, 14(3), 525–533.

19. Zhang, Z., Qin, Y., Cheng, X., Zhu, L., Kou, L., Li, J., & Sun, F. (2015). Metro Station Safety Status Prediction Based on GA-SVR. In International Conference on Electrical and Information Technologies for Rail Transportation (pp. 57–69). Berlin: Springer.

20. Andrade, A. R., & Teixeira, P. F. (2015). Statistical modelling of railway track geometry degradation using Hierarchical Bayesian models. Reliability Engineering and System Safety, 142, 169–183.

21. Moridpour, S., Mazloumi, E., & Hesami, R. (2017). Application of Artificial Neural Networks in Predicting the Degradation of Tram Tracks Using Maintenance Data. In Applied Big Data Analytics in Operations Management (pp. 30–54). Delhi, India: IGI Global.

22. Falamarzi, A., Moridpour, S., Nazem, M., & Hesami, R. (2018). Rail Degradation Prediction Models for Tram System : Melbourne Case Study. Journal of Advanced Transportation, 2018, 8.

Empirical Study on Maximum Traffic Throughputs at Intersections

Patikhom Cheevarunothai[1] and Ruangsak Kaewpikul[1]

[1] *Department of Civil and Environmental Engineering, Thammasat University, Pathumthani 12121, Thailand*

Abstract. Traffic congestion in big cities of Thailand has been a major problem seriously deterring economy growth. An average travel speed has become lower than 15 km/hour during peak-hour periods. One of the current major questions is how to find the most optimum green time that minimizes vehicle delay at an intersection. Since the identification of this optimum green time significantly depends on the knowledge of how fast vehicles can pass through an intersection, the information of the maximum traffic throughputs at intersections at different locations is indispensable. In this study, therefore, traffic counts were extracted manually from the videos recorded at 4 major intersections in Saensuk city. After the data were analyzed, the maximum traffic throughputs for different lane numbers and configurations were calculated and summarized. A better understanding of this maximum traffic throughput at intersections will lead to more accurate estimation of optimal green time.

1 Introduction

Traffic congestion in Thailand has been a major problem slowing down economy growth. With the number of vehicles exceeding the available roadway capacity in the network, traffic condition during rush hours becomes extremely congested. For example, at present, there are about 7 million registered vehicles in Bangkok but the current roadway capacity is only available for 2.5 million vehicles. The rush-hour periods expand longer each year mainly because of the increasing number of personal vehicles.

For last 20 years, to minimize traffic congestion, several new roads and expressways were built. However, there has been no sign of improvement on travel time in the area. Under bad economy for about a decade and the unsuccessful outcomes from previous traffic congestion mitigation strategies, most of major cities have limited budget and less support for new road construction projects from the communities. Therefore, an improvement on the current operations of existing traffic systems becomes more feasible option.

In 2015, after experiencing severe traffic congestion on major roads during peak hours, Saensuk city started a new traffic mitigation program. This new program's first phase started the work on major intersections with the highest congested level. The program emphasizes on the application of Intelligent Traffic Systems (ITS) to lessen the traffic congestion level. At the selected intersections, there was an attempt to use vehicle data collected from recorded videos to enhance an understanding of traffic congestion and identify the optimum green time that results in high intersection capacity. To be able to select the optimum green time, a good understanding of possible maximum traffic throughput at intersection is essential. As a result, the data on maximum traffic throughput for different lane numbers and configurations were collected and summarized in this study.

2 Literature review

There has been many researches with the effort to minimize vehicle delay at intersections. The work is related to important traffic parameters such as traffic capacity and saturation flow rate at an intersection. Samples of the researches can include the following works. In 1968, Miller [1] conducted a research to estimate capacity at signalized intersections in Australia. After that, Akcelik [2] analyzed not only capacity but also timing at signal intersections. Kimber et al. [3] developed a procedure to predict the saturation flow rate at an intersection that is controlled by traffic signals. Bester and Meyers [4] also worked on saturation flow rate at an intersection.

In recent years, non-motorized mode has become more popular. As a result, Li et al. [5] conducted a research on the saturation flow rate of shared non-motorized lane at intersections. Applying a better understanding of the traffic throughput at an intersection, many researchers had developed efficient ways to improve traffic throughputs, such as Liu and Chang who worked on an arterial signal optimization for intersections [6] and Li and Elefteriadou who worked on maximizing the throughput of turn bays at a signalized intersection [7]. Additional works may include the work from Zhou et al. [8], Wu et al. [9], and Chunhui et al. [10]. As advanced traffic-related and automobile technologies are rapidly growing and increasingly play an important role in our present society, new research related to the reduction of vehicle delay at intersections has become broader.

3 Data source

Intersections with high traffic congestion level in the Saensuk city were selected for this study, which include 4 major intersections that are 1) Galaxy, 2) Burapa, 3) Kao Larm, and 4) Lam Tann. These intersections are on the major arterials in the city. Each intersection has different number of arms and lanes and/or lane configurations. For example, the Galaxy intersection is a three-way junction as shown in Fig. 1. The Southbound (SB) approach has 4 lanes. Similarly, the Northbound (NB) approach has 4 lanes. However, the Eastbound (EB) approach has 3 lanes. The lane type is also displayed by white arrows in Fig. 1.

Figure 1. Galaxy Intersection

The Burapa intersection is also a three-way junction with 2 lanes EB, 3 lanes WB, and 3 lanes NB (shown in Fig. 2). The special thing at this intersection is the rightmost u-turn lane of the WB approach. The last two intersections are Kao Larm and Lam Tann intersections, which are four-way junctions (Fig. 3 and 4, respectively). The Kao Larm intersection has 2 lanes EB, 4 lanes WB, 2 lanes NB, and 3 lanes SB. Similarly, for the Lam Tann intersection, there are 1 lane EB, 2 lanes WB, 2 lanes NB, and 3 lanes SB.

To collect traffic count data, the CCTV cameras were installed at these 4 major intersections (displayed in Fig. 2). Videos were recorded for 24 hours for the days with high traffic congestion. After the investigation of traffic condition, the time interval 16:00-20:00 has been identified as the time period with highest traffic volume. Therefore, the manual traffic count was conducted for this time period and the results are summarized in this study.

Figure 2. Installation of CCTV cameras and a snapshot of recorded videos

4 Maximum traffic throughout

To ensure the accuracy of the collected traffic data, vehicle counts were extracted manually for three signal cycles in every half hour between 16:00-20:00 and shown in Tables 1-4. Lane 1 is always the rightmost lane in each approach. According to Tables 1-4, the maximum number of vehicles that can pass an intersection on the left lane is always smaller than its adjacent right lane. For example, based on the median values of traffic counts on the NB Lanes 2 and 3 at the Galaxy intersection, the number of vehicles that could pass the intersection in the same time period on the left lane can be as low as 45% of the number of vehicles on the adjacent right lane on the same approach. Similarly, Lane 4 SB is the leftmost lane and the traffic count is only 548 vehicles compared to 1078 vehicles on Lane 3 SB.

Table 1. Maximum traffic throughput at Galaxy intersection

time	cycle	Galaxy								
		NB			SB				EB	
		lane 1	lane 2	lane 3	lane 1	lane 2	lane 3	lane 4	lane 1	lane 2
16:00	1	1,770	1,368	544	1,424	1,503	1,298	1,398	1,205	165
	2	1,536	1,587	1,204	1,425	1,329	1,567	643	1,283	935
	3	1,466	1,168	205	1,225	1,240	1,187	981	1,480	1,132
16:30	1	1,557	1,349	544	1,276	1,331	1,240	539	1,366	1,087
	2	1,651	1,531	1,360	1,422	1,271	1,326	702	760	1,072
	3	1,473	1,349	530	1,481	1,459	1,109	339	1,073	1,020
17:00	1	1,391	1,495	984	1,422	1,537	736	351	1,513	901
	2	1,705	1,643	1,138	1,305	1,439	1,320	307	1,295	619
	3	1,717	1,506	932	1,526	1,548	1,321	1,187	1,615	1,018
17:30	1	1,692	1,551	363	1,196	1,525	995	541	1,407	1,194
	2	1,807	1,200	843	1,313	1,288	1,091	386	1,194	1,520
	3	1,633	1,339	836	1,242	1,176	1,106	225	1,493	1,364
18:00	1	1,470	1,618	1,064	1,270	1,498	775	776	1,197	1,202
	2	1,014	825	585	1,461	1,037	881	561	1,494	1,017
	3	1,812	1,790	437	1,461	966	966	938	1,201	1,319
18:30	1	1,628	1,301	661	1,452	1,450	1,078	1,041	1,380	1,235
	2	1,623	1,483	610	1,172	1,298	894	348	1,220	1,298
	3	1,610	1,320	682	1,439	1,126	980	801	1,321	1,317
19:00	1	1,467	1,383	338	1,452	1,051	1,078	271	1,381	1,044
	2	1,659	1,304	406	1,172	1,350	894	397	1,220	1,282
	3	1,333	1,370	1,056	1,439	1,338	980	1,011	1,321	1,083
19:30	1	1,100	1,299	683	1,258	1,643	772	896	1,283	1,145
	2	963	1,128	702	1,418	1,389	886	974	1,428	1,470
	3	1,463	1,409	624	1,342	1,367	1,307	312	1,444	646
20:00	1	1,255	895	389	1,263	1,683	1,064	483	1,446	1,036
	2	735	808	230	1,386	1,541	1,226	477	1,268	1,847
	3	1,545	1,206	452	1,290	1,380	1,018	312	1,137	1,238
Average		1,477	1,355	646	1,359	1,402	1,079	701	1,337	1,283
Median		1,545	1,383	624	1,428	1,399	1,078	548	1,321	1,245

At the Burapa intersection, since Lane 1 is for through and U-turn traffic, its maximum traffic throughput is less than that of Lane 2, which is through-left-turn lane. At this intersection, traffic demands on Lane 3 NB, Lane 2 EB, and Lane 3 WB are significantly less than the capacity, so those lanes were not included in Table 2.

Table 2. Maximum traffic throughput at Burapa intersection

time	cycle	Burapa				
		NB		EB	WB	
		lane 1	lane 2	lane 1	lane 1	lane 2
16:00	1	1,559	556	1,345	1,147	1,134
	2	1,001	300	1,107	848	1,274
	3	1,179	468	983	1,173	1,152
16:30	1	952	528	1,559	642	1,084
	2	1,208	1,369	1,248	948	1,114
	3	1,207	—	328	1,110	1,291
17:00	1	1,289	944	1,137	678	1,063
	2	1,066	372	1,072	592	1,380
	3	1,312	1,533	1,250	835	1,346
17:30	1	1,228	—	719	1,124	1,251
	2	1,002	936	834	1,026	1,508
	3	1,009	1,318	208	881	1,388
18:00	1	681	506	1,231	626	1,395
	2	983	516	478	584	1,288
	3	1,267	375	454	918	1,232
18:30	1	1,127	—	1,044	826	866
	2	1,092	912	955	889	1,147
	3	1,279	574	1,227	569	1,020

https://doi.org/10.1051/matecconf/201925902008

19:00	1	1,211	—	822	877	1,272
	2	1,167	1,083	987	661	1,363
	3	1,159	554	608	823	1,407
19:30	1	985	—	1,024	929	1,075
	2	1,071	680	1,020	885	1,225
	3	1,071	570	1,137	777	699
20:00	1	1,246	879	569	566	1,100
	2	721	1,513	1,547	817	957
	3	1,084	1,057	1,391	831	869
ave.		1,168	1,057	1,134	836	1,181
median		1127	627	1024	835	1225

Figure 2. Burapa Intersection

Table 3. Maximum traffic throughput at Kao larm intersection

time	cycle	Kao Larm								
		NB		SB		EB		WB		
		lane 1	lane 2	lane 1	lane 2	lane 1	lane 3	lane 1	lane 2	lane 3
16:00	1	1,381	1,721	1,283	1,088	2,232	1,305	1,803	2,313	1,165
	2	1,305	1,297	1,101	983	855	—	809	740	1,397
	3	1,449	1,028	866	1,044	2,466	—	1,213	2,271	1,451
16:30	1	1,386	1,134	740	—	1,264	—	778	429	1,037
	2	2,074	1,236	1,304	715	1,957	—	1,231	1,610	—
	3	2,777	1,383	1,082	—	1,898	—	1,161	1,477	1,280
17:00	1	1,484	842	744	—	1,130	—	1,458	1,326	914
	2	1,893	1,282	1,958	—	1,656	—	1,383	1,802	1,163
	3	1,888	933	951	—	418	—	1,509	1,926	1,224
17:30	1	1,402	1,571	1,578	—	1,638	—	1,143	1,152	1,088
	2	1,822	1,483	1,081	1,111	1,154	—	1,619	1,638	737
	3	1,301	1,375	261	588	1,339	—	1,212	1,398	1,089
18:00	1	1,270	1,074	1,880	—	826	—	894	1,686	460
	2	1,367	774	1,110	1,105	1,270	—	1,087	1,477	547
	3	1,288	1,407	849	1,720	1,707	—	1,127	1,746	1,228
18:30	1	1,514	747	589	1,440	1,667	—	840	1,714	738
	2	1,512	1,388	1,487	2,001	1,708	—	1,044	1,469	465
	3	1,346	1,389	1,329	—	2,279	—	903	1,423	1,572
19:00	1	1,008	1,277	1,217	1,339	1,283	358	1,179	955	1,493
	2	1,956	420	929	—	446	—	1,458	1,593	954
	3	1,477	896	587	543	1,289	—	976	1,330	1,258
19:30	1	641	1,020	1,576	1,246	1,249	852	1,409	1,177	1,334
	2	1,698	1,533	606	639	1,546	496	1,224	1,001	1,284
	3	2,362	941	669	—	1,434	527	1,101	1,088	1,575
20:00	1	1,808	1,120	1,710	—	648	1,335	1,167	1,858	681
	2	966	1,223	1,852	473	1,906	—	746	542	—
	3	1,977	1,131	1,302	1,254	445	—	753	1,488	—
ave.		1,642	1,398	1,363	1,066	1,398	846	1,140	1,382	1,097
median		1312	1223	1317	1096.5	1434	705	1167	1469	1164

Figure 3. Kao Larm Intersection

Table 4. Maximum traffic throughput at Lam Tann intersection

time	cycle	Lam Tann					
		NB		SB		EB	WB
		lane 1	lane 2	lane 1	lane 2	lane 1	lane 1
16:00	1	1,403	1,165	1,353	562	1,572	525
	2	1,118	1,245	1,371	654	1,021	1,419
	3	1,494	1,601	1,387	782	1,447	1,347
16:30	1	1,372	1,611	—	—	1,089	1,745
	2	1,513	820	632	—	1,079	1,881
	3	1,289	585	—	—	758	1,148
17:00	1	1,320	1,365	1,527	—	1,789	1,511
	2	1,586	1,301	—	—	1,102	1,558
	3	1,406	790	1,191	—	1,579	1,747
17:30	1	1,305	628	1,356	—	820	1,749
	2	1,610	1,557	1,341	—	1,448	1,690
	3	1,345	1,429	706	—	1,316	1,705
18:00	1	1,267	1,061	1,085	—	435	1,596
	2	1,349	1,281	673	—	1,165	2,170
	3	1,364	1,163	600	—	1,990	1,802
18:30	1	1,368	945	1,462	634	1,619	1,730
	2	1,028	1,316	1,572	1,103	1,995	1,612
	3	1,645	1,581	1,531	886	1,184	2,504
19:00	1	649	719	1,504	—	1,906	971
	2	1,300	1,200	2,236	—	2,394	1,908
	3	1,258	890	1,314	—	2,154	1,125
19:30	1	1,163	1,273	1,498	—	2,772	861
	2	1,247	1,267	1,696	234	1,224	1,300
	3	932	1,221	1,117	1,155	2,587	716
20:00	1	1,403	1,165	1,353	562	1,572	525
	2	1,118	1,245	1,371	654	1,021	1,419
	3	1,494	1,601	1,387	782	1,447	1,347
		1,353	1,287	1,425	886	1,573	1,686
		1,345	1,245	1,363.5	654	1,447	1,558

Figure 4. Lam Tann Intersection

Because each traffic lane allows different kinds of movements (e.g., through, left turn, right turn) and maximum speed for each movement is different, we separated traffic lanes into 6 types: 1) through lane, 2) right-turn lane, 3) through-right-turn lane, 4) through-left-turn lane, 5) through-right-left-turn lane, and 6) through-U-turn lane. The possible maximum traffic throughputs for each lane type is summarized in Table 5.

The surprising finding is that the traffic throughputs on Type 1 lane is never higher than 1,500 vehicles/hour and the traffic throughputs for Types 2 and 3 lanes are always higher than that of Type 1 lane. According to Table 5, the medians of vehicle numbers passing through the selected intersection on Types 2 and 3 lanes are 1,310 and 1,402 vehicles/hour, but Type 1 lane has only 1,088 vehicles/hour.

 https://doi.org/10.1051/matecconf/201925902008

Table 5. Traffic throughput by lane type

Intersection	Direction	Lane#	Type 1	Type 2	Type 3	Type 4	Type 5	Type 6
Galaxy	NB	1	1,477					
		2	1,355					
		3	846					
	SB	1	-	1,369				
		2	-		1,402			
		3	1,079					
		4	701					
	EB	1		1,337				
		2		1,283				
Burapa	NB	1	1,168					
		2	1,057					
	EB	1			1,134			
	WB	1						1,140
		2				1,181		
Kao Larm	NB	1			1,642			
		2				1,398		
	SB	1			1,365			
		2	1,066					
	EB	1			1,396			
		2				846		
	WB	1		1,140				
		2			1,662			
		3	1,097					
Lam Tann	NB	1			1,353			
		2	1,287					
	SB	1			1,425			
		2				886		
	EB	1					1,573	
	WB	1			1,686			
median			1,088	1,310	1,402	1,034	1,573	1,140

5 Conclusions and recommendations

In this study, the information of possible maximum traffic throughputs at major intersections was collected using the traffic count data extracted from the videos recorded by deploying CCTV cameras. Four major intersections with high traffic congestion in Saensuk city were chosen and afterward new CCTV cameras were installed at these intersections. Traffic throughput data were manually extracted from the recorded videos for the time period of 16:00-20:00 because, after the visual observation, this time period covers peak traffic hours.

The maximum traffic throughputs for different number of lanes and lane configurations at the chosen intersections were collected and analyzed. There are two 3-way and two 4-way intersections. Traffic throughputs for different lane types should be treated separately. With a better understanding of possible maximum traffic throughputs, the identification of optimum green time to minimize vehicle delay would become more accurate. In the future work, a statistical analysis to identify a relationship among important parameters, such as green time vs. maximum throughput will be conducted.

References

1. A.J. Miller, The Capacity of Signalized Intersections in Australia. Australian Road Research Board, Bull. No. **3** (1968)
2. R. Akcelik, Traffic Signal: Capacity and Timing Analysis. Australian Road Research Board, Research Report ARR No.**123** (1981)
3. R.M. Kimber, M. McDonald and N.B. Hounsell, The Prediction of Saturation Flows for Road Junctions Controlled by Traffic Signals. Transport and Road Research Laboratory, Research Report **67** (1986)
4. C.J. Bester, and Meyers W.L. Saturation Flow Rates. *The 26th Southern African Transport Conference*. Pretoria, South Africa (2007)
5. L. Li, D. Lin, X. Chen and W. Ma, Saturation Flow Rate of Shared Non-Motorized lane at intersections. *Transportation Research Board 94th Annual Meeting*, Washington, D.C., USA (2015)
6. Y. Liu and G. L. Chang, An Arterial Signal Optimization Model for Intersections Experiencing Queue Spillback and Lane Blockage, *Transportation Research Part C: Emerging Technologies*, Vol. **19**, pg. 130-144 (2011)
7. Z. Li and L. Elefteriadou, Maximizing the Traffic Throughput of Turn Bays at a Signalized Intersection Approach. *Journal of Transportation Engineering*, Vol. **139**, Issue 5 (2013)
8. H. Zhou, H. G. Hawkins Jr., and Y. Zhang, Arterial Signal Coordination with Uneven Double Cycling, *Transportation Research Part A: Policy and Practice*, Vol. **103**, pg. 409-429 (2017)
9. J. Wu, D. Ghosal, M. Zhang and C. N. Chuah, Delay-Based Traffic Signal Control for Throughput Optimality and Fairness at an Isolated Intersection, *IEEE Transaction on Vehicular Technology*, Vol. **67** Issue 2 (2018)
10. Y. Chunhui, F. Yiheng, L. Henry X, M. Wanjing, and Y. Xiaoguang, Integrated Optimization of Traffic Signals and Vehicle Trajectories at Isolated Urban Intersections, Transportation Research Part B: Methodological, Vol. **112**, pg. 89-112 (2018)

MATEC Web of Conferences **259**, 02009 (2019) https://doi.org/10.1051/matecconf/201925902009
ICTLE 2018

An enhanced routing algorithm using ant colony optimization and VANET infrastructure

Noussaiba Melaouene and Rahal Romadi

Team of information Research and indexing documents, texts and multimedia, ENSIAS, Mohammed V University, Rabat, Morocco

Abstract. For the last fifty years, finding efficient vehicle routes has been studied as a representative logistics problem. In the transportation field, finding the shortest path in a road network is a common problem. VANET presents an innovation opportunity in the transportation field that enables services for intelligent transportation system (ITS) especially communication features. Because of VANET features [1] and despite road obstacles, a route for the shortest path can be established at a given moment. This paper proposes an enhanced algorithm, based on ACO Ant Colony Optimization and related to VANET infrastructure that aims to find the shortest path from the source to destination through the optimal path; in addition, a storage on static nodes is installed in each intersection in a VANET environment and for a specific time.

1. Introduction

A vehicular ad hoc network (VANET) joins features in vehicles in order to make them smarter [1]. Vehicular Communication in VANETs considers two major architectures: Vehicle-to-vehicle communication **(V2V)** that enables communication between vehicles, and Vehicle-to-infrastructure **(V2I)** that is the wireless exchange of information between vehicles and roadside. Figure 1 shows a Vehicular Ad Hoc Network characterized by a highly dynamic topology that results in a frequent disconnected network. Thanks to unlimited battery storage, nodes in VANETs do not experience power limitation. To improve the safety of the transportation area, VANETs network enables the development of many applications such as safety applications to avoid accidents, efficiency applications to maximize road capacity and avoid traffic congestion, and commercial applications to access internet.

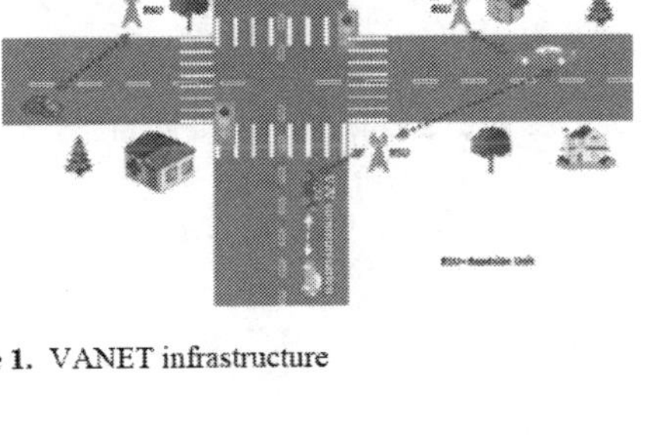

Figure 1. VANET infrastructure

Ants in nature manage amazing feats like building nests and finding food. The swarm intelligence of natural environmental agent called ant colony optimization (ACO) was developed by Dorigo [2, 3] as a kind of meta-heuristic optimization method. The ACO is based on the communication system used by ants when they look for food sources [4]. Ants search food starting from their nest and taking different paths to reach food. In this situation, the only necessary information is the path taken to food. Nevertheless, the complete topology of the environment is that not clear for a single ant which deposits the pheromone along its path. Subsequent ants use an indirect communication, which consists of tracing the pheromone left by the previous ants. The more the ants pass through a particular path, the more the concentration of the pheromone increases along that path. The pheromone acts as significant stimuli since other ants can sense the pheromone deposited by each other, and they generally take the path with maximum pheromone concentration [5]. Ants follow this method to progressively converge on a single optimum path between their nest and food.

Initially, the important goal is to provide the shortest and the most optimal path for a vehicle from source S to its destination D. This can be realized through the swarm intelligence techniques that usea behavior similar to the living organisms in nature (explained above). These techniques form a part of artificial intelligence and have a wide range of applications that exists in the literature. Some of these techniques are being used in vehicular networks that implement routing or vehicles behavior control.

In a real VANET network, obstacles between two or more vehicles may exist. These obstacles weaken the transmitted signal and hinder communication. Therefore, it is necessary to find alternative communication routes for vehicles in these conditions expand it through all Road Side Units (RSU) in routes in a given time for better quality services, increase the waiting time and improve information centralization. To meet this objective, this paper proposes an enhanced algorithm based on Ant Colony Optimization (ACO); a nature-inspired algorithm with fast response time, low computational effort, and that is particularly useful for routing problems.

This paper is organized as follows: section 2 outlines a literature survey of many proposed algorithm for routing problems in VANETs, section 3 presents the proposed work and its main goal, while the detailed steps are explained in section 4. Some assumptions and the method to implement the algorithm are presented in section5. Section 6 concludes and offers a future outlook.

2. Literature survey

A large number of routing protocols have been proposed for VANETs [6, 7]. The main goal of these routing protocols is to provide the best path through multiple wireless communications. Some ACO-based algorithms for the routing problems in VANETs have been proposed. In this section, most representative literature on vehicular routing areas using ACO algorithm (Ant Colony Optimization) are provided.

R.Silva et al in [8] propose an ACO-based algorithm for multi-objective routing in VANET. The proposed algorithm, called AntRs, find routes considering the best commitment between shortest path (number of nodes/vehicles in a route) and the lowest probability of disconnection. Simulations are implemented in three different scenarios namely static routing, static routing with obstacles, and dynamic routing. Results are very promising showing a good adaptability. Nevertheless, the algorithm still has to be improved to fit high-density and dynamic networks.

G.Caro in [9] describes a hybrid algorithm, named AntHocNet, combining both reactive and proactive elements. After a reactive path setup, the algorithm probes, maintains and improves path in a proactive way. The algorithm is based on the nature-inspired Ant Colony Optimization framework. The advantage exists in terms of packet delivery ratio, average end-to-end delay and average jitter, and increases in larger, sparser and more dynamic environments. However, AntHocNet is less efficient in terms of routing overhead.

Toklu in [10] proposes a robust multiple ant colony system to solve the uncertain travel cost problem. To generate a collection of solutions with different levels of protection against the uncertainty, the author uses multiple ant colonies in parallel; each of which focuses on a different conservativeness degree. This method is handled by incorporating linear formulations from the field of robust optimization into the metaheuristic approach.

G.Li in [11] proposes a new adaptive multi-criteria VANET routing protocol, named VACO (Vehicular routing protocol based on Ant Colony Optimization). The proposed algorithm combines both reactive and proactive components to respectively establish and maintain best routing paths. The simulations indicate that VACO shows better performance than reference protocols (GPSR and CAR). The algorithm still has to be improved to fit with realistic vehicular traffic.

3. Proposed work

The proposed algorithm aims to provide the best path from source to destination. First, an area with randomly scattered N vehicles is considered. The area contains multiple intersections and obstacles. To reduce the probability of disconnection, the communication features in VANETs are required to establish a connection between vehicles (V2V) and between vehicle and infrastructure (V2I). The route between the source S and the destination D is divided into sub-routes to make research easier and to reduce the waiting time.

At the beginning, the source vehicle S starts by checking if a route already exists in the neighborhood or not. If the answer is affirmative, the vehicle S follows the given routing path. If not, the search of the best path to destination starts using Ant Colony Optimization; this is called the reactive path exploration where the exploration of the path is established on demand. When a *forward* ant arrives to destination D, respecting the less end-to-end delay, a reactive *backward ant* is generated to return to the source S and deposit the pheromone on the optimal road; the deposited pheromone describes the quality of the road. Once the route is established, the optimal route found is recorded and sent to the roadside units at intersections, where the information is updated for a definite time interval. The goal of the information centralization at intersections is to reduce the movement of vehicles individually on the routing path. Finally, a process of pheromone evaporation is necessary in each time interval.

Choosing VANET environment was necessary for its network characteristics. First, the VANET network is homogeneous; vehicles have the same physical capacity and the same radio power [1, 8].Second, there is no hierarchy between vehicles thus the network can be described as Flat. Finally the dynamic topology of the VANETs network enables communication during movement.

It is important to mention that network congestion is not considered in this paper. The optimal path will be the one with the best commitment between the less end-to-end delay and the lowest hops number between the source and the destination.

4. The algorithm description

In this section, a detailed description of the different components of our proposed algorithm is given.

4.1 The path exploration

The aim of the algorithm is to find the best path from source S to destination D. To do so, N vehicles are randomly scattered in the area. The source vehicle S sends route request to the two neighboring intersections to check whether a route to destination is already recorded or not. If the route exists and is updated, the intersection sends an affirmative message to S, and then the source vehicle S directly implements data packet forwarding [12]. Otherwise, S starts to search the best path between Source and destination utilizing the Ant Colony Optimization by sending *forward ant*.

At the beginning, the source S sends *ants* randomly to the neighboring intersections searching candidate routing path to destination D. When arriving at intersection I, the task of the *forward ant* is to find the path connecting S to D. At each intersection, according to whether or not the intersection has the routing information to D, an *ant* is either unicast or broadcast. The next intersection j of the forward ant is chosen probabilistically according to equation (1) [8, 13] based on the global pheromone deposited at intersection (i).

$$\mathbf{Pi,j} = \mathbf{\tau i,j} \Big/ \sum_{k=1}^{n} \mathbf{\tau k} \quad (1)$$

where **Pi,j** is the probability to choose the intersection(j) from the intersection(i), **τi,j** is the pheromone amount on the road segment between intersection i and j, $\sum_{k=1}^{n} \mathbf{\tau k}$ is the sum of the pheromone in the whole path from intersection (i).

The *forward ant K* records every intersection it has passed. Every intersection has an identifier ID. Therefore, by each past intersection, the number of hops increases. Each *forward ant* keeps a list of the intersections it has visited.

4.2 The path establishment

Upon arrival at the destination D, the *forward ant* is converted into a *backward ant*. Using unicast transmission, the backward ant travels back to the source vehicle S following the reverse path. The backward ant sets up a path towards the destination D and then deposits and updates the pheromone in each go through intersection.

4.3 The pheromone deposit update

After the route exploration and establishment, the pheromone deposit is necessary for the other ants to follow the routing path easier. Through the return path from destination D to the source S by the backward ant, the pheromone value is updated or created. The process of pheromone deposit follows the equation (2) [14] below:

$$\mathbf{\tau i,j} \leftarrow \rho \times \mathbf{\tau i,j} + \sum_{K} \mathbf{\Delta\tau i,j^{h}} \quad \rho\epsilon\,(\mathbf{0,1}) \quad (2)$$

where **τi,j** is the pheromone amount on the road segment (i, j), **ρ** is an evaporation coefficient, **k**>0 is the number of road segments in the chosen path, $\mathbf{\Delta\tau i,j}^{h}$ are the increased pheromones on the road segment (i,j) of route h found by the backward ant. The pheromone increment-updating rule uses the ant-weight strategy presented by [15].

The iteration is repeated until all *backward ants* arrive to the source S. The route possessing the highest pheromone value is considered as the optimal path between source S and destination D.

4.4 The pheromone evaporation

A process of pheromone evaporation is applied at the trails/roads to avoid a rapid convergence to a local optimum. To imitate the real ant's pheromone, the mechanism of evaporation consists in decreasing the level of the pheromone value on all road segment links. As known, the choice of a path follows a probabilistic equation thus wrong decision can happen. Therefore, to make ants forget the low-quality routes, the pheromone should be evaporated according to the equation (3) [13] below:

$$\mathbf{\tau i,j} = (\mathbf{1} - \mathbf{\rho})\mathbf{\tau i,j} \quad (3)$$

where **τi,j** is the pheromone amount on the road segment between intersection i and j and **ρ** is an evaporation coefficient.

4.5 Updating the traffic information

If the path establishment is successful, a number of good paths between source and destination are made available. The intersections are updated with the information of the best path between the source S and the destination D. Otherwise, if no *backward ant* returns to the source S after a certain amount of time (1s), data are temporarily buffered and the whole process is restarted.

The main steps of the proposed algorithm are given in the flowchart illustrated in Figure 2.

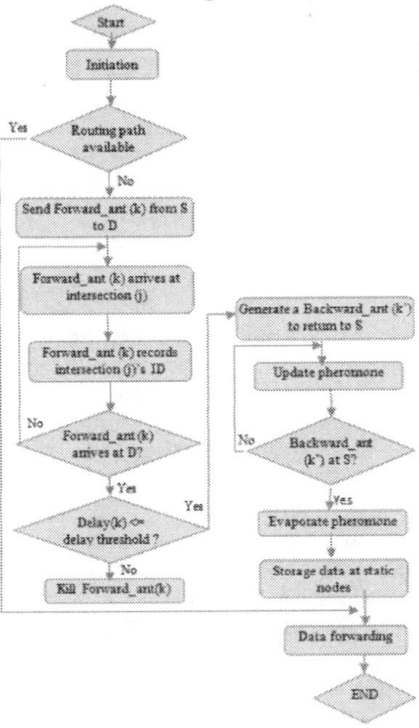

Figure 2. Flowchart of the proposed algorithm

5. Method

Considering the VANET infrastructure and features, this paper assumes that each vehicle is equipped with GPS facility, digital map and navigation system [16, 17]. With these equipments, the vehicles will have the ability to know their own speed, their geographical position and the position of intersections. With the help of the location service [18], the source vehicle is supposed to obtain the geographic location of its respective destination. To make the storage and the update of traffic information possible, a static node in each intersection is implemented. The VANETs have the same radio power that makes the data loss null. Moreover, it is supposed that the urban street map is abstracted as a square area with N vehicles scattered and distributed randomly. Obviously, the route between the source S and the destination D can be defined as a succession of intersections; supposing that M intersections in the area are connected by a set of road segments. Note that the intersection (1) is the first intersection connected to the source S, and the intersection (M) is the last one connected to the destination D. The reference vehicle is randomly positioned in the network. The proposed algorithm aims to find the best and shortest path between source and destination considering the best commitment between the delay and the minimum number of hops. To do so, it is assumed that for each road segment between two intersections, a delay threshold exists. For each hop, the delay made by the forward ant and the delay threshold is compared so that the shortest path in terms of time can be chosen.

The figure below illustrates and describes the proposed algorithm. The bases for implementation are as follows:

- First, after the random distribution of vehicles, neighboring intersections are to be found.
- If the routing path is available, data packet forwarding is implemented. Otherwise, forward ants are created.
- Forward ants start being sent. The first ant chooses the route haphazardly. The next ants will choose the route according to the probability given in equation 1.
- *For Every forward ant sent. Pheromone evaporation is applied according to the equation* 3.
- The ID of each intersection is recorded on each hop before choosing the next destination corresponding to equation 1.
- *When forward ant reaches destination D,* a comparison is done between the delays made by the ants and the delay threshold, in order to choose the shortest route in terms of time.
- If the ant's delay is bigger than the threshold, the forward ant is killed. Otherwise, a backward ant is generated.
- When returning to the origin, the pheromone table is updated according to equation 2, where the pheromone amount recorded on trails is updated.
- Notice that the obstacles and the radio power limit the search of neighbors for a given vehicle, therefore it would be wise to divide the simulation into three parts: static routing, static routing with obstacles and dynamic routing.

```
set running parameters
spread vehicles randomly

for i= 0 -> N (vehicles number)
        find neighboring intersection of vehicle(i)

    if routing path available at neighboring intersection
    then vehicle(i) implements data packets forwarding

    else  create forward_ant
end for

for k= 0 -> N (vehicles number)
for j= 0 -> M (Intersections number)

    if  forward_ant(k) arrive at intersection(j) then
                forward_ant(k) records the ID of intersection(j)
                choose next destination intersection (j+1)
                send forward_ant(k)

    else if forward_ant(k) arrive at vehicle(N)  then
        ( if delay <= delay_threshold then
                forward_ant(k) <- backward_ant(k')
                backward_ant(k') return to origin S
                    (if backward_ant(k') at vehicle(1) then
                        evaporate pheromone
                    )
          else
                destroy forward_ant(k)
        )
    else if  forward_ant(k') in return then
            update pheromone
    end if
end for
end for
```

Figure 3. The proposed algorithm

6. Conclusion

The choice of VANET routing research field is explained by the advantages and features it offers, such as the homogeneous network that results in the same physical capacity and the same radio power. Still, many challenges characterize the VANET environment including high dynamic topology, frequent disconnect network and others. This paper discusses the problem of routing protocols in VANETs; the proposed algorithm aims to find the shortest and the most optimal path considering the best commitment between the number of hops and the delay, thanks to the swarm intelligence techniques usea behavior similar to the living organisms in nature. This analogy with nature enables the use of the Ant Colony Optimization. The algorithm starts with a search of an existing routing path in static nodes, when a negative feedback is signaled. The proposed algorithm usesthe ACO-algorithm to start a reactive search from source to destination. After the path setup phase, the algorithm probes the path by the level of the pheromone amount in the trails. The higher the pheromone amount is, the best routing path itis. Finally, an evaporation pheromone operation is implemented to avoid the convergence of a local optimum. The advantage exists in terms of average end-to-end delay, shortest path computed by the number of hops between source and destination, and the best routing path information centralized in all static nodes at all intersections in the area. Nevertheless, the algorithm needs more performance in terms of routing overhead. As for future works, a simulation in three parts will be made, starting with static routing, static routing with obstacles and dynamic routing. In addition, congestion during rush hours in urban areas will be taken into consideration.

7. Future work and perspectives

To investigate the performance evaluation of our routing protocol, a simulation in three parts will be made as future work, starting with static routing, static routing with obstacles and dynamic routing. In the simulation experiments, we will use the Vehicular Ad Hoc Networks Mobility Simulator (VanetMobiSim [19]) with the network simulator NS-2 to generate vehicles mobility. In addition, congestion during rush hours in urban areas will be taken into consideration.

8. Acknowledgment

This paper was fully supported by the team of Information Research and Indexing Documents Texts and Multimedia at National School of Computer Science and Systems Analysis of Mohammed V University, Rabat, Morocco.

Refrences

1. Melaouene, N., and Romadi, R. 2017. VANET's Applications and its systems based on RFID: state of the art. In *proceeding of CoDIT 4th Interational conference on control, decision and information technologies* (Barcelona, Spain, April 5-7, 2017).
2. Dorigo, M., Maniezzo, V., and Colorni A. 1996. Ant System: Optimization by a colony of cooperating Agents. *IEEE transactions on sysems, man, and cybernetics*. Vol. 26, NO. 1, (Feb. 1996)
3. Dorigo, M., and stützle, T. 2004. Ant colony optimization. *MIT press, (Cambridge, USA).*
4. Mantilla-Gaviria, I. A., Diaz-Morcillo, A., and Tejedor, Jv. B. 2013. An Ant Colony Optimization Algorithm for Microwave Corrugated Filters Design. *Hindawi Publishing Corporation Journal of Computational Engineering*. Volume2013.
5. Kaur, S. Er. 2013. Shortest Path Finding Algorithm Using Ant Colony Optimization. *International Journal of Engineering Research & Technology*. Vol. 2 Issue 6 (June 2013), 2278-0181.
6. Kumar, R., and Dave, M. 2011. A comparative study of various routing protocols in VANET. *IJCSI international journal of computer science issues (July 2011). Vol. 8, Issue 4. 1694-0814.*
7. Lee, K.C., Lee, U., and Gerla, M. 2010. Survey of routing protocols in vehicular ad Hoc Networks. *Chapter 8 in Advances in vehicular Ad-Hoc Networks: developments and challenges*. University of Wollongong, UK.
8. Silva, R., Lopes, H. S., and Junior, W. G. 2013. A heuristic algorithm Based on Ant Colony Optimization for Multi-objective Routing in Vehicle Ad-Hoc Networks. In *proceedings of the BRICS Congress on Computational Intelligence & 11th Brazilian Congress on Computational Intelligent* (Recife, Brazil, September 08-11, 2013).
9. Caro, G.D., Ducatelle, F., and Gambardella, L.M. 2005. AntHocNet: an adaptive nature-inspired algorithm for routing in mobile ad hoc networks. *Euro. Trans. Telecomms. (2005). 16. 443-455.*
10. Toklu, N.E., Montemanni, R., and Gambardella, L.M. 2013. A Robust multiple Ant Colony System for the capacitated vehicle routing problem. *IEEE international conference on systems, man, and cybernetics. DOI= 10.1109/SMC.2013.322.*
11. Li, G., and Boukhatem, L. 2013. Adaptive Vehicular Routing Protocol Based on Ant Colony Optimization. In *ProceedingVANET'13 (Taipei, Taiwan, June 25, 2013).*
12. Li, G. 2015. *Adaptive and Opportunistic QoS-based Routing Protocol in VANETs*. Doctoral Thesis. University of Paris SUD.
13. Blum, C. 2005. Ant Colony optimization: Introduction and recent trends. *Physics of life reviews 2.* (2005),353-373.
14. Bin, Y., Zhong-Zhen, Y., and Baozhen, Y. 2009. An improved ant colony optimization for vehicle routing problem. *European journal of operational Research*. 196, (2009), 171-176
15. Yang, Z.Z., Yu, B., and Cheng, C.T. 2007. A parallel ant colony algorithm for bus network optimization. *Computer-aided Civil and infrastructure engineering 22. 44-55.*
16. Song, X., Li, X., Tang, W., and Zhang, W. 2016. A fusion strategy for reliable vehicle positioning utilizing RFID and in-vehicle sensors. *Information fusion*. 31, (2016). 76-86.
17. Duchon, F., Hubinsky, P., Hanzel, J., Babinec, A., and T1ôlgyessy, M. 2012. Intelligent vehicles as the robotic applications. *ProcediaEngineering*. 48, (2012). 105-114.
18. Karagiannis, G., Altintas, O., Ekici, E., Heijenk, G., Jarupan, B., Lin, K., and Weil, T. 2011. Vehicular networking: A survey and tutorial on requirements, architectures, challenges, standards and solutions. *IEEE Communications Surveys & Tutorials. 13, 4 (Jul. 2011), 584-616.*
19. Harri, J., Filali, F., Bonnet, C., and Fiore, M. 2006. Vanetmobisim: generating realistic mobility patterns for vanets. In *proceeding of (VANET'06) IEEE international Workshop on Vehicular Ad Hoc Networks* (Los Angeles, USA, Sep.2006).

The safety analysis of transportation vehicles carrying prefabricated units under the action of gusts

Gang Yao[1,2], Mingpu Wang[1,2] and Yang Yang[1,2]

[1]*Key Laboratory of New Technology for Construction of Cities in Mountain Area, Ministry of Education, 400045 Chongqing, China*
[2]*School of Civil Engineering, Chongqing University, 400045 Chongqing, China*

Abstract. The prefabricated-unit industry has developed rapidly and the safety of vehicles carrying prefabricated units has received extensive attention. In order to evaluate the safety of vehicles carrying prefabricated units under the action of gusts, the roughness of road surfaces was simulated by the sinusoid superposition method. Furthermore, the wind load was simulated by the harmonic synthesis method. The analysis model of the 13-DOF vehicle was established, and the wind-vehicle-road system was formed. In addition, the safety of the vehicles carrying prefabricated units under gusts was analyzed. It is demonstrated that when a vehicle encounters a sudden gust during driving, lateral deviation of the body will occur. The contact force on the windward side wheel will drop sharply and the wheel is easy to get off the ground, which decides whether a rollover accident happens or not. It sheds light on the safety analysis for ultra-wide and overweight vehicles under gust action.

1 Introduction

The factory site of prefabricated units takes up a large area and is generally selected in a relatively remote area for cost reduction. It has a certain distance from the construction site and hence the number of relative transportation vehicles has been increasing in recent years. The prefabricated-unit plant has a reasonable haul distance of 70.0 km and the number of the transportation vehicles for prefabricated units in the radiating area increases. Prefabricated units are self-contained and large, and need ultra-wide and over-weight vehicles to transport. Therefore risk during driving is rapidly increased. Traffic safety is related to many factors such as vehicle type, road conditions, various loads acting on vehicle, and drivers' controllability [1]. At the same time, gusts have a greater impact on the vehicle safety [2].

Under the action of gust, if the speed limit of the vehicles carrying prefabricated units is too low or the passage is prohibited, there will be a negative impact on construction sites and prefabricated-unit plants. On the one hand, the original material entry plan of a construction site will be delayed, and then the entire progress of the construction site will be affected. What's more, if the transport path where the material enters the site involves traffic control, the road usage time must be re-applied, adding a lot of extra work. On the other hand, if the prefabricated units can't be transported out of the finished product yard of a prefabricated-unit factory in time, the production schedule of the factory will be delayed, resulting in the lag of subsequent production plan.

If there is no speed limit on the vehicles carrying prefabricated units under the action of gusts, the probability of traffic accidents would greatly augment. The transportation of prefabricated units is a newly arising issue from the development of prefabricated buildings. At present, the specifications for prefabricated-unit transport in China are incomplete, and there are few studies on the transportation safety of prefabricated units.

In summary, it is necessary to analyze the safety of vehicles carrying prefabricated units under the action of gusts. In this paper, the roughness of road surface was simulated by the sinusoid superposition method. The wind load was simulated by the harmonic synthesis method. The analysis model of the 13-DOF vehicle was established, and the wind-vehicle-road system was formed. Additionally, the safety of the vehicles carrying prefabricated units was analyzed with gusts.

2 Wind-Vehicle-Road system

2.1 Wind load model

At present, the Monte-Carlo random variable simulation method is widely used to simulate fluctuating wind speed, among which the wave superposition method is extensively applied because of its small error and good stability [3-7]. Hence, harmonic synthesis method was performed in this paper to simulate wind load where the cut-off frequency was 4.0 rad/s, the frequency equivalent was 1024, the sampling time interval was 0.25 s, and the sampling time was 6.5s. Moreover, the point of wind load was the mass center of vehicle. The wind load paralleling to the direction of bodywork was negligible, and the wind load perpendicular to the bodywork was simulated. Wind speed samples are shown in Figure 1. It can be seen from Figure 2 that the power spectral density of simulated wind speed shows a high agreement with that of standard wind speed.

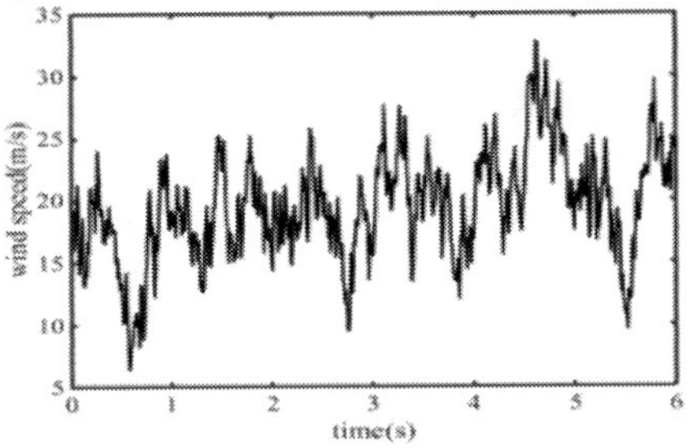

Figure 1. Samples of fluctuating wind speed

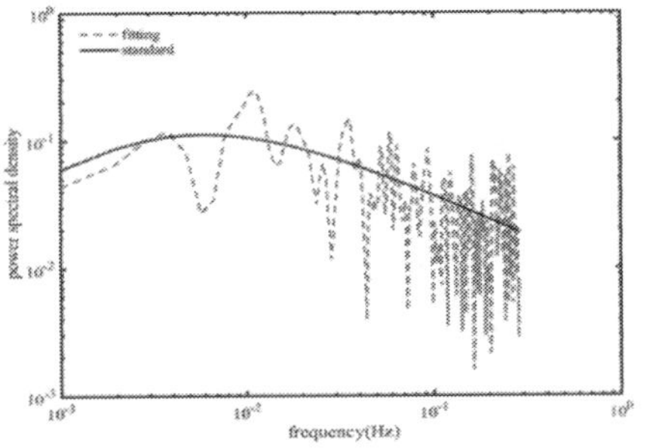

Figure 2. Comparison of power spectral density

2.2 Road roughness

On usual, road space frequency ranges from 0.011 m-1 to 2.83 m-1, and time frequency ranges from 0.44 Hz to 28.3 Hz including suspension quality natural frequency 1.0~2.0 Hz and non-suspension quality natural frequency 10.0~15.0 Hz [8]. Road roughness is divided into eight categories and A, B, C road roughness are the most common to see [9]. Three-dimensional road roughness of B level was simulated on the condition that time frequency was divided into 200 portions and sampling frequency was taken as 0.01s. Road roughness with 40 m both in width and length was simulated by sinusoid superposition method.

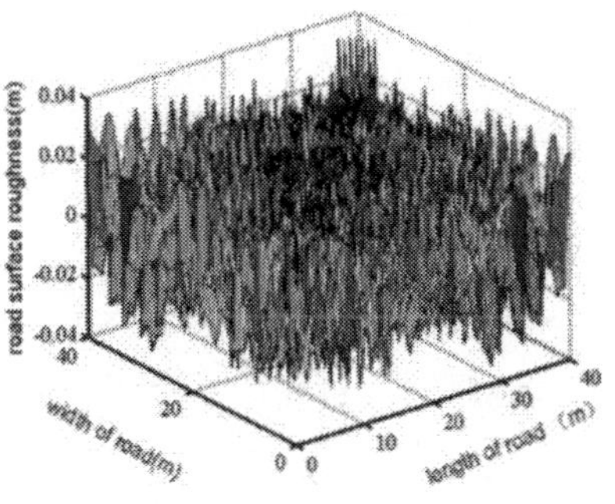

Figure 3. Road roughness generated by sinusoid superposition method

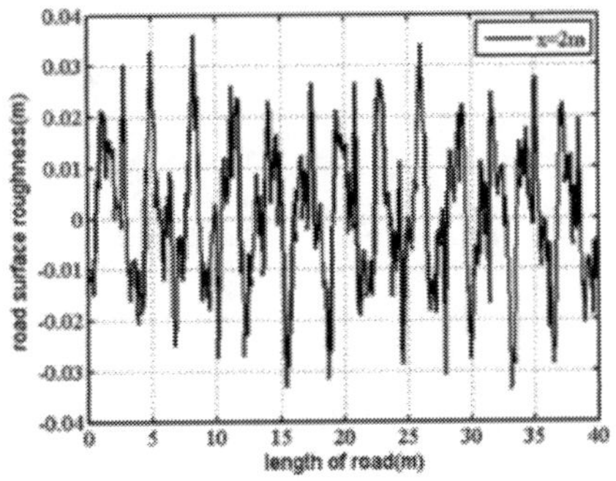

Figure 4. Roughness along the road generated by sinusoid superposition method at 2 m width of road

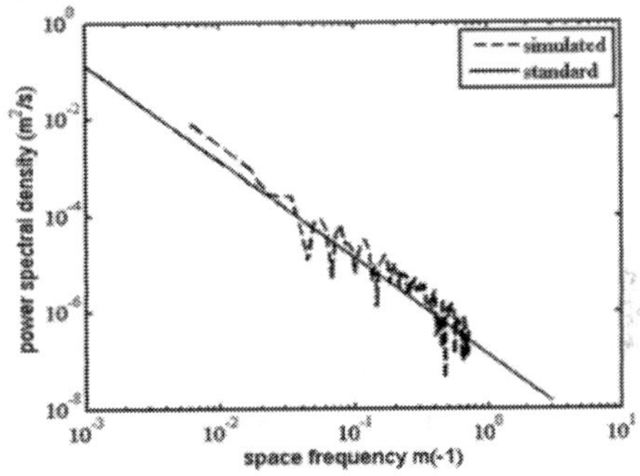

Figure 5. Simulated power spectral density by sinusoid superposition method and the standard one

Three-dimensional road roughness simulated by sinusoid superposition method is given in Figure 3 and 4. It can be seen that variation of road roughness ranged from -0.03 m to 0.03 m. Power spectral density (Figure 5) was simulated to confirm the correctness of road roughness generated by sinusoid superposition method.

2.3 Vehicle model

In order to establish the equation of vehicle motion, the following assumptions were made: (1) Vehicle moved in the longitudinal direction of road with constant speed, and longitudinal dynamic effects were not considered. (2) Spring constitutive models were linear. (3) Damping was viscous. (4) All vehicle units were symmetrical. (5) Vehicle units only made vibrations of small displacement [10-12].

There were five parameters in the equation of vehicle motion. The specific meaning is as follows:

2.3.1 Displacement parameters

The total freedom degrees for four wheels were:

$$Z_{s1}, Y_{s1}, Z_{s2}, Y_{s2}, Z_{s3}, Y_{s3}, Z_{s4}, Y_{s4}$$

Lateral independence freedom degrees:

$$Y_{c1}, Y_{c2}, Y_{c3}, Y_{c4}$$

Bodywork of five freedom degrees:

$$Z_V, Y_V, \theta_V, \phi_V, \varphi_V$$

 https://doi.org/10.1051/matecconf/201925903001

2.3.2 Quality parameters

The 13-DOF vehicle system mass and mass inertia moment parameters were defined as follows:

Four wheels:

$$M_{s1}, M_{s2}, M_{s3}, M_{s4}$$

Mass at the contact site of wheel and road:

$$M_{c1}, M_{c2}, M_{c3}, M_{c4}$$

Body:

$$M_v, J_{xv}, J_{yv}, J_{zv}$$

Among them, M, J—represented the mass and mass moments of the units; the subscripts *x, y, z* indicated the horizontal axis direction, the vertical axis direction, and the vertical direction, separately.

2.3.3 Geometric parameters

The relative positions (geometric dimensions) of various units in vehicle system model with damping elements and springs are given in Table 1.

Table 1. Geometric parameters of suspension system

Parameters	Meanings
L_1	The distance from the rear axle to the mass center of bodywork
L_2	The distance from the front axle to the mass center of bodywork
b_1	Half horizontal distance of wheels
h_1	Distance between body center and lateral spring
h_v	The height of body center from ground

2.3.4 Stiffness parameters

For two-axis suspension vehicle model, the stiffness parameter consisted both of the upper (*u*) and the lower (*l*) suspension systems.

Upper suspension system:

$$K_{uz1}, K_{uz2}, K_{uz3}, K_{uz4}, K_{uy1}, K_{uy2}, K_{uy3}, K_{uy4}$$

Lower suspension system:

$$K_{lz1}, K_{lz2}, K_{lz3}, K_{lz4}, K_{ly1}, K_{ly2}, K_{ly3}, K_{ly4}$$

2.3.5 Damping parameters

Damping parameters corresponding to the stiffness parameters of suspension system:

Upper suspension system:

$$C_{uz1}, C_{uz2}, C_{uz3}, C_{uz4}, C_{uy1}, C_{uy2}, C_{uy3}, C_{uy4}$$

Lower suspension system:

$$C_{lz1}, C_{lz2}, C_{lz3}, C_{lz4}, C_{ly1}, C_{ly2}, C_{ly3}, C_{ly4}$$

3 Safety analysis

3.1 Initial conditions

The wind direction and the driving direction were assumed to be vertical, and at this point, the external excitation of the vehicle included road surface roughness and wind load. The moment when the vehicle started to experience road surface friction was 0.0 s. When analyzing the sudden wind load, firstly, the vehicle travelled in a windless area for 6.0 s. Within 6.0~6.5 s, the average wind speed was input in program, and wind speed samples were generated. The force was added to the bodywork. When studying the influences of pulsating wind, different wind speeds and vehicle speeds, the road surface roughness and wind load excitations were given at the same time in 0.0 s. In this article, whether an accident occurs or not was assessed according to Baker's vehicle accident criteria [13-15].

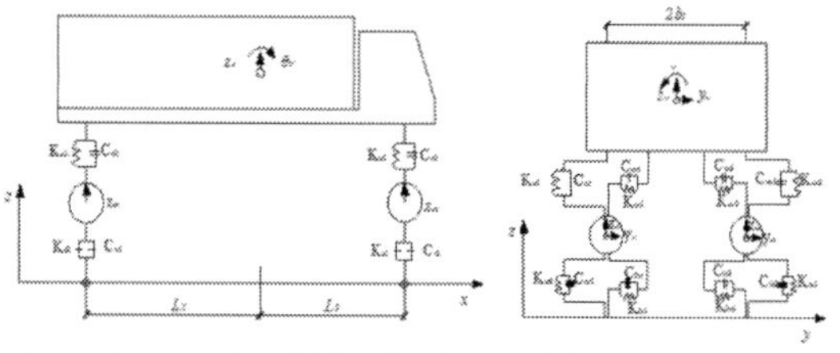

Figure 6. Two axis vehicle vibration model

3.2 Influences of gusts

The vehicle speed was 60.0 km/h, and the average wind speed was 20.0 m/s. The road surface roughness ranked Grade B, and the time integration step in the program was 0.01 s, at the same time, the air density was 1.29 kg/m^3. The vertical displacement and lateral displacement of vehicle driving in the absence of wind for 6.5 s are shown in Figure 7. The vertical displacement and lateral displacement of vehicle driving in the absence of wind for 6.0 s and followed by 0.5 s in sudden gust are shown in Figure 8.

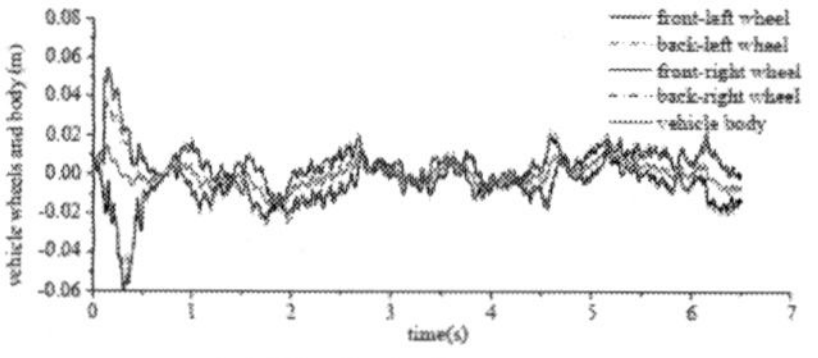

(a) Vertical displacement

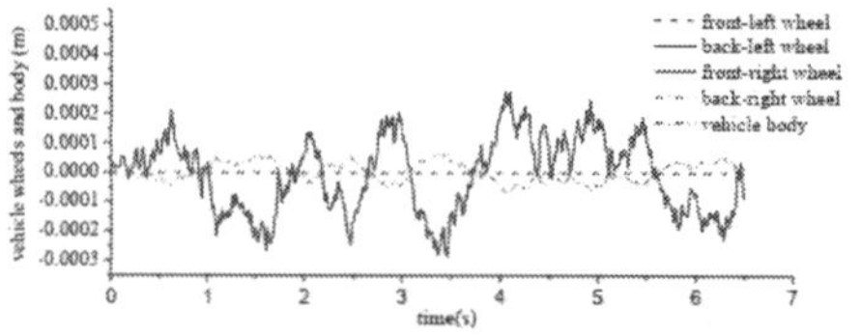

(b) Lateral displacement

Figure 7. Displacement of vehicle without gust

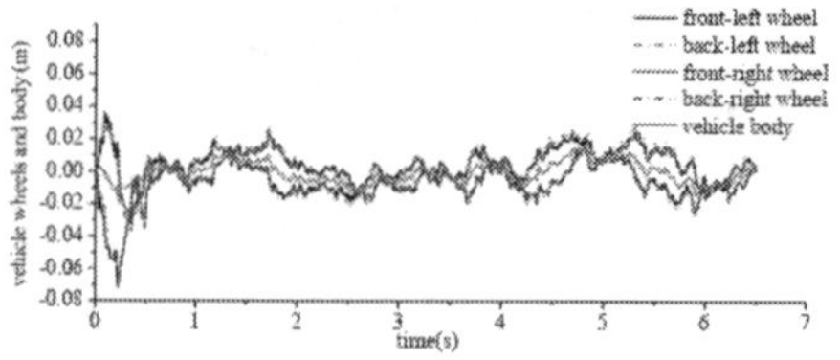

(a) Vertical displacement

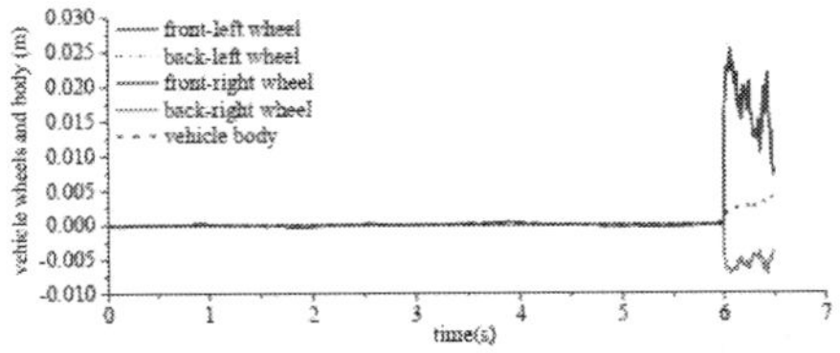

(b) Lateral displacement

Figure 8. Displacement of vehicle with gust

The followings can be derived from Figure 7 and Figure 8. (1) No matter there is gust or not, the responses of vehicle to sudden change are all completed within 1.0 s. The excitation inputs of three-dimensional road surface roughness to the wheels on right and left sides are different, so the vertical displacement curve of right wheels is different from that of left wheels. (2) Under Grade B of road surface roughness, the vertical displacement range of wheels after stability is -0.03~0.04 m. (3) The vertical displacement range of bodywork is -0.01~0.02 m. The vertical displacement ranges of bodywork and wheels in Figure 7(a) and Figure 8(a) are consistent, so the vertical displacements are basically not affected by gust. (4) There are only three lateral displacement curves in Figure 7(b) and Figure 8(b). The front left and right wheels as well as the rear left and right wheels are both coincident. This result accords with the assumption in the establishment of vehicle motion equation that the connection between the left and right wheels is rigid, which illuminates that the procedure runs correctly. (5) Without wind, the lateral displacement of wheels is tiny and fluctuates within 0~0.2 mm. The lateral displacement changes of bodywork are at the magnitude of 10-22 m that can be ignored, hence there was no lateral deviation of bodywork. (6) The vehicle experienced gust in 6th second, and it can be seen from Figure 8(b) that the lateral displacement of bodywork took place within the next 0.5 s. In Baker's accident model, this offset is the cause of a side-slip accident. Under this operating condition, the cumulative displacement of bodywork within 0.5 s was 0.132 m, and 0.5 m is the limit of sideslip. It can thus be seen that no sideslip accident happened.

4 Discussion

In this study, the wind-vehicle-road system was established. It is found that the vertical displacement of a vehicle body will not generally be affected under the action of gusts. However, the lateral deviation of the body will occur, which is the direct cause of a sliding accident. Wang et al. [16] built a road vehicle-bridge-wind (RVBW) system and found that vehicles entering a sharp crosswind gust would have a significant risk of overturning, while the focus of the research was the impact of tower and bridge dynamic response on vehicles. Moreover, William et al. [17] indicated that the probability of overturning accidents was relatively high in various kinds of wind-induced vehicle accidents. In another study [18], it was manifested that the lift oscillations of high sided road vehicles were dominated by the large scale instability caused by wake vortices. In this paper, the data shows that the contact force on the rear wheel of windward side is minimal and the wheel is easy to get off the ground, which decides whether a rollover accident happens or not. This provides a reference for the future studies on the more accurate safety thresholds.

5 Conclusion

In order to assess the safety of transportation vehicles carrying prefabricated units under the action of gusts, the roughness of road surfaces was simulated by the sinusoid superposition method, and the wind load was simulated by the harmonic synthesis method. The analysis model of the 13-DOF vehicle was established, and the wind-vehicle-road system was formed. Additionally, the safety of transportation vehicles was analyzed with the relevant parameters. The results of this study are summarized as follows.

When a vehicle encounters a sudden gust during driving, the vertical displacements of its body and wheels are basically not influenced, but lateral deviation of the body will occur. The deviation will augment as the wind speed increases, and this is the direct cause of a sideslip accident. When the body is subjected to sudden gusts, the contact force on the left (lee) side wheel will rise, while the one on the right (windward) side wheel will drop. Therefore, it can be assumed that the contact force on the rear wheel of windward side is minimal and the wheel is easy to get off the ground, which decides whether a rollover accident happens or not.

In this article, the wind-vehicle-road system was formed, and the safety of transportation vehicles was analyzed with fixed parameters. However, the working conditions considered in this study are limited. Other impact parameters and safety thresholds need further study.

6 Acknowledgment

This study is financially supported by the national 13th five-year key research and development program of China (2016YFC0701909-1), Fundamental Research Funds for the Central Universities (106112017CDJXY200009, 106112016CDJRC000101).

References

1. S.Q. Wu, S.S. Law, Vehicle axle load identification on bridge deck with irregular road surface profile. Eng. Struc., **33**, 591-601(2011)
2. H. Kozmar, K. Butler et al., Downslope gusty wind loading of vehicles on bridges. J. Bridge Eng., **20**(11), 04015008(2015)
3. R.N. Meroney, R. Derickson, Virtual reality in wind engineering: the windy world within the computer. J. Wind Eng., **11**(2), 11–26(2014)

4. B. Blocken, 50 years of Computational Wind Engineering: Past, present and future. J. Wind Eng. Ind. Aerodyn., **129**, 69–102(2015)
5. D.G. Vernay, B. l. Raphae, I.F.C. Smith, Improving simulation predictions of wind around buildings using measurements through system identification techniques. Build. Environ., **94**(2), 620–631(2015)
6. Y Tominaga, Flow around a high-rise building using steady and unsteady RANS CFD: Effect of large-scale fluctuations on the velocity statistics. J. Wind Eng. Ind. Aerodyn., **142**, 93–103(2015)
7. P. Patlakas, G. Galanis, N Barranger, Extreme wind events in a complex maritime environment: Ways of quantification. J. Wind Eng. Ind. Aerodyn., **149**, 89-101(2016)
8. P.F. Li, S.G. Feng, X.L. Deng, Three-dimensional virtual road's harmonic reconstruction. Auto Engnieer, **11**(8), 24–27(2014)
9. K. Bogsjö, K. Podgórski, I. Rychlik, Models for road surface roughness. Vehicle System Dynamics, **50**(5), 725–747(2012)
10. H.L. Zhang, W.Q. Yang, Evaluation method of pavement roughness based on 5-DOF human-vehicle- road vibration model. Journal of Traffic and Transportation Engneering, **10**(4), 16–22(2010)
11. M. Agostinacchio, D. Ciampa, S. Olita, The vibrations induced by surface irregularities in road pavements–a Matlab approach. Europe Transportation Research Review, **11**(6), 267–275(2014)
12. A. Pazooki, S. Rakheja, D. Cao, Modeling and validation of off-road vehicle ride dynamics. Mechanical Systems and Signal Processing, **28**, 679–695(2012)
13. C.J. Baker, J. Jones, F. Lopez-Calleja et al., Measurements of the cross wind forces on trains. J. Wind Eng. Ind. Aerodyn., **92**, 547-563(2004)
14. C.J. Baker, S. Reynolds, Wind-induced accidents of road vehicles. Accident Analysis & Prevention, **24**(6), 559-575(1992)
15. M. Sterling, A.D. Quinn, D.M. Hargreaves, et al. A comparison of different methods to evaluate the wind induced forces on a high sided lorry. J. Wind Eng. Ind. Aerodyn., **98**, 10-20(2010)
16. B. Wang, Y.L. Xu, Safety analysis of a road vehicle passing by a bridge tower under crosswinds. J. Journal of Wind Engineering and Industrial Aerodynamics; Journal of Wind Engineering & Industrial Aerodynamics, **137**, 25-36(2015)
17. Y. William, W. Oraby, S. Metwally, Analysis of Vehicle Lateral Dynamics due to Variable Wind Gusts. J. SAE International Journal of Commercial Vehicles, 7(2), 666-674(2014)
18. S.A. Coleman, C.J. Baker, D.M. Hargreaves et al., 1994. An experimental study of the aerodynamic behaviour of high sided lorries in cross winds. J. Journal of Wind Engineering and Industrial Aerodynamics, **53**(3), 401-429(1994)

Risk evaluation for the navigation environment of the LNG ship based on the cloud model

Shiyuan Zhang[1], Lianbo Li[2] and Xinze Li[3]

[1]*Navigation College, Dalian Maritime University, 212, 3800 laboratory, 1 Linghai Road, Ganjingzi District, Dalian 116029, Liaoning, China, 86-15941196261*
[2]*Navigation College, Dalian Maritime University, 405, Administration Building, 1 Linghai Road, Ganjingzi District, Dalian 116029, Liaoning, China, 86-18900982638*
[3]*Navigation College, Dalian Maritime University, 212, 3800 laboratory, 1 Linghai Road, Ganjingzi District, Dalian 116029, Liaoning, China, 86-13709875988*

Abstract. In order to enhance water transport safety of the LNG ship and improve the objectivity of risk assessment of LNG ships navigation environment, the cloud model which can realize the conversion between qualitative and quantitative evaluation indexes is used to evaluate the LNG ships navigation environment. Combined with the Delphi method, cloud weight and cloud model of indexes at all levels in the evaluation system are determined. At the same time, simulation is carried out by MATLAB software and cloud chart is obtained. And then we analyze the risk level of different factors in LNG ships navigation environment to get the evaluation results, which provides decision-making basis for navigation safety management. Overall, it is testified that the cloud model can reduce the randomness and fuzziness in the process of evaluating and the irrationality caused by the subjective assumption of weight distribution, which makes the results more reasonable and credible.

1. Introduction

In recent years, natural gas has already become the major energy consumption, while the storage of natural gas is inadequate in China. Importing natural gas through waterway transportation is the main method for people in China to acquire. Therefore, LNG transportation has become an indispensable part of shipping. In view of the flammability of LNG, an increasing number of people concentrate on navigation safety of LNG ship [1-2].

Nowadays, there are many kinds of methods which are used to evaluate the risks of navigation safety. In order to ensure the maritime navigation safety of Xiamen port and improve the efficiency of port operations, Huang Pengfei uses AHP-Fuzzy comprehensive evaluation to evaluate the navigation safety of Xiamen port. They propose corresponding countermeasures and suggestions of various risks control [3]. Zhu Jingjun uses a comprehensive evaluation method to evaluate several ports. Case studies indicate that the model uses fewer indexes and facilitates the weighing of factors [4]. Zheng Yibin uses quantitative analysis on the evaluation index to get the overall actual evaluation on the channel maintenance level of the region [5]. Fan Yaotian uses the modified model which combines BP neural network with cluster analysis to reduce effects of experts' subjectivity when evaluating the severity of marine traffic flow conflicts [6]. Li Lianbo uses two kinds of calculation methods of grey connection degree to comparatively analyze marine accidents between 2009 and 2013 in Liaoning waters and predict the traffic situation in the future [7]. There is a certain reference for the outcome of navigation evaluation by using the above methods, but they ignore the fuzziness of navigation safety to some degree. What's more, during the evaluating process, the feedback from different experts according the same factor is random and the measuring data which are limited by the lack of equipment are also uncertainty. Therefore, we should not only consider the fuzziness of navigation safety, but also consider the randomness and uncertainty of various factors during evaluating the navigation safety.

On the basis of fuzzy mathematics and probability statistics, the cloud model [8-9] combines the fuzziness, randomness and uncertainty of the qualitative concepts in the natural language together, so as to realize the natural transformation between the qualitative and the quantitative. Considering the fuzziness of navigation safety, this paper combines the advantages of the cloud model and calculates the weight of the evaluation index of navigation safety at all levels to reduce the influence of subjective factors on the weight of the evaluation indexes at all levels. In addition, the establishment of the cloud model can fulfill the transformation between the qualitative and quantitative evaluation indexes, which can reasonably solve the problem of randomness and uncertainty in the traditional methods, and make the evaluation result more reasonable.

2. Construction of risk assessment index for navigation environment of jiangsu binhai lng vessel

LNG industry has always believed that natural gas leakage is the biggest potential risk of the receiving station. SIGTTO (The Society of International Gas Tanker and Terminal Operators) believes that one of the main links of the LNG leakage risk caused by the LNG receiving station is the navigation process of the LNG ship in the channel.

Therefore, this paper, based on the specific conditions of the Jiangsu seashore and the characteristics of the navigation environment in the waters of Jiangsu, evaluates the environmental risks of the navigation of Jiangsu coastal LNG ships from human factors, ship factors, environmental factors (including hydrology, meteorology and traffic environment) and management factors, and establishes an evaluation index system, see Table1.

Table 1. Transportation environmental risk assessment index system for LNG ship of Jiangsu Province

Primary index	Secondary index	Third-level index
LNG navigation environment of ships of Jiangsu BinhaiC	Human factorC_1	Competence C_{11}
		Education C_{12}
		Safety consciousness C_{13}
	Ship factorC_2	Ship scale C_{21}
		Maneuverability C_{22}
		Ship equipment C_{23}
	Environmental factorC_3	Wind and wave C_{31}
		Visibility C_{32}
		Channel length and width C_{33}
		Navigation obstruction C_{34}
		Traffic flow C_{35}
	Management factorC_4	Tug C_{41}
		Traffic service C_{42}
		Maritime supervision C_{43}
		Emergency resources C_{44}

Without effective control of factors affecting navigation safety, navigation accidents may occur. Risk assessment is on the basis of determining the risk and knowing its distribution, analyzing various factors affecting the degree of risk, finding key risk factors through the method of primary and secondary arrangement, providing the decision basis for the managers and ensuring the navigation safety of the waters.

3. The theory of cloud model

3.1 The definition of cloud and its digital features

Let C be a qualitative concept on the quantitative domain U represented by exact data. If quantitative value $x \in U$, $x$is a random implementation on C and the membership degree $\mu(x) \in [0,1]$ of C to x is a stable random number of $\mu: U \to [0,1], x \in U, x \to \mu(x)$, the distribution of $x$on the domain U is called cloud, and each X is called a cloud droplet$(x, \mu(x))$.

The three numerical characteristics of cloud are meanEx, entropy$En$and super entropyHerespectively, which can be recorded as $A = T(Ex, En, He)$. They distribute mainly between $[Ex - 3En, Ex + 3En]$ .The qualitative concept is reflected by quantitative concept, and the mean Ex is the most representative one. Entropy$En$is used to counter the fuzziness of the model, reflecting the range of the samples contained in the concept. The super entropyHerepresents the uncertainty of entropy En , which reflects the randomness of qualitative concepts and the degree of dispersion of models. Figure 1 is a cloud picture of $(Ex = 20, En = 3, He = 0.2)$ respectively.

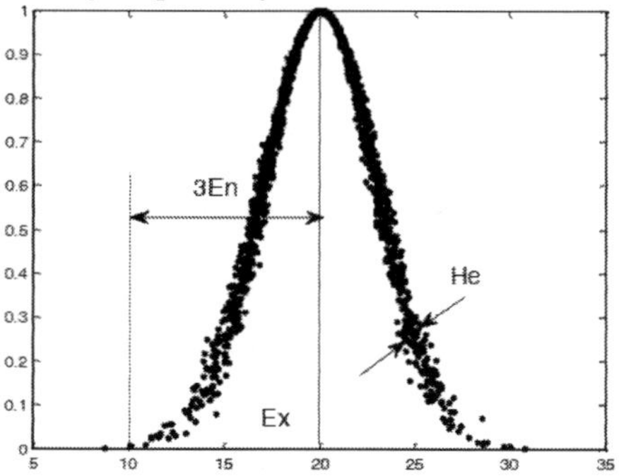

Figure1. The digital features of the cloud.

The X axis represents a value corresponding to C, and the Y axis represents the corresponding membership degree.

3.2 Cloud generator

The basic operations of the cloud model include the normal cloud generator and the reverse cloud generator [10].

(1)The principle of the normal cloud generator is that after the input of three digital featuresEx、En、$He$of the cloud, an example of random occurrence under this concept can be obtained, as follows.

Input: The three eigenvalues of the qualitative conceptEx、En、$He$and the number of cloud droplets required$N$.

Process: Generating a normal random number$E'n$, which satisfies$En' \sim N(En, He^2)$; Generating a normal random number $xi \sim N(Ex, E'n^2)$; Carrying on a calculation to$yi$with $yi = exp[-(xi - Ex)^2/2(E'n)^2]$; Producing cloud droplets(xi, yi) and repeating the above steps until the expected number of cloud droplets is obtained.

Output: A number of cloud droplets and their determination are obtained.

(2)The principle of the inverse cloud generator is to transform the actual data into qualitative concepts of (Ex, En, He) and the specific algorithm is as follows.

Input: A number of cloud droplets and their determination.

Process: Calculating mean of sample with $\bar{X} = \frac{1}{n}\sum_{i=1}^{n} xi$, absolute center distance of first order sample with $\frac{1}{N}\sum_{i=1}^{n}|xi - \bar{X}|$ and sample variance with $S^2 = \frac{1}{n-1}\sum_{i=1}^{n}(xi - \bar{X})^2$;

The expectation of sample:

$$Ex = \frac{1}{n}\sum_{i=1}^{n} xi \tag{1}$$

The entropy of sample:

$$En = \sqrt{\frac{\pi}{2}} \times \frac{1}{N}\sum_{i=1}^{n}|xi - \bar{X}| \tag{2}$$

The super entropy of the sample:

$$He = \sqrt{S^2 - En^2} \tag{3}$$

Output: Qualitative concept of samples and expectedEx, entropyEn and super entropyHe.

3.3 Basic operation of cloud model

The cloud computing whose essence is the operation of cloud computing to form a virtual cloud is necessary for risk assessment of LNG vessel navigation environment [11]. If there are two cloud models $C1(Ex1, En1, He1)$ and $C2(Ex2, En2, He2)$ on the same domain, the resulting virtual cloud is $C(Ex, En, He)$. The calculation formula is followed, see Table 2.

Table 2. The basic operation of the cloud

Operation rule	*Ex*	*En*	*He*
+	$Ex_1 + Ex_2$	$\sqrt{En_1{}^2 + En_2{}^2}$	$\sqrt{En_1{}^2 + En_2{}^2}$
−	$Ex_1 - Ex_2$	$\sqrt{En_1{}^2 + En_2{}^2}$	$\sqrt{En_1{}^2 + En_2{}^2}$
×	$Ex_1 \times Ex_2$	$\lvert Ex_1 Ex_2\rvert\sqrt{\left(\frac{En_1}{Ex_1}\right)^2 + \left(\frac{En_2}{Ex_2}\right)^2}$	$\lvert Ex_1 Ex_2\rvert\sqrt{\left(\frac{He_1}{Ex_1}\right)^2 + \left(\frac{He_2}{Ex_2}\right)^2}$
÷	$Ex_1 \div Ex_2$	$\left\lvert\frac{Ex_1}{Ex_2}\right\rvert\sqrt{\left(\frac{En_1}{Ex_1}\right)^2 + \left(\frac{En_2}{Ex_2}\right)^2}$	$\left\lvert\frac{Ex_1}{Ex_2}\right\rvert\sqrt{\left(\frac{He_1}{Ex_1}\right)^2 + \left(\frac{He_2}{Ex_2}\right)^2}$

3.4 Integrated operation

In our daily work, we have different views on the same thing. Therefore, it produces more than two clouds of the same type, and the integrated operation is to integrate these same types of clouds with a certain algorithm to get a more generalized concept. The calculation formula is as follows:

$$Ex = \frac{Ex_1En_1 + Ex_2En_2 + \cdots + Ex_nEn_n}{En_1 + En_2 + \cdots + En_n} \tag{4}$$

$$En = En_1 + En_2 + \cdots + En_n \tag{5}$$

$$He = \frac{He_1En_1 + He_2En_2 + \cdots + He_nEn_n}{En_1 + En_2 + \cdots + En_n} \tag{6}$$

4. Comprehensive evaluation of cloud model

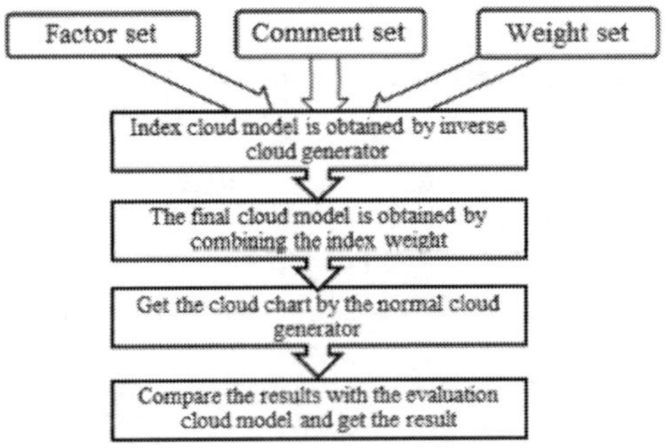

Figure 2. Cloud model comprehensive evaluation process.

This paper applies the cloud model to the comprehensive evaluation process as shown in Figure 2.

(1)According to the target object, some influential indicators are identified. The first level index to the two level index is $U = \{U_1, U_2, \cdots, U_n\}$. The two level index Ui can be divided into several three level index $U_i = \{U_{i1}, U_{i2}, \cdots, U_{in}\}$, $i = 1,2, \cdots, n$. The set of the above indexes is the factor set.

(2)After evaluating the above indicators, the set of commentary of all evaluation indicators is called the comment set, which is $V = \{V_1, V_2, \cdots, V\}$.

(3)The weight of each index of the above factor sets is calculated by cloud model. First, through many experts score the importance of the lower level indicators in the higher level, this paper divides the importance into five grades, and establishes a suitable cloud scale according to the grade of evaluation. The evaluation level and the matching cloud scale are shown in Table 3 and Table 4. Then the obtained fraction is processed by the reverse cloud generator to get the cloud model. If the weight cloud droplet distribution is scattered, it should be fed back and corrected until a more stable weight cloud is obtained. Finally, the weight is normalized to get the weight of the upper level of each index [12-13].

Table 3. Indicator importance level

Weight class	Assignment interval
Very important	[0.8,1]
Important	[0.6,0.8)
Less important	[0.4,0.6)
General important	[0.2,0.4)
Unimportant	[0,0.2)

Table 4. Indicator importance level cloud scale

Weight class	Cloud model
Very important	(1,0.103,0.013)
Important	(0.691,0.064,0.008)
Less important	(0.500,0.039,0.005)
General important	(0.309,0.064,0.008)
Unimportant	(0,0.103,0.013)

(4)According to the actual situation of the navigation environment of LNG ships, the expert group has found the highest evaluation value and the minimum evaluation value for its evaluation index. The evaluation index is divided into five grades, which is shown in Table 5. Then the two groups of evaluation values are obtained, using the reverse cloud generator to generate the highest evaluation cloud model and the lowest evaluation cloud model, and then use the two group of cloud models to get a more generalized cloud model.

Table 5. Rating of risk indicators

Evaluation grade	Assignment interval
Lowest	[0.8,1]
Lower	[0.6,0.8)
Ordinary	[0.4,0.6)
Higher	[0.2,0.4)
Highest	[0,0.2)

(5)The weight of all levels of the LNG navigation environmental risk assessment is combined with the evaluation values at all levels. According to the formula $C_i = \sum_{j=1}^{m}(C_{ij} \times W_{ij})$, the cloud model of all levels of indicators can be calculated. After the final cloud model C is obtained, the cloud map generated by the cloud is compared with the cloud map of the navigation environment evaluation results. The evaluation result is the evaluation level which is the closest to the final cloud map. The results of the assessment of the risk of navigable environment at all levels are shown in Table 6.

Table 6. Risk assessment results level cloud scale

Evaluation grade	Cloud model
Lowest	(1,0.103,0.013)
Lower	(0.691,0.064,0.008)
Ordinary	(0.500,0.039,0.005)
Higher	(0.309,0.064,0.008)
Highest	(0,0.103,0.013)

5. Example analysis based on jiangsu seashore

In order to meet the demand of natural gas in Jiangsu Province, improve the capacity of natural gas supply, improve the supply pattern and ensure the safety of gas supply, the LNG project is to be set up in the coastal area of Jiangsu province. The construction of this project can effectively solve the problems of energy supply security and environmental protection. Therefore, the risk assessment can be carried out with the cloud model.

5.1 Evaluation index of cloud weight

Ten experts of the marine environmental safety assessment are invited to score the indicators at all levels of Jiangsu coastal LNG navigation environmental risk assessment. It takes the proportion of competency in human factors as an example, see Table 7. The score of competency is processed by inverse cloud generator. The results are fed back and corrected until the stable cloud map is obtained. Finally, the weight cloud $w_{11} = (0.708,0.024,0.002)$ of the competence is obtained, and the method of obtaining the weight cloud of other indexes is also the same. The weight cloud is normalized to get the weight cloud at all levels, see Table 8.

Table 7. Competency of human factors in the weight of the score

Expert	1	2	3	4	5
Weight scoring	0.68	0.72	0.71	0.75	0.72
Expert	**6**	**7**	**8**	**9**	**10**
Weight scoring	0.70	0.71	0.72	0.69	0.72

Table 8. The weights of the indicators at all levels

Secondary index weight	Weight cloud model	Third-level index weight	Weight cloud model
w_1	(0.268,0.024,0.003)	w_{11}	(0.429,0.024,0.002)
		w_{12}	(0.258,0.021,0.002)
		w_{13}	(0.314,0.022,0.003)
w_2	(0.237,0.014,0.003)	w_{21}	(0.308,0.021,0.002)
		w_{22}	(0.368,0.020,0.002)
		w_{23}	(0.324,0.021,0.002)
w_3	(0.279,0.019,0.002)	w_{31}	(0.151,0.019,0.002)
		w_{32}	(0.162,0.018,0.002)
		w_{33}	(0.230,0.021,0.003)
		w_{34}	(0.198,0.023,0.004)
		w_{35}	(0.259,0.022,0.002)
w_4	(0.216,0.025,0.002)	w_{41}	(0.274,0.025,0.005)
		w_{42}	(0.264,0.022,0.002)
		w_{43}	(0.234,0.018,0.003)
		w_{44}	(0.227,0.019,0.003)

5.2 Third-level index cloud model

Taking the capacity of three level indicators as an example, the maximum and minimum scores are scored, see Table 9. The maximum cloud model and minimum cloud model are obtained by using the type (1) ~ (3). The integrated cloud model is $C_{11} = (0.875,0.061,0.016)$, and other three levels of cloud model can be obtained by using the type (4) ~ (6), see Table 10.

Table 9. The score of the competency

Expert	1	2	3	4	5
Maximum	0.90	0.95	0.90	0.95	0.85
Minimum	0.85	0.85	0.85	0.90	0.80
Expert	**6**	**7**	**8**	**9**	**10**

Maximum	0.85	0.95	0.90	0.95	0.90
Minimum	0.80	0.90	0.85	0.85	0.87

5.3 Other index cloud models

According to the index value and its weight of the third-level index which has been calculated, we can calculate the cloud model of the upper level index. Taking the determination process of the two level index artificial factor cloud model as an example, the cloud models of the third-level index corresponding to human factors are $C_{11} = (0.875,0.061,0.016)$, $C_{12} = (0.862,0.043,0.019)$, $C_{13} = (0.835,0.036,0.021)$ respectively. The calculation of cloud models for human factors is as follows:

$$C_1 = \begin{pmatrix} w_{11} \\ w_{12} \\ w_{13} \end{pmatrix}^T \circ \begin{pmatrix} C_{11} \\ C_{12} \\ C_{13} \end{pmatrix}$$
$$= \begin{bmatrix} Ex_{w_{11}} & En_{w_{11}} & He_{w_{11}} \\ Ex_{w_{12}} & En_{w_{12}} & He_{w_{12}} \\ Ex_{w_{13}} & En_{w_{13}} & He_{w_{13}} \end{bmatrix}^T \circ \begin{bmatrix} Ex_{C_{11}} & En_{C_{11}} & He_{C_{11}} \\ Ex_{C_{12}} & En_{C_{12}} & He_{C_{12}} \\ Ex_{C_{13}} & En_{C_{13}} & He_{C_{13}} \end{bmatrix}$$
$$= \begin{bmatrix} Ex_{w_{11}}Ex_{C_{11}} & +Ex_{w_{12}}Ex_{C_{12}} & +Ex_{w_{13}}Ex_{C_{13}} \\ En_{w_{11}}En_{C_{11}} & +En_{w_{12}}En_{C_{12}} & +En_{w_{13}}En_{C_{13}} \\ He_{w_{11}}He_{C_{11}} & +He_{w_{12}}He_{C_{12}} & +He_{w_{13}}He_{C_{13}} \end{bmatrix}^T$$
$$= \left(Ex_{C_1}, En_{C_1}, He_{C_1}\right) \quad (7)$$

According to the basic operation of clouds, the cloud model of human factors can be calculated to be $C_1 = (0.860,0.065,0.022)$. The calculation method of cloud models of other indicators at various levels is the same as above, as shown in Table 10.

Table 10. Risk assessment index cloud model

Secondary index cloud model		Third-level index cloud model	
C_1	(0.860,0.065,0.022)	C_{11}	(0.875,0.061,0.016)
		C_{12}	(0.862,0.043,0.019)
		C_{13}	(0.835,0.036,0.021)
C_2	(0.897,0.048,0.025)	C_{21}	(0.923,0.059,0.026)
		C_{22}	(0.905,0.049,0.026)
		C_{23}	(0.863,0.052,0.023)
C_3	(0.756,0.045,0.022)	C_{31}	(0.854,0.033,0.015)
		C_{32}	(0.752,0.035,0.023)
		C_{33}	(0.736,0.039,0.016)
		C_{34}	(0.728,0.023,0.018)
		C_{35}	(0.741,0.025,0.022)
C_4	(0.582,0.043,0.018)	C_{41}	(0.697,0.043,0.017)
		C_{42}	(0.675,0.025,0.016)
		C_{43}	(0.386,0.034,0.014)
		C_{44}	(0.539,0.031,0.015)

5.4 Final cloud model establishment and index analysis at all levels

According to the above conclusion, the cloud model of Jiangsu coastal LNG navigation environment risk assessment is $C = (0.780,0.080,0.023)$. The cloud map of the final result is compared with the cloud model at all levels in the Table 6 cloud scale. The result, as shown in Figure 3, is between "lowest" and "lower" and closer to the "lower" risk assessment of the navigation environment of the ship's navigable ships in the coastal area of Jiangsu. Therefore, it can be considered that its rating is "lower".

By analyzing Figure 4 and Figure 5, it shows that the evaluations of artificial factors and ship factors are between "lowest" and "lower" and is closer to "lowest". Therefore, the evaluation levels of human factors and ship factors are "low"; As shown by Figure 6, the environmental factor index is between "lowest" and "lower", and Closer to "lower", so it can be considered that the evaluation of environmental factors is "lower"; As shown in Figure 7, the management factor index is between "ordinary" and "lower", and closer to "ordinary", so it is considered that the evaluation level of management factors is "ordinary".

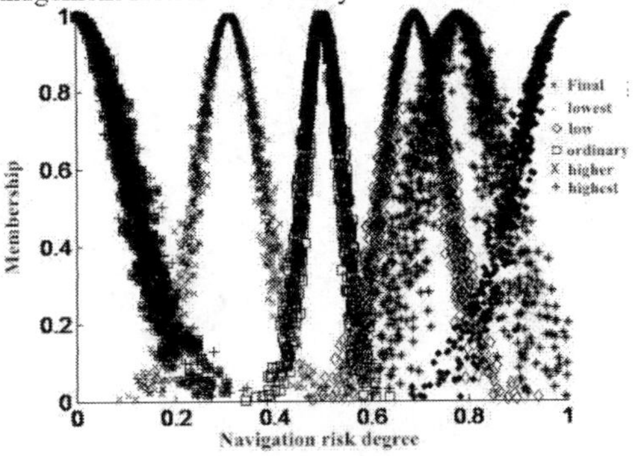

Figure 3. Target risk assessment cloud.

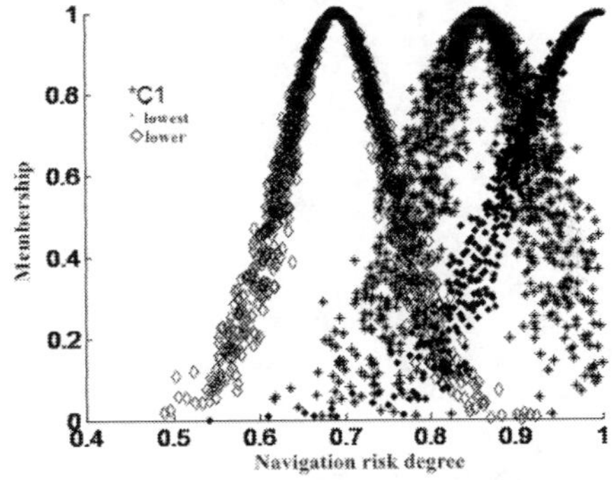

Figure 4. The cloud of human factors indicators.

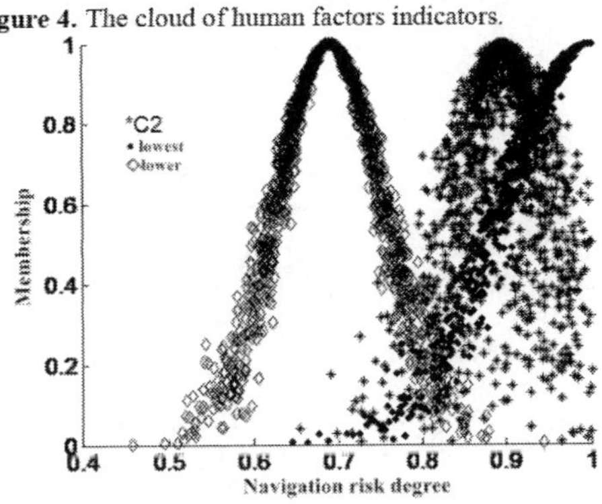

Figure 5. The cloud of the ship factors indicators.

LNG crew are trained professionally and LNG ships are built well. Therefore, the risk of accidents is lower in the two aspects of human factors and ship factors. With the continuous development of the port, the traffic in the main channel is becoming more and more busy and the traffic flow in the channel is increasing, so the safety of navigation is affected adversely. Therefore, environmental factors will have a certain impact on the navigation of Jiangsu seashore. Because of the new construction of the wharf, the lag of navigation supervision equipment and the particularity of LNG, involving many aspects and large indeterminacy, the impact of management factors on navigation is relatively large. In the future, the local relevant department is likely to improve wharf on the two aspects of environmental factors and management factors.

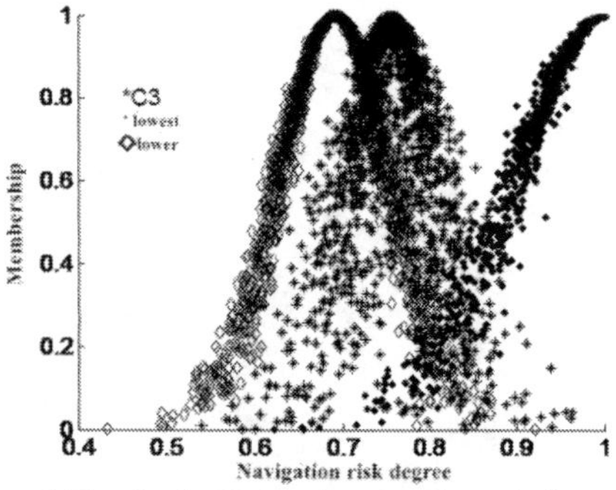

Figure 6. The cloud of the environment factors indicators.

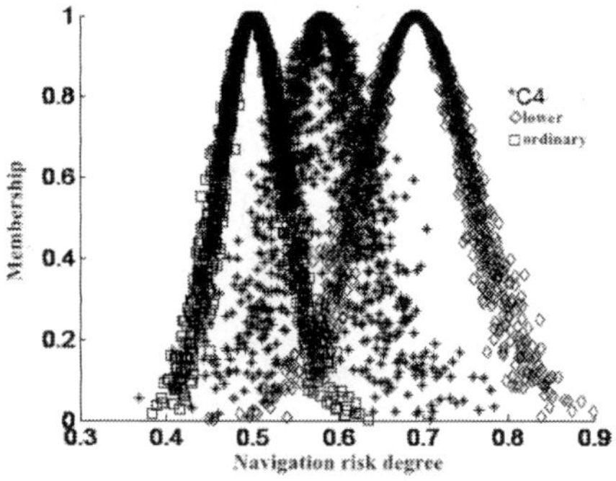

Figure 7. The cloud of the management factors indicators.

6. Conclusion

In this paper, we make full use of the advantages of the cloud model which can be used to transform the qualitative problem into the quantitative problem and take full consideration of the randomness and fuzziness of the sample, and combine the Delphi method to determine the cloud weight and cloud model of all levels of the LNG navigation environmental risk assessment system. Applying the above theories to analyze the navigation environment of the coastal LNG ships in Jiangsu. In this way, the MATLAB software is used to simulate the cloud map, which directly reflects the risk assessment level of the Jiangsu littoral indicators, and then we summarize and analyze the evaluation results. It provides a reasonable direction for the follow-up development of the navigation environment of Jiangsu coastal LNG ships, and can also be used as a reference for other regions for the navigation safety of LNG ship, which provides support of theory and practice for the development of shipping safety.

7. Acknowledgment

This work is supported by National Natural Science Foundation (NNSF) of China under Grant 51709032 and the Fundamental Research Funds for the Central Universities under Grant 3132018153 and Grant 3132018282.

References

1. Ma jinjing, Huangpu lixia, Han li, et al. Current status and prospect of LNG in land water transportation safety standards at home and abroad[J]. Natural Gas Industry, 2015,(12):117-123.
2. Walter Chukwunonso Ikealumba,Hongwei Wu. Effect of atmospheric and sea stability on liquefied natural gas (LNG) dispersion: Implications to Australian LNG marine transport[J]. Fuel,2017,197.
3. Huang P, Du Z, Cheng C. Navigation Safety Evaluation of Xiamen port Based on AHP-Fuzzy[A]. Proceedings of 2015 Joint International Mechanical, Electronic and Information Technology Conference [C]. 2015:10.
4. Zhu jingjun, Lan peizhen, Xu shenghao. Comprehensive port navigation safety evaluation[J]. Navigation of China, 2015,38(04):79-82. [2017-08-04].
5. Zheng yibin, Liu yu. Evaluation on maintenance level of provincial inland waterway based on factor analysis[J]. Port& Waterway Engineering, 2013, (06):143-148.
6. Fan yaotian, Wang chi. Evaluation methods of Severity level of water traffic flow conflict based on BP neural network[J]. Journal of Transport Information and Safety,2017, (02):23-29.
7. Li lianbo, Liu junpo, Niu jiawei, et al. Weighted grey relative analysis and prediction for marine accidents in Liaoning waters[J]. Journal of Jiangsu University of Science and Technology (Natural Science Edition) ,2016,(06):605-609.
8. Zhang S B, Chun-Xiang X U, Yu-Jun A A. Study on the Risk Evaluation Approach Based on Cloud Model[J]. Journal of University of Electronic Science & Technology of China, 2013, 42(1):92-97+104
9. Zhou Y, Wang L. Multi Level Fuzzy Comprehensive Evaluation Method Based on Cloud Model Theory[J]. Computer Simulation,2016,33(12):390-395.
10. Niu erxuan, Meng bin, Shen siwei. Evaluation model and empirical study of port enterprise green logistics based on cloud model[J]. Journal of Dalian Maritime University,2017, (02):67-74.
11. Zhang yong, Zhao dongning, Li deyi. The Similar Cloud and the Measurement Method[J]. Information and Control, 2004,02:129-132.
12. Liu cuilian, Yang liu, Liu jianmei. Application of cloud model to evaluation on port low-carbon green

development[J]. Journal of Shanghai Maritime University,2015,04:31-36.
13. Xu Zheng-jie, Zhang You-peng, Su Hong-sheng. Application of risk assessment on fuzzy comprehensive evaluation method based on the cloud model[J]. Journal of Safety and Environment, 2014,14(2):69-72.

MATEC Web of Conferences **259**, 03003 (2019) https://doi.org/10.1051/matecconf/201925903003
ICTLE 2018

Study on the takeoff landing number of emergency rescue aircraft demand after earthquake

Zhang xiuyan[1,2] and Chi hong[1]

[1]*University of Chinese Academy of Sciences, Institutes of Science and Development, Beijing, China,100190*
[2]*Civil Aviation University of China, Economics and Management College, Tianjin, China, 300300*

Abstract: Due to the lack of information after earthquake, the demand for takeoff landing number of emergency rescue aircraft is not clear. The model for takeoff landing number of emergency rescue aircraft demand after earthquake is constructed. Firstly, the total task amount of emergency rescue is determined according to the number of injured persons and the number of victims after the earthquake. Then, according to the damage degree of the ground roads in the earthquake area, the amount of aviation emergency rescue mission is determined. Finally, based on the information of emergency rescue aircraft fleet, the actual takeoff landing number of emergency rescue aircraft is calculated. The rationality of the model is verified by the data of 5 actual earthquake events from 2008, to provide guidance for effective aviation emergency rescue after earthquake.

1 INTRODUCTION

China is located between the two major earthquake belts of the Pacific earthquake belt and the Eurasian earthquake belt, which are squeezed by the Pacific plate, the India plate and the Philippines Sea plate. The earthquake source is shallow, the frequency is high, the magnitude is large and the distribution is wide. From the Wenchuan earthquake in 2008, the Ya'an earthquake in 2013 to the Jiuzhaigou earthquake in 2017, serious loss of life and property was caused to our country. Emergency rescue work after earthquake is key to reduce earthquake losses. With the progress of the society and the development of the economy, the aviation emergency rescue force is increasingly involved in the emergency rescue work after earthquake, and it has become an organic supplement to the ground emergency rescue force. Compared with other traditional rescue methods, the aviation emergency rescue force has the characteristics of flexible, rapid and efficient and not restricted by geographical space. It plays a more and more important role in the emergency work after earthquake.

However, we could not quickly and accurately obtain information about casualties, housing and road damage after earthquake, which led to the failure to determine the specific needs of aviation emergency rescue force for disaster area immediately. The indistinct demand directly lead to the mismatch between the dispatched aviation emergency rescue force and the actual rescue needs. During emergency rescue after earthquake, time is life. If the dispatching emergency rescue force is lower than the actual demand, it will increase the rescue time and reduce the rescue efficiency. Otherwise, it will cause the waste of rescue resources and increase the command load of rescue site.

At present, most of the research on emergency demand after earthquake is focused on the demand of relief materials, namely how to calculate the needed relief materials according to the basic information obtained after earthquake [1-5]. Based on decades of earthquake data, Nie summed up the relationship between the demand of more than 80 kinds of materials and the number of victims [6]. Guo constructed the BP network using 6 factors including earthquake occurrence time, magnitude and so on as the input data, trained the network according to the actual data of 16 earthquakes, and finally completed the prediction of the number of casualties [7]. Lu uses the IDF-IKF method to study the quantitative relationship between number of survival and material demand [8]. Xu used time series method to predict the change rule of material demand after the earthquake [9]. The literature about emergency rescue force after earthquake is usually focused on how to allocate task for rescue forces and to plan rescue routes under the assumption that the demand is known [10-14]. Du studied the optimal allocation of ambulance helicopters during the flow process of the wounded in the tent hospital [15]. Liu studied the transportation plan for sending relief personnel and injured persons from different transfer points by helicopters according to the different demand levels of disaster areas [16]. On the basis of the daily demand for various materials of each affected area, Zdamar has studied how to complete the delivery of materials from the supply point to the demand point by using various modes of transportation [17]. Lu studied helicopter scheduling problem after earthquake by building a bi-level multi-objective programming model [18].

Few literature is related to the needs of the rescue force. The demand for emergency rescue force after earthquake is still relied on expert experience, without mature method yet. Therefore, it is necessary to study the influence factors of the emergency rescue task after earthquake, and build the takeoff landing number model of the emergency rescue aircraft after earthquake, so as to provide the accurate demand information for the follow-up emergency rescue aircraft scheduling and improve the efficiency of the rescue.

2 Study framework

According to the literatures mentioned above, the key factor to determine the size of relief needs is the loss information caused by earthquakes. Similarly, the factors determining the demand for emergency rescue aircraft after earthquakes are also the loss information after earthquakes, including (1) the number of injured people and the total number of affected people after earthquakes: determine the total amount of emergency rescue tasks; (2) the degree of damage to the ground roads after earthquake: determine the level of demand for aviation emergency rescue forces.

A model for calculating the takeoff landing number of aircraft after earthquake is constructed, which propose a quantitative method for calculating the takeoff landing number of emergency rescue aircraft based on the demand information, as shown in Figure 1.

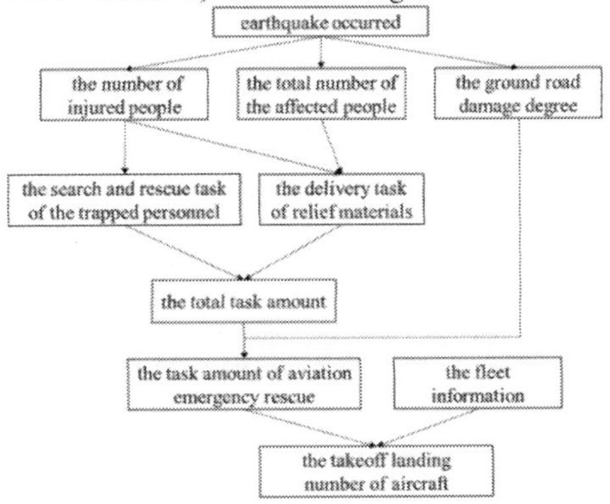

Figure 1. The study framework

3 The takeoff landing number model of emergency rescue aircraft demand after earthquake

The aviation emergency rescue task A after earthquake depends on the total rescue task T and the demand degree of the aviation emergency rescue work, that is the damage degree r of the ground road in the earthquake area. But even all the ground roads are not damaged, there will be certain aviation emergency rescue demand due to timeliness requirements, which is the datum value of aviation emergency rescue after earthquake δ.

Accordingly, the takeoff landing number model of emergency rescue aircraft demand after earthquake is put forward, as follows:

$$\mathbf{A} = \frac{S_d}{S_e}\left[n_w/\alpha + (\mu_l * n_e + \mu_m * n_w)/\beta \right] + \delta \quad (1)$$

S_d: The damage range of the road;
S_e: The scope of the earthquake wave;
n_w: The number of injured people;
n_e: The total number of the affected people;
μ_l: The coefficient of living material, meaning the average tonnage of living materials needed for each person;
μ_m: The coefficient of medical material, meaning the average tonnage of medical materials required for each injured person;
α: The average number of passengers of the rescue aircraft;
β:The average tonnage of the rescue aircraft.

3.1 Calculation of total task amount of emergency rescue after earthquake

Because of the difficulty in defense against earthquake disasters beforehand, the requirement for rescue afterwards is relatively high. In the event of an earthquake, the most important thing is to save people. We must rescue the trapped personnel with limited resources as soon as possible. The second is to provide necessary materials for people in earthquake area to ensure they can survive. Generally speaking, the two most important things in the emergency rescue process after earthquake are search and rescue of trapped personnel and relief supplies. Search and rescue trapped personnel are mainly in the initial rescue phase after the earthquake, searching for the injured and trapped people in the earthquake area and sending them to a nearby hospital or emergency medical center. The delivery of relief materials refers to the provision of the necessary materials, including living materials and medical materials, to the personnel in the earthquake area, which usually runs through the whole process of emergency rescue. The total task T of emergency rescue after earthquake can be expressed as follows:

$$\mathrm{T} = T_S + T_P \quad (2)$$

T_S is the search and rescue task of the trapped personnel; T_P is the delivery task of relief materials.

In general, the higher the magnitude, the greater the intensity, the higher the population density in the earthquake area, the more the number of disaster victims is and the greater the search and rescue task T_S of the trapped personnel is. The amount of relief supplies, that is, the amount of living materials and medical materials needed in the earthquake area, directly depends on the number of affected population and the number of injured population in the earthquake area. That is to say, the search and rescue task of the trapped personnel T_S and the delivery task of relief materials T_P are determined by the number of the affected population, so the total task T of emergency rescue after earthquake can be estimated by the number of the affected population in the earthquake area.

After the earthquake, we need to rescue immediately according to the basic information such as magnitude, intensity and the number of injured persons, so as to minimize the loss of casualties. It is assumed that, the basic information including the number of injured people n_w, the total number of the affected people n_e, and the material demand can be obtained immediately after earthquake, and updated in real time with the development of the earthquake. Then the total task amount T of emergency rescue after earthquake can be calculated directly according to n_w and n_e, and the calculation formula is as follows

$$T_S = n_w \tag{3}$$

$$T_P = \mu_l * n_e + \mu_m * n_w \tag{4}$$

The search and rescue task of the trapped personnel T_S refers to the number of people who need to be transported out of the earthquake area, directly got from the number of injured people n_w. The delivery task of relief materials T_P refers to the total tonnage of relief materials that need to be transported to the earthquake area, including living materials and medical materials.

3.2 Calculation of task amount of aviation emergency rescue after earthquake

The biggest advantage of air transportation is fast speed. The longer the distance is, the more time the air transportation can save, the more obvious the characteristics are. The destruction of the road after earthquake will not affect the route condition. It is an essential transportation mode for the emergency rescue after earthquake. The greater the degree of damage to the ground roads after earthquake, the higher the demand for aviation emergency rescue forces is. The aviation emergency rescue task after earthquake A increases while r increasing and vice versa.

Because the structure of ground road is a space network which has a long distribution in space, the damage degree will be affected by different factors such as earthquake intensity. After the large-scale earthquake disaster, the ground road is the main escape channel of the disaster area and the input channel of the rescue materials. It plays a very important role in the emergency rescue of the earthquake disaster. After the earthquake, the ground roads will be damaged to varying degrees, resulting in reduced road capacity, and personnel escape and rescue supplies are affected.

Generally speaking, the ground roads are damaged in large area or even interrupted during the early stage after earthquake, and r is close to 1. In the mid-term of earthquake rescue, the ground roads gradually recover, but they can be damaged by the aftershocks again, so r is spirally reduced. During the recovery period, the ground roads are basically completely restored, and r is close to 0.

The demand for emergency materials has a very strong timeliness, so we need to analyze and judge the degree of road damage at the first time. The higher the road damage, the more difficult the ground vehicles is, even impassable, so in order to implement the rescue as soon as possible, it is necessary to transport the wounded and materials first by the rescue aircraft.

After earthquake, it is usually impossible to know the extent of the road damage in real time and accurately, but it is possible to know about the scope of the earthquake wave and the damage range of the road in this area. Then the ground road damage degree r can be equivalent to the road damage ratio within the scope of the disaster.

$$r = S_d/S_e \tag{5}$$

Then the search and rescue task of the trapped personnel by aircraft A_S and the delivery task of relief materials by aircraft A_P can be calculated separately from the following formula:

$$A_S = r * T_S \tag{6}$$

$$A_P = r * T_P \tag{7}$$

3.3 Calculation of the takeoff landing number of emergency rescue aircraft

The calculation of the takeoff landing number of emergency rescue aircraft bases on the search and rescue task of the trapped personnel by aircraft A_S and the delivery task of relief materials by aircraft A_P. Assuming that the size and type of rescue aircraft are known, the average number of passengers α and the average tonnage β of the rescue aircraft can be calculated. The takeoff landing number of emergency rescue aircraft A can be calculated from the following formula:

$$A = A_S/\alpha + A_P/\beta + \delta \tag{8}$$

Considering that a part of the aircraft that carries out the task of searching and rescuing the trapped personnel can carry out the mission of relief supplies at the same time, the actual needs of the aviation emergency rescue force should be within the following range:

$$[max(A_S/\alpha, A_P/\beta) + \delta, A] \tag{9}$$

4 Example analysis

It is more and more common for aviation emergency rescue forces to participate in emergency rescue work after earthquake since the Wenchuan earthquake in 2008. Therefore, considering the data verifiability, the Wenchuan earthquake in 2008, the Ya'an earthquake in 2013, the Dingxi earthquake in Gansu in 2013, the Xinjiang earthquake in 2014, and the Jiuzhaigou earthquake in 2017 are used as an example to carry out the model verification and analysis. Specific data is shown in Table 1.

Table 1. Table of earthquake case data

Earthquake name	The injured people n_w	The affected people n_e	The damage degree of road r	The actual takeoff landing number
the Wenchuan earthquake in 2008	374643	6437000	0.95	450
the Ya'an earthquake in 2013	11470	383000	0.9	53
the Dingxi earthquake in Gansu in 2013	582	440000	0.1	20
the Xinjiang earthquake in 2014	0	7838	0.8	2
the Jiuzhaigou earthquake in 2017	525	176492	0.9	127

Data source: China earthquake Information Network

The purpose of this paper is to calculate the total demand for delivering materials, so it is not necessary to distinguish the types of goods and materials. According to the corresponding relationship between the demand for all kinds of materials after earthquake and the number of affected people by Nie [6], the value of μ_l is 2kg per person, and the value of μ_m is 1kg per person. Based on the type of the rescue aircraft participating in the five earthquake, the average number of passengers α was determined to be 10 persons per frame, and the average tonnage β was 2 tons per frame. The datum value of aviation emergency rescue after earthquake δ was 5. The calculating result is shown in Table 2.

Table 2. Comparison table of model calculation results

Earthquake name	Actual result	Calculating result
the Wenchuan earthquake in 2008	450	[6298, 41889]
the Ya'an earthquake in 2013	53	[1037, 1387]
the Dingxi earthquake in Gansu in 2013	20	[49, 55]
the Xinjiang earthquake in 2014	2	[11,11]
the Jiuzhaigou earthquake in 2017	127	[111, 143]

From the above results, it can be seen that the actual takeoff landing number of rescue aircraft in 2008 Wenchuan earthquake is quite different from the calculation results. This is because the aviation emergency rescue force had not yet been universally involved in emergency rescue work after earthquake, and the rescuing aircraft did not meet the needs of the rescue. But after 2008, with the continuous development of the general aviation industry, the emergency rescue force of civil aviation is becoming more and more perfect. Many general aviation enterprises even take the initiative to participate in earthquake rescue. Therefore, the aircraft involved in the rescue gradually improved in meeting the needs of rescue. Until the Jiuzhaigou earthquake in 2017, the actual rescue takeoff landing number had fallen within the rescue requirement interval, basically meeting the needs of rescue.

5 Conclusions

By analyzing the main task demand of emergency rescue after earthquake and the influence factors of the emergency rescue demand for aviation, this paper calculates the takeoff landing number of rescue aircraft after earthquake, and draws the following conclusions:

(1) The number of injured people after earthquake and the number of affected people in the earthquake area are the key factors for determining the amount of emergency rescue tasks

(2) The damage degree to the ground roads in the earthquake area is the key factor for determining the demand for the aviation emergency rescue force after earthquake.

(3) The actual takeoff landing number of aircraft involved in rescue is different with the type of rescue aircraft, which is expressed using interval representation. How to dispatch aircraft belongs to mission assignment problem of rescue aircraft, which is also the research direction of the next step. That is, how to optimize the rescue aircraft scheduling after determining the demand for aviation emergency rescue forces after earthquake, so as to improve the efficiency of rescue.

(4) In the case analysis, the input data of the model are the total number of injuries and the number of affected people of each earthquake, so the calculation results are the total demand for the emergency rescue force of the earthquake. The next step is to use the number of injuries and the number of affected people per day after earthquake as input data, to calculate the change of the daily aviation emergency rescue force demand, and to realize the dynamic scheduling of the rescue aircraft after earthquake.

6 Acknowledgment

Thanks to the support of National key research and development plan 2016YFC0802601.

References

1. Kang Lijun. Application Research of Particle Swarm Optimization BP Neural Network on Emergency Resource Demand Forecasting [D], 2013, Lanzhou Jiaotong University
2. Li Lili. Research on demand forecasting for large-scale earthquake emergency rescue materials based on Grey Modeling Technology [D], 2013, Industrial and Commercial University Of Chongqing
3. Wang Zhengxin, Liu Sifeng. Forecasting Demand of the Disaster Emergency Supplies Based on Fourier-GM(1,1) Model, Systems Engineering, 2013(08): 60-64
4. Shuzhen Li, Alateng Tuya. Disaster Risk Research Literature on Statistics Analysis in China Journal

Net, Journal of Risk Analysis and Crisis Response, Vol. 5, No. 2 (July 2015), 129-140

5. Fusheng Yu. Decision-Making Model in the Environment of Complex Structure Data, Journal of Risk Analysis and Crisis Response, Vol. 3, No. 2 (August 2013), 103-109
6. Nie Gaozhong. Models on rapid jugement for the emergent rescue needs during earthquake – by analysis on post-earthquake events, Resources Science, 2001(01): 69-76
7. Guo Jinfen, Zhou Gang. Research on Emergency Material Demand Forecast Method under Large-scale Earthquakes, Value Engineering, 2011. 30(22): 27-29
8. Lu, C., K. Ying and H. Chen, Real-time relief distribution in the aftermath of disasters – A rolling horizon approach. Transportation Research Part E: Logistics and Transportation Review, 2016(93): p. 1-20.
9. Xu, X., Y. Qi and Z. Hua. Forecasting demand of commodities after natural disasters. Expert Systems with Applications, 2010. 37(6): p. 4313-4317.
10. Li Meixia, Che Haitao. Emergency Resource Dispatching Model and Algorithm, Operations Research and Management Science, 2011(03): 72-76
11. Wei Guoqiang, Wu Liang, Yang, Yongqing. Fuzzy programming model of emergency resources dispatching. Computer Engineering and Applications,2011,47(33):222-225
12. Shi Biao. A Two-stage Vehicle Scheduling Model of Transportation of Emergency Resources, Systems Engineering, 2012(07): 105-111
13. Junping Yan. The Spatial Symmetry Axis of Earthquake Hazard in China, Journal of Risk Analysis and Crisis Response, Vol. 3, No. 1 (May 2013), 59-64
14. Zhi Zhang, et al. Research Framework for Emergency Rescue of Disasters and Crises, Journal of Risk Analysis and Crisis Response, Vol. 2, No. 3 (November 2012), 173-177
15. Du Haijian. Study on Optimization of Allocation of Ambulance Helicopter in Earthquake Medical Relief. Helicopter Technique, 2014(01): 16-21
16. Liu, Y., et al. Robust optimization for relief logistics planning under uncertainties in demand and transportation time. Applied Mathematical Modelling, 2018(55):262-280.
17. Zdamar, L., E. Ekinci and B.K. Ükyazici, Emergency Logistics Planning in Natural Disasters. Annals of Operations Research, 2004(129): 217–245.
18. Lu, G.B., L. Zdamar and A. Evik, An interactive approach for hierarchical analysis of helicopter logistics in disaster relief operations. European Journal of Operational Research, 2002(140):118–133.

MATEC Web of Conferences **259**, 03004 (2019) https://doi.org/10.1051/matecconf/201925903004
ICTLE 2018

Research on Anomie Behavior of Drivers on Non-physical Isolated Urban Road Based on Game Theory

Meiye Li[1], Meiying Li[2], Xiaoxia Hu[1] and Ge Li[1]

[1] *School of Highway, Chang'an University, Middle of South 2rd Ring Road, Xi'an, Shaan'xi Province, China.*
[2]*School of Control Science and Engineering, Shandong Univiversity, Shandong, China.*

Abstract. In order to overtake motor vehicles ahead, some motor vehicles tend to enter non-motorized lane illegally on non-physical isolated urban road, which cause a negative influence on non-motor vehicles. In this paper, this anomie behavior is studied by headway distribution function, fluid dynamics model and game theory. Utilize headway distribution function to calculate the probability of critical gap required for motor vehicles entering non-motorized lanes, and fluid mechanics model to analysis delay. After that, considering critical gap, delay, management expense and fine for anomie behavior, the time value functions of drivers and cyclists are established and regarded as the payment functions in mixed strategy Nash equilibrium. Finally, according to the result of Nash equilibrium, two optimal probability models are put forward. Under different circumstances of non-motor vehicle flow, motor vehicle speed, management expense and fine for anomie behavior, the first model quantified the optimal probability that drivers commit anomie behavior and the second model quantified the optimal probability that traffic department implements management. Furthermore, models can provide references for the management of anomie behavior on non-physical isolated urban road.

1 Introduction

Traffic anomie behavior on non-physical isolated urban road refers to the behavior that travelers drive into the non-motorized lane illegally for the fake of overtaking the slow-speed vehicle ahead, which affects traffic safety and efficiency adversely. In China, mixed traffic is a typical characteristic on urban roads, especially non-physical isolated urban roads. However, due to the lack of traffic administration and existing management methods are unreasonable, the traffic anomie behaviors are very common on urban roads, causing a serious impact on the traffic safety and efficiency.

In the aspect of traffic anomie behavior. In the aspect of anomie behavior. The present study focuses on the misbehavior detection, analysing the influencing factors of the traffic anomie behavior. In term of the misbehavior detection, the data, which was collected through simulating a vehicle crash in different traffic scenarios, was used to design a misbehaviour detection scheme considering two inputs of emergency electronic brake light and lane change[1]. Considering the development of connected car, Kamel J. et al. put forward a state-of-the-art of misbehavior detection method based on the theory of exchanging Vehicle–to–X communication (V2X) messages[2]. In term of the factor analysis, factors of the traffic anomie behavior, such as age, gender, travel purpose, education background and vehicle type, were analysed to built the regression model of traffic anomie behavior[3]. Based on the theory of planned behavior, combined with external stimulus, traffic environment and cost pressures a theoretical model of urban traffic anomie behavior on the basis of the theory of planned behavior was constructed[4]. According to the difference of time, place and character, the urban traffic anomie behavior is classified and a logistic regression model was established based on traffic participant's physiology, traffic psychology characteristics and traffic environment characteristics[5].

In the aspect of mixed traffic on road section. The present study mainly focuses on simulating the behavior of drivers and analysing the traffic characteristics under different mixed traffic conditions. In term of longitudinal movements of vehicles, several studies has been carried out to evaluate the different car following models under mixed traffic conditions and compare the performance of these car following models[6,7]. In term of lateral movements of vehicles, based on the cellular automata theory, some studies has put forward different lane changing models to simulate lane changing behavior under mixed traffic conditions. Considering different types of vehicles take different lane changing rules, a multi-speed lane traffic flow model, which includes motorized vehicle and electric bicycle, was established based on mixed traffic flow of existing NaSch research[8]. Due to the car and the truck have different limited deceleration capabilities, Li X. et al. put forward a single lane safety driving (SD) model with limited deceleration capability to two-lane SD model, in which car–truck heterogeneous traffic is considered[9]. In addition, the traffic characteristics, such as delay, headway, speed, acceleration and deceleration, were analysed under different mixed traffic conditions. Vehicle trajectory data in mixed traffic was extracted from videos, and the mixed traffic flow characteristics, such as speed, acceleration, deceleration and longitudinal spacing, were analysed[10]. Traffic volume and speed data was collected for various sections of six lanes and four lane roads, Shah P M. et al. provided an analytical approach to speeds of heterogeneous system of traffic of the various sections[11]. Due to the variety of vehicles with their different static and dynamic characteristics. An attempt was made to study the effect of static and dynamic PCU values on the capacity analysis of four lane divided urban roads and to study the effect of proportion of two-wheeler on PCU values[12]. Under mixed traffic conditions, the car-following models such as Gipps, Intelligent Driver Model (IDM), Krauss Model and Das and Asundi were selected to analyse the capacity, PCU estimation and Level of Service (LOS) [13].

In the aspect of anomie behavior and mixed traffic, present studies mainly focus on the present situation description , reason exploration and influence analysis. The significance of these studies is to improve the traffic condition. However, the measures of controlling the anomie behavior and improving the mixed traffic are insufficient. In order to put forward reasonable suggestions, the analysis of behaviors of traffic participants are needed. In fact, anomie behavior on non-physical isolated urban road was a two-party game process. In this two-party game process, players are motor vehicles and non motorized vehicles.

In this paper, the anomie behavior, motor vehicles drive into non-motorized lane illegally for the sake of overtaking, was studied by headway distribution function, fluid dynamics model and game theory. Considering critical gap, delay, management expense and fine for anomie behavior, the time value functions of drivers and cyclists were established and regarded as the payment functions in mixed strategy Nash equilibrium. Finally, under different circumstances of motor vehicle speed, management expense, fine for anomie behavior and time value of two players, two optimal probability models were put forward. The first model quantified the optimal probability that drivers committed anomie behavior and the second model quantified the optimal probability that the traffic department implements the management. Moreover, the models could provide references for the management of anomie behavior on non-physical isolated urban road.

2 Analysis of anomie behavior of motor vehicle

2.1 Description of driving process

In order to overtake motor vehicles ahead, some motor vehicles tend to enter non-motorized lane illegally on non-physical isolated urban road. This process is showed in Figure 1 and described as follows: ① The road form is a single lane, and there is no physical isolation facilities between the motor lane and the non motorized lane; ② Distribution of non-motor vehicle is discrete and there is some low-speed motor vehicles on the motor lane. In order to overtake the low-speed motor vehicles ahead, rear motor vehicle waits for the suitable non-motor vehicle spacing to enter the non-motorized lane. If the distance between group k non motorized vehicles and group $k+1$ non motorized vehicles is suitable, the rear motor vehicle will drive into the non-motorized lane for the sake of overtaking the vehicle ahead and drive back to motor lane. The distance that just meets the need of overtaking is defined as critical gap in this paper; ③ When the rear motor vehicle drove into the non-motorized lane, considering the driving safety, group $k+1$ non motorized vehicles will brake until the longitudinal distance between motor vehicle and reders does not meet the safety requirements, and the aggregated wave will be formed; ④ After that, group $k+1$ non motorized vehicles will speed up, and dissipation wave will be formed.

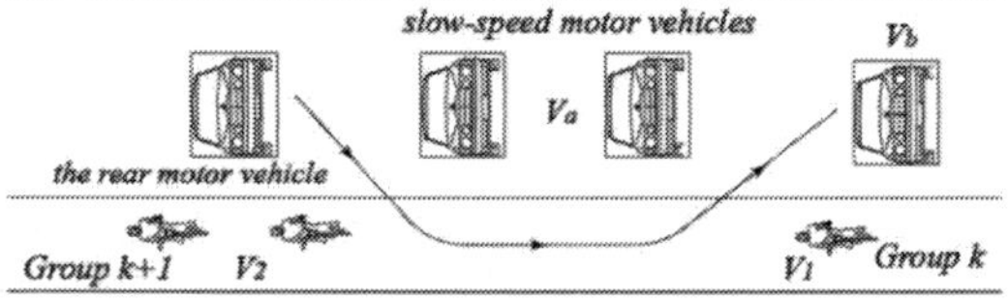

Figure 1. Process of anomie mixed traffic.

In this paper, group $k+1$ non motorized vehicles and the rear motor vehicle are selected as research objective, the change process of traffic flow can be abstracted as: deceleration (aggregated wave) —— following —— acceleration (dissipation wave).

In reality, the rear motor vehicle adopts anomie behavior or not is related to the management measures of traffic department. It's easy to know that there is a game between the motor vehicle and non motorized vehicles, so the game model of motor vehicle and non motorized vehicles is needed. There are two players in the game, include motor vehicle and non motorized vehicles. However, considering that slow-moving traffic is advocated by urban policy and protecting the "non-motor vehicle right" is an important principle in traffic management, so traffic management department can represent the interests of non-motor vehicles. Finally, the players in the game are motor vehicle and traffic management department. The purpose of motor vehicles is passing through the road as fast as possible, which means time saving, and the purpose of traffic management department is supervising the anomie behavior of motor vehicle drivers and reducing the delay of non-motor vehicles.

2.2 Critical gap

Headway is the time interval between two adjacent vehicles in car-following, which represents the time available to the driver of the following vehicle to reach the same level of deceleration as the lead vehicle in case it brakes [14].In the past, the research on the headway was mostly based on the motor vehicle, the research on the headway of non-motor vehicle was inadequately. In this paper, the distribution function of the headway of non motorized vehicle can be obtained by fitting with actual survey data, and the distribution function can be described as equation(1).

$$P(h<t)=\frac{\int_0^t f(t)dt}{Q} \quad (1)$$

where, P is the probability that non-motor vehicle headway h is less than t, $f(t)$ is time headway function of non-motor vehicle, and can be obtained by fitting with actual survey data; Q is the number of non-motor vehicles, which passed the survey line during the investigation period.

Assuming that the time just required for a motor vehicle to overtake lower motor vehicles ahead is τ, and τ is defined as critical gap. Therefore, h larger than τ is a necessary condition for drivers to overtake lower motor vehicles ahead by way of entering non-motorized lane illegally. According to the equation(1), the probability of occurrence of critical gap can be calculated as follows:

$$P_0(h\geq\tau)=1-\frac{\int_0^\tau f(t)dt}{Q} \quad (2)$$

2.3 Fluid dynamics model

Lighthill and Whitham put forward fluid dynamics model by comparing traffic flow to fluid, the model regarded the change of traffic density as the water wave and abstract as a traffic wave. When the traffic conditions change, the changes of traffic density will produce a propagation of traffic wave [15].

In this paper, the delay of group $k+1$ non motorized vehicles was studied when a motor vehicle driver committed anomie behavior. As shown in Figure 1, while the anomie motor vehicle entering the non-motor vehicle lane, assuming that the speed of it is V_c. After it entering the non-motor vehicle lane, the motor vehicle will overtake the front car and the acceleration is a. At the same time, if non-motorized riders recognizing that the longitudinal distance between motor vehicle and reders does not meet the safety requirements, the non-motorized riders of group $k+1$ will brake until the longitudinal distance meets the safe longitudinal distance L_0, after that the speed of non motorized vehicles will increase and dissipation wave is formed. In addition, due to the braking and acceleration time of non motorized vehicle is quite short, we can be assumed that the speed of non-motor vehicle reduce from V_1 to V_2 immediately after braking, and increase from V_2 to V_1 immediately after acceleration. The time tA that non-motor vehicle follows motor vehicle can be calculated by the equation(3).

$$V_c t_A+\frac{1}{2}at_A^2-V_2 t_A=L_0 \quad (3)$$

The distance x_A that non-motor vehicle follows motor vehicle can be obtained.

$$x_A=V_2 t_A \quad (4)$$

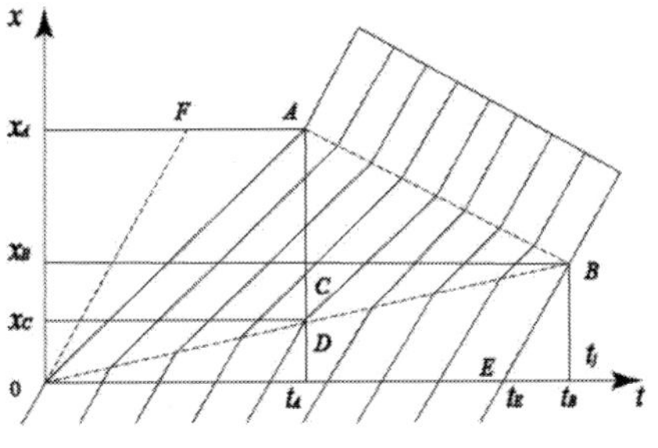

Figure 2. Time and space of motor vehicle.

The time-space trajectory of the group $k+1$ non motorized vehicles is shown in Figure 2. As the speed of non motorized vehicles slow down, the density of group $k+1$ increases from K_1 to K_2 and the volume decreases from Q_1 to Q_2. At the same time, the aggregated wave is formed and its speed equals to the slope of dashed line OB, it can be obtained as follows:

$$w_1=\frac{Q_2-Q_1}{K_2-K_1} \quad (5)$$

When the longitudinal distance between motor vehicle and non motorized vehicles reaches L_0, non motorized vehicles will speed up, the density of group $k+1$ increases from K_2 to K_3 and the volume decreases from Q_2 to Q_3. At the same time, the dissipation wave is formed and its speed equals to the slope of dashed line AB, it can be obtained as follows:

$$w_2=\frac{Q_3-Q_2}{K_3-K_2} \quad (6)$$

The congestion time is from t_0 to t_A. The most congested time is at t_A and complete dissipation time is at t_B, so the difference between t_B and t_A is the dissipation time t_s. The congestion duration t_j and dissipation time t_s can be calculated as follows:

$$w_1(t_A+t_s)=x_A+w_2 t_s \quad (7)$$

$$t_s=\frac{x_A-w_1 t_A}{w_1-w_2} \quad (8)$$

$$t_j=t_A+t_s \quad (9)$$

The flow of non motorized vehicles, skipped by the aggregated wave in unit time, is Q_{W1}. The total number of non motorized vehicles, skipped by the aggregated wave, is N. Q_{W1} and N can be calculated as follows:

$$Q_{w1}=\frac{V_2-V_1}{\frac{1}{K_2}-\frac{1}{K_1}} \quad (10)$$

$$N=Q_{w1}\cdot t_j \quad (11)$$

Finally, the total delay time of group *k+1* can be calculated by the equation(12).

$$D=\frac{1}{2}\cdot(\frac{x_A}{V_2}-\frac{x_A}{V_1})\cdot N \quad (12)$$

According to the equation(3) to (11), the equation(12) can be converted as follows :

$$D=\frac{1}{2}\cdot\frac{V_2-\frac{Q_3-Q_2}{K_3-K_2}}{\frac{Q_2-Q_1}{K_2-K_1}-\frac{Q_3-Q_2}{K_3-K_2}}\cdot(\frac{V_2-V_c+\sqrt{(V_2-V_c)^2+2aL_0}}{a})^2\cdot(1-\frac{V_2}{V_1})\cdot\frac{V_2-V_1}{\frac{1}{K_2}-\frac{1}{K_1}}$$

(13)

3 Game model

3.1 Model hypothesis

Assuming that motor vehicle drivers know that the anomie behavior will be fined, and the penalty for each anomie vehicle is *b*. The probability of traffic management is α and the average cost of supervision and management of each vehicle is *a*. Anomie behavior is bound to be fined once traffic management was implemented. The probability that motor vehicle drivers doesn't commit anomie behavior is *β*. In order to simplify the influence of anomie behavior on the payment function of non motor vehicles, the anomie behavior of motor vehicle drivers would be stopped immediately when it was found and fined, so the anomie behavior wouldn't cause a interference to non motor vehicles. If the driver follow the car ahead, the speed of motor vehicle is equal to the vehicle ahead and the speed is denoted as V_a. To the contrary, if the driver take the strategy to commit anomie behavior, the motor vehicle driver can adopt expected speed and it is denoted as V_b.

3.2 Description of players in game

There are two players in the game, motor vehicle and non-motor vehicle. Considering that slow-moving traffic is advocated by urban policy and protecting the "non-motor vehicle right" is an important principle in traffic management, so traffic management department can represent the interests of non-motor vehicles.

The purpose of motor vehicles is passing through the road as fast as possible. And the purpose of traffic management department is supervising the anomie behavior of drivers and reducing the delay of non-motor vehicles. Their strategy sets are showed as follows:

(1) Strategy set of traffic management department:

G={*G1, G2*}={management, no management}

(2) Strategy set of motor vehicle:

H={*H1, H2*}={anomie, no anomie}

3.3 Game strategy analysis

There are two types of payment functions in the game, including time cost and economic cost. Time cost refer to the time of travel and economic cost contains the cost of management and fine. In order to construct the payment function of each player in the game reasonably, the time cost and economic cost are unified quantification into economic costs by average time value . The game matrix is shown in Table 1.

Table 1. Game matrix of motor vehicle and traffic management department.

traffic management department	motor vehicle	
	no anomie	anomie
management	(C_1, C_2)	(C_3, C_4)
no management	(C_5, C_6)	(C_7, C_8)

When the traffic management department implements the management and the motor vehicle doesn't adopt anomie behavior. The payment function of traffic management department only includes management fees.

$$C_1=a \quad (14)$$

The payment function of the motor vehicle is shown as follows.

$$C_2=\frac{L}{V_a}VOT_1\cdot K \quad (15)$$

where *L* indicates the distance between the end of the road and the site where the rear vehicle meets the front motor vehicle with low speed, or the following distance when the rear motor vehicle driver doesn't adopt anomie behavior. V_a indicates following speed and it equals to the speed of front vehicle. VOT_1 is the average time value of motor vehicle driver. *K* is the average number of passengers on a motor vehicle in the survey area.

When the government implements the management and the motor vehicle commits anomie behavior. The motor vehicle driver would commit anomie behavior once the critical gap occurred. The payment function of traffic management department includes the cost of supervision and management of each vehicle and fine for anomie behavior when the critical gap occurred.

$$C_3=a-P_0b \quad (16)$$

If the critical gap occurred, the payment function of motor vehicle includes the fine when the motor vehicle

https://doi.org/10.1051/matecconf/201925903004

planed to commit anomie behavior but was stopped and the time cost that the motor vehicle following the vehicle ahead and driving through the road after it was fined. If the critical gap didn't occur, the payment function of motor vehicle equals to the time cost that the motor vehicle following the vehicle ahead and driving through the road. So the total payment function of motor vehicle can be described as:

$$C_4 = P_0\frac{L}{V_a}VOT_1\cdot K + P_0b + (1-P_0)\frac{L}{V_a}VOT_1\cdot K$$
$$= \frac{L}{V_a}VOT_1\cdot K + P_0b$$

(17)

When the government doesn't implement the management and the motor vehicle doesn't adopt anomie behavior. In this condition, non-motor vehicles can pass through the road at the expected speed and no regulatory fees are required. There is no payment for traffic management department.

$$C_5 = 0 \qquad (18)$$

The payment function of motor vehicle equals to the time cost that the rear motor vehicle following the vehicle ahead and driving through the road.

$$C_6 = \frac{L}{V_a}VOT_1\cdot K \qquad (19)$$

When the government doesn't implement the management and the motor vehicle commits anomie behavior. In this condition, the non-motor vehicles is influenced by the anomic motor vehicle. The payment function of traffic management department can be represented by the delay of non-motor vehicles.

$$C_7 = D\cdot VOT_2 \qquad (20)$$

where VOT_2 is average time value of non-motor vehicle driver.

Due to the traffic management department doesn't implement the management, there is no fine for anomie behavior. The payment function of motor vehicle equals to the time cost that the rear motor vehicle following the vehicle ahead and driving through the road.

$$C_8 = \frac{L}{V_b}\cdot VOT_1\cdot K \qquad (21)$$

Finally, the payoff matrix of the game can be summarized in the table 2.

Table 2. Payoff matrix of the game.

traffic management department	**motor vehicle**	
	no anomie	anomie
management	(a , $\frac{L}{V_a}VOT_1\cdot K$)	($a-P_0b$, $\frac{L}{V_a}VOT_1\cdot K+P_0b$)
no	(0,	($D\cdot VOT_2 , \frac{L}{V}\cdot VOT_1\cdot K$)
management	$\frac{L}{V_a}VOT_1\cdot K$)	

4 Result

Given the probability β that motor vehicle driver doesn't adopt anomie behavior, if the expected utilities of the traffic management department are equal whether it implements the management or not, the following equation can be obtained.

$$\beta a + (1-\beta)\cdot(a-P_0b) = \beta\cdot 0 + (1-\beta)D\cdot VOT_2$$

(22)

According to the equation(22), the optimal probability β^* that motor vehicle driver doesn't adopt anomie behavior can be obtained.

$$\beta^* = 1 - \frac{a}{P_0\cdot b + D\cdot VOT_2}$$

(23)

Given the probability α that the traffic management department implements the management, if the expected utilities of the motor vehicle drivers are equal whether they adopt anomie behavior or not, the following equation can be obtained.

$$\alpha\frac{L}{V_a}VOT_1\cdot K + (1-\alpha)\cdot\frac{L}{V_a}VOT_1\cdot K =$$
$$\alpha\cdot(\frac{L}{V_a}VOT_1\cdot K + P_0b) - (1-\alpha)\cdot\frac{L}{V_b}VOT_1\cdot K$$

(24)

According to the equation(24), the optimal probability α^* that the traffic management department implements the management can be obtained.

$$\alpha^* = \frac{1}{\dfrac{P_0b}{(\frac{1}{V_a}-\frac{1}{V_b})\cdot L\cdot VOT_1\cdot K}+1} \qquad (25)$$

The solution of the mixed strategy Nash equilibrium is ($\alpha^* = \frac{1}{\dfrac{P_0b}{(\frac{1}{V_a}-\frac{1}{V_b})\cdot L\cdot VOT_1\cdot K}+1}$, $\beta^* = 1-\frac{a}{P_0\cdot b+D\cdot VOT_2}$),

where $P_0(h\geq\tau) = 1-\frac{\int_0^\tau f(t)dt}{Q}$, $D = \frac{1}{2}\cdot\frac{V_2-\frac{Q_3-Q_2}{K_3-K_2}}{\frac{Q_2-Q_1}{K_2-K_1}-\frac{Q_3-Q_2}{K_3-K_2}}\cdot$

$$(\frac{V_2-V_c+\sqrt{(V_2-V_c)^2+2aL_0}}{a})^2\cdot(1-\frac{V_2}{V_1})\cdot\frac{V_2-V_1}{\frac{1}{K_2}-\frac{1}{K_1}}\cdot$$

5 Conclusion

5.1 Optimal probability model of the Traffic department

On the non-physical isolated urban road, the optimal probability(α^*) that traffic department implements the management is:

$$\alpha^* = \frac{1}{\dfrac{(1-\dfrac{\int_0^{\tau} f(t)dt}{Q})b}{(\dfrac{1}{V_a}-\dfrac{1}{V_b})\cdot L\cdot VOT_1\cdot K}+1} \tag{26}$$

Under different circumstances of time value of motor vehicle driver, motor vehicle speed, management expense , fine for anomie behavior, non-motor vehicle flow and the passenger number of motor vehicles, the equation(26) quantified the optimal probability that the traffic management department implements management.

For the traffic management department. On one hand, with the increase of average time value of motor vehicle drivers(VOT_1) and average passenger number of motor vehicles(K), the optimal probability (α^*) that the traffic management department implements management is increasing, which means α^* is closely related to the development of regional economy and the travel habits of residents. One the another hand, with the increase of the fine for anomie behavior(b), the the optimal probability (α^*) that the traffic management department implements management is decreasing, which means motor vehicle drivers tend to comply with the traffic rules if the fine is big enough.

5.2 Optimal probability model of the motor vehicle

On the non-physical isolated urban road, the optimal probability(β^*) that the rear motor vehicle driver doesn't adopt anomie behavior is:

$$\beta^* = 1 - a / [(1-\frac{\int_0^{\tau} f(t)dt}{Q})\cdot b + \frac{1}{2}\cdot\frac{V_2-\dfrac{Q_3-Q_2}{K_3-K_2}}{\dfrac{Q_2-Q_1}{K_2-K_1}-\dfrac{Q_3-Q_2}{K_3-K_2}}\cdot$$

$$(\frac{V_2-V_c+\sqrt{(V_2-V_c)^2+2aL_0}}{a})^2\cdot(1-\frac{V_2}{V_1})\frac{V_2-V_1}{\dfrac{1}{K_2}-\dfrac{1}{K_1}}VOT_2]$$

(27)

Under different circumstances of law enforcement cost of each motor vehicle , average time value of non-motor vehicle riders, motor vehicle speed, management expense , fine for anomie behavior, non-motor vehicle flow and the average passenger number of motor vehicles, the equation(27) quantified the optimal probability that motor vehicle driver doesn't adopt anomie behavior.

For the motor vehicle driver. On one hand, with the increase of law enforcement cost of each motor vehicle(a), the optimal probability that motor vehicle driver doesn't adopt anomie behavior is decreasing, which means motor vehicle drivers are inclined to commit anomie behavior and drive into the non-motorized lane illegally for the fake of overtaking the low-speed vehicle ahead when the law enforcement cost was at a low level. One the another hand, with the increase of the fine for anomie behavior(b), the optimal probability that motor vehicle driver doesn't adopt anomie behavior is increasing.

References

1. A. A. Ardo, A. Zainal, F. A. Ghaleb. Indian J. Sci. Technol. **9** (2016)
2. J. Kamel, A. Kaiser, I. Jemaa. VTC2018-Spring. (2018)
3. R. W. van der Heijden, S. Dietzel, T. Leinmüller. arXiv preprint arXiv. (2016)
4. Li Gaofeng. CNKI. (2016)
5. Yi Zige. CNKI. (2016)
6. C.H. Naveen Kumar. M. Tech Project Report (2013)
7. V. Kanagaraj, G. Asaithambi, C.H.N. Kumar. Procedia-Social and Behavioral Sciences. **104** (2013)
8. You Shiguang, Xiao Wen, Chen Yuguang. Highway Engineering. **3** (2017)
9. Li Xin, Li Xingang, Xiao Yao, Jia Bin. Phys. A. **451** (2016)
10. V. Kanagaraj, G. Asaithambi, T. Toledo. Transp. Res. Rec. NO.2491 (2015)
11. P.M. Shah, N. Gupta. Indian J. Sci. Technol. 9 (2016)
12. C. Chauhan, S. Gupte, K. Bhatt. IJSRSET. **3** (2017)
13. G. Asaithambi, V. Kanagaraj, K.K. Srinivasan. Transportation letters. **10** (2018)
14. Xu Jiqian, Chen Xuewu. Fundamental Traffic Engineering (*Highway Traffic Press*. 2008)
15. Chen Kuanmin, Yan Baojie. Highway Capacity Analysis (*China Communications Press*, 2011)

Utilizing Bicycle Compatibility Index and Bicycle Level of Service for Cycleway networks

Qiang Liu[1], Riken Homma[1] and Kazuhisa Iki[1]

[1] *Graduate School of Science and Technology, Kumamoto University, Japan*

Abstract. Bicycle is widely used in Japan by people of all age groups in daily usage, which may significantly ease traffic congestion. Responding to the cyclists increasing, the method to assess the quality of bicycle travel become necessary. Previous studies reported several approaches to obtain evaluation methods. However, cycleway evaluation in Japan is still far behind the evaluation methods developed in Europe or America. This paper concentrated on familiarizing readers with two methods for evaluating the quality of bicycle facilities and then presenting some proposals of cycleway evaluation in Japan referencing to these two methods. The first method, Bicycle Compatibility Index (BCI), is used to evaluate the road environment for cycling according to the road characteristics by statistical analysis. The second method, Bicycle Level of Service (BLOS), also represents an evaluation of safety for bicyclists. Both of the above methods offered equations of comfort and safety perceptions of bicyclists according to cycling environments. By introducing these methods in combination, this paper enables the readers to maximize the comparative advantages of both BCI and BLOS. The comparison includes sensitivity of variables and the development of both methods. Then we applied BCI and BLOS to evaluate the target roadways in Kumamoto, Japan.

1 Introduction

The use of bicycles is increasing and is being promoted as a low cost, healthy, efficient, and non-polluting way of transportation for all age groups. Responding to the increase in the number of cyclists, the bicycle is now considered as an important mode of urban transportation in Japan. To create a more comfortable cycling environment for cyclists, the Japanese government elaborated the Cycleway Improvement Plan to improve the cycleway network. In response to this situation, the method to assess the quality of bicycle travel become necessary. However, there was no methodology widely accepted by engineers and planners to evaluate the environment of the cycleway in Japan. As yet, several researchers from Europe and America have proposed several approaches to obtain evaluation methods. However, the road geometry in Japan is a little different from that in Europe and America. The objective of this research is to familiarize readers with two popular methods for evaluating the quality of bicycle facilities and then compared these two methods. At last, we will present some proposals of cycleway evaluation in Japan referencing to these two methods.

Previous studies reported several approaches to obtain evaluation methods for bicycle. In the 1960s, the Highway Capacity Manual (HCM) presented the level of service (LOS) to assess the roadway of motorized vehicles [1]. In 1997, on the base of LOS, Landis and his team proposed the method of bicycle level of service (BLOS) to assess the perception of safety and comfort in bicycle facilities [2]. The Florida Department of Transportation (FDOT) concluded that the BLOS is the best method. It has been applied to the roadways in America and Canada. In 2007, also based on the LOS, Søren developed a rigorous methodology that can describe the bicyclist level of service in Denmark [3]. For the bicycle compatibility index (BCI), in 1998, it was developed by the Federal Highway Administration (FHWA) in America; it was developed for urban and suburban areas to measure the cycling environment of the roadway segments [4]. The BCI can clearly reflect the impact on bicycle friendliness by traffic and roadway factors such as curb lane width, number of curb lanes, bike lane width, traffic volume, traffic speed and type of roadside development. In 2003, Jones developed a rural bicycle compatibility index (RBCI) for rating of rural roadway segments [5]. And Based on BCI, Ilie present a model to evaluate the gaps in bicycle network designed for India [6]. In Japan, some previous studies have also been working on the evaluation methods for bicycle and BCI was also used in Tokyo to measure the comfort of cyclists. [7,8]. However, there was still no method widely accepted in Japan. Planners and engineers in Japan urgently need efficient methods to evaluate the cycleway.

In this paper, it begins by introducing BCI and BLOS respectively to familiarizing readers with them. Next, we compared BCI and BLOS in some aspects so that we can

https://doi.org/10.1051/matecconf/201925903005

enable the readers to maximize the comparative advantages of these two methods. Then we applied BCI and BLOS to evaluate the target roadways in Kumamoto, Japan. According to the results obtained, vehicle volume, curb lane width, and pavement condition of a street segment approach have the most significant influences on its service quality. Finally, some proposals of cycleway evaluation in Japan are presented referencing to comparison of BCI and BLOS.

2 Bicycle Compatibility Index

The comfort of cyclists is one of the most common measures used to assess the cycleway environment. In 2005, Hasan used the BCI to generate the most comfortable bicycle routes and, based on the BCI score, a color-coded map was elaborated [9]. In this study, based on the roadway and traffic factors, we used the calculation of BCI to evaluate the cyclists' comfort on each street segment. The comfort index in BCI aims to capture the sentiments of more than 200 participants by letting them watch videos in which the road can be clearly seen. Then, the FHWA let them assess the "bicycle friendliness" of these videos (road segments) according to the cycling environments and through a regression analysis obtained the equation below [10].

$$BCI=3.67-0.966BL-0.41BLW-0.498CLW+0.002CLV+0.0004OLV+0.022SPD+0.506PKG-0.264AREA+AF \quad (1)$$

where:
BL: presence of a bicycle lane or paved shoulder ≥ 0.9 m
YES: BL=1; NO: BL=0;
BLW: bicycle lane (or paved shoulder) width;
CLW: curb lane width;
CLV: curb lane volume;
OLV: other lane(s) volume;
SPD: legal speed limit +15 km/h;
PKG: presence of a parking lane with more than 30% occupancy YES: PKG=1; NO: PKG=0;
AREA: type of roadside development
Residential: AREA=1; Other type: AREA=0;
AF: $f_r+f_u+f_i$, f_r: adjustment factor for truck volumes,
f_u: adjustment factor for parking turnover,
f_i: adjustment factor for right turn volumes.

The concept of compatibility of a roadway with cyclists is an important factor in determining whether the cycleway needs to be improved; thus, we used the BCI to evaluate the compatibility of the roadway environment with cyclists in Kumamoto. The above equation requires various data that would take a lot of times but here we used the data of the traffic census survey (TCS). The TCS website provided the characteristics of the roadways with all the attributes that the BCI needed.

Through equation (1) and the data of TCS, the authors could calculate the compatibility of each of the targeted main road in Kumamoto city. Then we could classify them into several compatibility levels, from high to low level. The compatibility level is also called the BCI level of service (LOS), which is classified into six levels [11]. Different with the LOS defined for motor vehicles, the BCI LOS was developed to describe cyclists [3]. In this study, according to the BCI LOS, the compatibility levels were classified from level A to level F, as presented in Table 1. The breakpoints between each of the level were defined as the 5th, 25th, 50th, 75th, and 95th percentiles of the total scores. Level A means that the road segment is extremely comfortable for the cyclist, while level F indicates the worst condition. Poor scores suggest that the roadway segments may need some improvements to enhance the comfort of cyclists. After statistical analysis, using the geographic information system (GIS), we put the calculated results into the map on each segment of the road and the output option is a color-coded map based on the BCI LOS of target roadways in Kumamoto city. Figure 1 shows the BCI results for the target roadways.

Table 1. Bicycle Compatibility Index score.

Level	BCI Range	Compatibility Level
A	0−1.50	Extremely High
B	1.51−2.30	Very High
C	2.31−3.40	Moderately High
D	3.41−4.40	Moderately Low
E	4.41−5.30	Very Low
F	> 5.30	Extremely Low

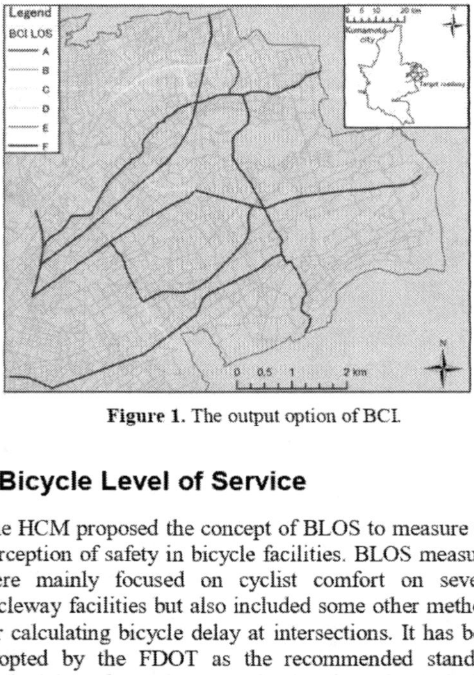

Figure 1. The output option of BCI.

3 Bicycle Level of Service

The HCM proposed the concept of BLOS to measure the perception of safety in bicycle facilities. BLOS measures were mainly focused on cyclist comfort on several cycleway facilities but also included some other methods for calculating bicycle delay at intersections. It has been adopted by the FDOT as the recommended standard methodology for cycleway evaluation throughout Florida. An important finding of BLOS was the importance of pavement surface condition which is frequently dismissed by some researchers. In this study, it confirmed that pavement surface condition plays an important role in cycleway evaluation. This method placed 150 participants in actual urban traffic and roadways to obtain feedback on real-time perceptions [2], rather than cyclist perceptions to video tapes. According to the perceptions of participants and several relevant variables, the following model was developed through the stepwise regression analysis:

$$BLOS=0.507\ln(Vol_{15}/L)+0.199SP_t(1+10.38HV)^2 + 7.066(1/PR_5)^2 - 0.005\ We^2 + 0.760 \qquad (2)$$

where:
Vol_{15} = volume of directional traffic in 15-minute time period;
L = total number of through lanes;
SP_t = effective speed limit = 1.1199 $ln(SP_p-20)$ + 0.8103, SP_p is posted speed;
HV = percentage of heavy vehicles;
PR_5 = FHWA's 5-point surface condition rating (5=best);
W_e = average effective width of outside through lane = $W_t + W_l - \Sigma W_r$,
W_t = total width of outside lane and shoulder/parking pavement,
W_l = width of paving from outside lane stripe to pavement edge,
ΣW_r = width reduction due to encroachments in outside lane.

Widespread use of the original form of the BLOS has led to several refinements. In this study, we used Version 2.0 of BLOS model. The same as BCI, equation (2) also need various data and we used the data of TCS to evaluate the compatibility of the roadway environment with cyclists by BLOS. The TCS website provided the characteristics of the roadways with the attributes that the BLOS needed. However, for the Kumamoto city, complete data were not available for all streets. The data were lacking for pavement surface conditions, which were collected in the field and rating by the authors. Table 2 shows the rating example for pavement surface conditions.

Table 2. Pavement condition rating.

Example	Pavement condition rating
	5(very good)
	4(good)
	3(medium)
	2(poor)
	1(very poor)

Equation (2) produces a BLOS score, which is also stratified into 6 service categories from level A to level F, as presented in Table 3. Level A means that the cyclist is best satisfied for the road segment, while level F indicates the worst. We used the BLOS calculator (created by League of Illinois Bicyclists) to calculate BLOS score of the roadway segment. The last output option is also a color-coded map based on the BLOS score by using GIS. Figure 2 shows the BLOS results based on BLOS calculator which is the same roadways with BCI.

Table 3. Bicycle Level of Service score.

Level	BLOS Range
A	0—1.50
B	1.51—2.50
C	2.51—3.50
D	3.51—4.50
E	4.51—5.50
F	> 5.51

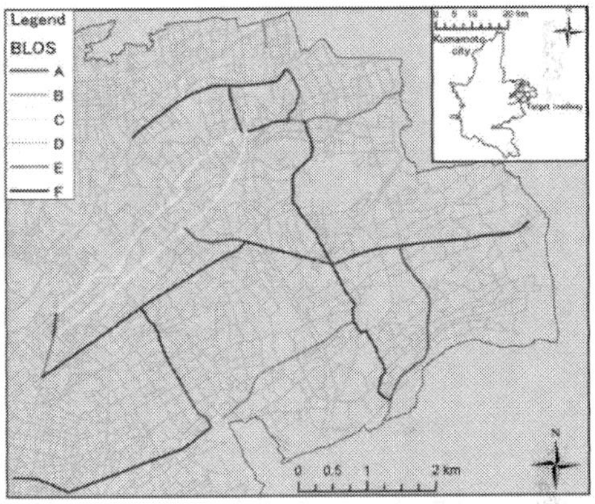

Figure 2. The output option of BLOS

4 Comparison of BCI and BLOS

BCI and BLOS have incorporated the variables that cyclists usually use. Both methods offered equations of comfort and safety perceptions of cyclists according to the roadway environments. Frequently, the results will provide greater detail by using more variables in the equation. However, more variables also let planners and engineers to spend more time and money because of time-consuming data collection. The last output option, color-coded map, can display both BCI and BLOS score of the target roadways and help planners to identify bicycle facilities that may need improvement (the intersections are not taken into consideration). However, there are also some differences between BCI and BLOS. In this section, we compared the basic information about BCI and BLOS which include the organizations, approaches, correlation coefficient (R^2), variables, etc. Then we described a sensitivity analysis to identify the variables that are sensitive.

4.1 The development

The BCI model has 9 variables (Table 4) in the equation with a very high correlation of R^2 = 0.89. BCI was developed by the FHWA based on the research of the Sorton et al. [12] and the Geelong Bikeplan Team [13]. The approaches used to develop BCI model was to get the sentiments of participants by letting them watch videos and then rate the "bicycle friendliness". A Pilot study comparing video survey and field survey was conducted and validated that the video survey is a valid technology to obtain real sentiments from participants. The video survey also can eliminate the risk to participants.

The BLOS was introduced in HCM as a procedure for assessing the comfort of bicycle facilities which include intersections. It has 9 variables (Table 4) and the R^2 is 0.77 (Version 2.0). The BLOS used a different approach with BCI. Instead of video survey, this method placed participants on actual field to obtain feedback because they considered that video survey cannot include all response stimuli.

Table 4. Variables for BCI and BLOS

Variable	BCI	BLOS
Curb lane width(CLW)	x	x
Bike lane width(BLW)	x	x
Paved shoulder width(PSW)	x	x
Traffic volume(VO)	x	x
Vehicle speed(VS)	x	x
Truck volume(TV)	x	x
On street parking(PARK)	x	x
Pavement condition(PC)		x
Lane number(LN)		x
Roadside development(RD)	x	
Turn right(TR)	x	

Note: An x indicates the attribute is used for the equation and a blank entry indicates not.

4.3 Sensitivity analysis

This section describes the sensitivity analysis for BCI and BLOS. According to the situation of Japan, here we chose 5 key variables to discuss the sensitivity include curb lane width (CLW), paved shoulder width (PSW), traffic volume (VO), vehicle speed (VS), and pavement condition (PC). Using OAT, the most common approach, to change one input variable and then to see what effect on the output. The changed variables and default variables were shown in table 5. For BCI, other default variables were: paved shoulder width=bike lane width=1, truck volume=80, on street parking=0, lane number=4, roadside development=1, turn right=0. For BLOS, other default variables were: paved shoulder width=bike lane width=1, percentage of truck=2%, on street parking=0, lane number=2, pavement condition=4.

Table 5 compared the sensitivities of BCI and BLOS to change in key variables. Less curb lane width tended to affect BCI by one level (from level D to level C) while extra curb lane width plays a lesser role. However, the BLOS was more sensitive to curb lane width that the level changed from level D to level B. The effect of paved shoulder width had less influence on BCI but affected BLOS by one level (from level D to level C). The effect of traffic volume on BCI was changed by four levels, which was more sensitive than BLOS changed by three. Both BCI and BLOS affected by vehicle speed by one level. Poor pavement condition had a significant effect on BLOS that dropped it to the worst category.

Table 5. Sensitivity analysis of BCI and BLOS

CLW (m)	PSW (m)	VO	VS (km/h)	PC	BCI	BLOS
Default variables						
3.7	1	4000	50	4	4.68(D)	3.56(D)
Effect of curb lane width						
2.7	1	4000	50	4	5.18(E)	3.95(D)
4.7	1	4000	50	4	4.18(D)	3.02(C)
5.7	1	4000	50	4	3.68(D)	2.4(B)
Effect of paved shoulder width						
3.7	2	4000	50	4	4.27(D)	3.26(C)
3.7	3	4000	50	4	3.86(D)	2.92(C)
3.7	4	4000	50	4	3.45(D)	2.54(C)
Effect of traffic volume						
3.7	1	500	50	4	2.84(C)	2.51(B)
3.7	1	2000	50	4	3.64(D)	3.21(C)
3.7	1	6000	50	4	5.73(F)	3.77(D)
3.7	1	8000	50	4	6.78(F)	3.91(D)
Effect of vehicle speed						
3.7	1	4000	30	4	4.24(D)	3.2(C)
3.7	1	4000	40	4	4.46(D)	3.43(C)
3.7	1	4000	60	4	4.9(E)	3.65(D)
Effect of pavement condition						
3.7	1	4000	50	1	-	10.1(F)
3.7	1	4000	50	3	-	3.9(D)
3.7	1	4000	50	5	-	3.4(C)

Note: BCI does not have a pavement condition term.

To identify the importance of variables more intuitively, we used equation (3) to calculate the effect index of each changed key variable to compare with default variables. According to results above, as the effect of paved shoulder width affected both method slightly, we didn`t calculate the effect index. The effect indexes of other variables are shown in table 6.

$$S_i=(V_i-V_{default})/V_{default} \quad (3)$$

Where:
S_i: effect index (%);
V_i: changed BCI/BLOS score;
V_{defaul}: default BCI/BLOS score.

Table 6 compared the effect indexes of BCI and BLOS. Extra curb lane width and less traffic volume strongly decreased the score of BCI and BLOS which means that it can strongly improve the comfort of the roadway segments. However, high traffic volume tended to strongly affect BCI and played less role in BLOS. For the pavement condition, poor conditions had an extremely significant impact on BLOS while good conditions had not. The critical variables can be referenced in the evaluation method in Japan.

Table 6. Variables for BCI and BLOS

CLW (m)	BCI	BLOS	VS (km/h)	BCI	BLOS
2.7	10.6%	10.9%	30	-9.4%	-10.1%
4.7	-10.6%	-15.1%	40	-4.7%	-3.6%
5.7	-21.3%	-32.5%	60	4.7%	2.5%
VO	BCI	BLOS	PC	BCI	BLOS
500	-39.3%	-29.4%	1	-	183.7%
2000	-22.2%	-9.8%	3	-	9.5%
6000	22.4%	5.8%	5	-	-4.4%
8000	44.8%	9.8%			

4 Conclusions

This paper introduced two popular methods for evaluating the quality of bicycle facilities and then compared these two methods. BCI and BLOS did the same thing. Both used various characteristics of roadway segments to calculate a BCI/BLOS score of comfort and safety for each roadway segment. After applying BCI and BLOS models in Kumamoto city, the output options indicated that both models are not significantly different.

From both methods we can learn about how to establish the evaluation system in Japan. The designs of BCI and BLOS models research were different. BCI suggested video survey and BLOS suggested field survey. Considering the safety of participants, we suggest that video survey is a better way to be held in Japan. Then we used sensitivity analysis to determine the sensitivity of each variable in BCI and BLOS. The sensitivity analysis exposed that some variables are more critical than other variables in affecting the BCI/BLOS formula. Therefore, the results of sensitivity analysis could help to determine which variables should be referenced in the evaluation method in Japan.

This study enables the readers to maximize the comparative advantages of both BCI and BLOS which will be helpful for researchers in Japan to establish a new evaluation system. It will be helpful not only for the research design but also for the variables chosen. As the results provide greater detail by using more variables. Our future research will continue to find suitable variables by studying on other evaluating methods. However, one challenge we face with is the data availability issue. In addition, although the suitable data may be available, it is only for the main road, not the branch way.

References

1. Federal Highways Administration (FHWA), *Highway Capacity Manual*, Washington, D.C., USA (2000).
2. B.W. Landis, V. Vattikuti, , and M. Brannick,. Transp. Res. Record, 1578, 119-126 (1997)
3. U.J. Søren, Transp. Res. Record, 2031, 43-51 (2007)
4. Federal Highways Administration (FHWA), Report No. FHWA-RD-98-072. Washington, D.C. (1998).
5. E.G. Jones, T.D. Carlson, Transp. Res. Record 1828 (2003)
6. A. Ilie, C. Oprea, D. Costescu, E. Roşca, O. Dinu, F. Ghionea, IOP Conference Series: Materials science and engineering **161** (2016)
7. S. Takanori, et al, Infrastructure Planning Review of JSCE **47**, ROMBUNNO.170 (2013) (in Japanese).
8. H. Nagano, master thesis, Chuo University (2013)
9. Z . Hasan, M. K. Mohammed, Transport, **20**, 4, 160-164 (2005)
10. D. Harkey, et al, Report No. FHWA-98-072 (1998)
11. Transportation Research Board (TRB), Highway Capacity Manual: Special Report 209, Washington, D.C. (1985)
12. Sorton, Alex and Thomas Walsh, Transp. Res. Record 1438 (1994).
13. Geelong Planning Committee, *Geelong Bikeplan* Geelong, Australia (1978)

Measuring Public Transport Accessibility for Elderly

Kaniz Fatima[1], Sara Moridpour[2]

[1] *PHD Student, RMIT University, Department of Engineering, Melbourne, Australia*
[2] *Senior Lecturer, RMIT University, Department of Engineering, Melbourne, Australia*

Abstract. The growth acceleration in the older population has been observed in many countries which creates a significant challenge to transportation. Public transport is an essential component of most people's lives. This specific group of population growth will require special attention to accessibility and mobility issues in combination with the transport system which undoubtedly can play a key role to support ageing. Good health, quality of life and independence are fundamental values for the life as well as mobility. For people such as senior citizens, public transport needs to be accessible and affordable in order to provide them with the opportunity to fulfil their medical, shopping and recreational needs. Providing access to transport for people with age 65 & over is not a marginal problem and it is expected to grow in the future. The most important concern in accessibility is the relatively short distance and high frequency movements from a permanent home. This study will identify the travel pattern & mode for elderly commuters in greater Melbourne, Australia. The study will also suggest strategies to improve the use of public transport system to promote ageing in place.

1 Introduction

Over the next few decades with the ageing of the baby boomer generation (those born between 1946 and 1964) combined with the increase in longevity and declining birth rates will produce a new aged group population all over the world [1]. By the middle of the next century roughly one in three citizens of the world's industrial nations will be over 65 [2]. At present people aged 65 years and over make up 13 per cent of Australia's population. By 2051, this proportion is projected to increase to more than 25 per cent, or more than one-quarter of the population. The over 85 age group is expected to 8.6 per cent by 2051 [3]. Figure 1 represents the aged population growth in a percentage all over the world. Good health, quality of life and independence are fundamental values for the life as well as mobility.

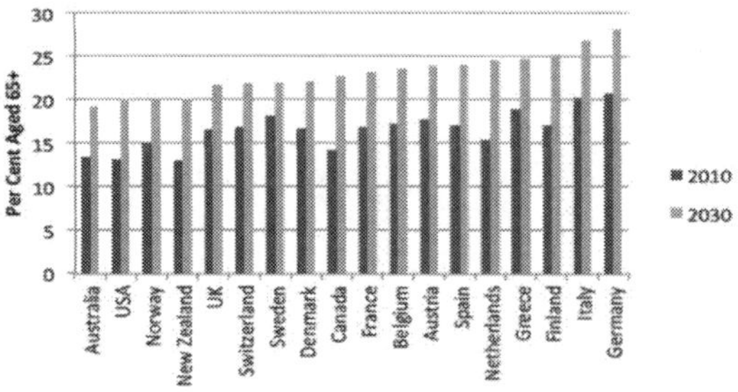

Figure 1. Aged people growth (%) in developed countries. (Source: United Nations, Department of Economic and Social Affairs)

The aim of this study is to identify Melbourne elderly travel pattern & public transportation accessibility. The study also proposes strategies to increase the quality of public transport usage among elderly commuters.

2 Literature review

The travel need of the older population may be even greater than the younger population since they more time for non-home activities and need more social services and health care [4]. Therefore, it is imperative to develop policies to help the increasing number of older people maintain necessary mobility in an aging society [5]. Mobility and accessibility are two important term for transport system. In brief, mobility concerns the way the individual moves, the conditions and infrastructure that are presented to him, the quality of transport, in relation to the ease with which the movement is made. Meanwhile accessibility is the objective of displacement itself and concerns issues related to distance, time and cost to reach a certain destination from a particular origin. The transport system cannot be claimed well enough without having mobility & accessibility for all age group of people. Travel by private transport is most comfortable & flexible way. Public transport accessibility policy should be a new better alternative to avoid dependency on private transport. [6] pointed out proper accessibility for elderly is not only important to older people themselves but also to those with who senior interact (especially where these others rely on tasks such as child care and voluntary work undertaken by senior citizens).

2.1 Elderly travel pattern & public transport accessibility

2.1.1 Ageing in Place

Many of senior people have chosen to remain living in their homes rather than in retirement villages or other forms of aged care accommodation that may have better access to services. This living style is known as "Aging in place". As a result they are adding numbers to the existing suburb population. In Australia it is now much more difficult to ensure their accessibility to both public & private transport facilities [7]. For older people who are no longer part of the main workforce, transportation is a way to maintain social bonds. As a result, the transport is strongly linked to independence, freedom of movement, choice, social activity and community involvement [8]. Most of the traditional transportation & environmental planning ignores the older people contribution [9]. The focus of this paper is to improve public transport accessibility expectations of seniors.

2.1.2 Elderly Trip Pattern

Mostly older people travel to meet their daily needs. Their trip-making primarily occurs during a mid-day peak and daylight hours. [10] studying New Zealanders showed that compared to younger adults (25-59), those aged 65 and over: do fewer trip per day; make shorter trips; travel fewer vehicle kilometres per day or year; make most trips with the purpose of social/recreational/personal business; have more 'at home' days than the general population; make a higher proportion of their trips as a passenger in a vehicle and fewer as a driver; make most of their trips between 9:30 a.m. and 3 p.m. [11].

Most common place to travel for the senior population group is shopping, health care centre & retirement recreation [12]. Some of the senior people are self-employed. It may also include some trips being made to undertake voluntary work, particularly formal volunteering [13]. Very few education trips are made by older people. A part of older population group is involved in 'Education escort' means taking children to school. Senior traveller also travel to visit retail services—banks, post office and chemists. One reason older people make fewer trips than younger people are because some of them find travelling through public transport is difficult. It can be seen that mobility difficulties increase with age, with only 4% of younger people having such difficulties, which increases to 17% for those aged 60–69 and to 39% for those aged 70 and over [6].

Using VISTA (Victorian Integrated Survey of Travel and Activity) 2012 travel pattern has been analysed by statistical software SPSS. Figure 2 represents the travel pattern for elderly commuters. Figure 3 represents elderly trip time using VISTA data 2012.

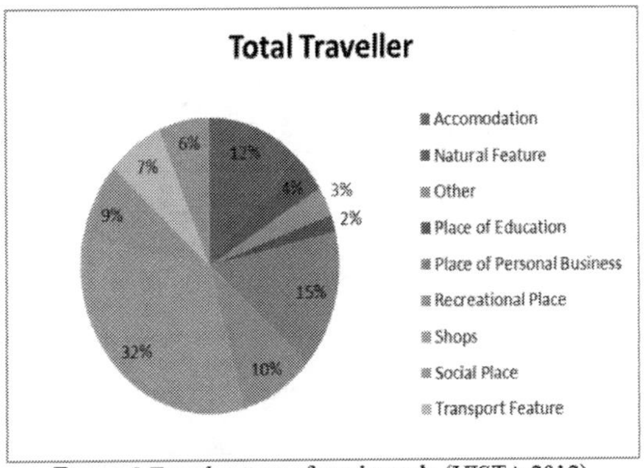

Figure 2. Travel pattern of aged people (VISTA 2012).

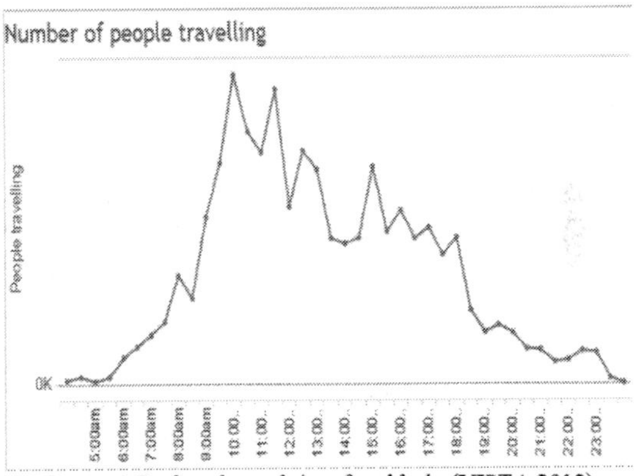

Figure 3. Preferred travel time for elderly (VISTA 2012).

2.1.3 Attitudes to public transport

A satisfactory journey using public transport will typically involve an acceptable walk to the bus stop along well maintained and well lit footways; a short, sheltered and secure wait for a service which arrives reliably, as expected from a readable timetable; a safe and comfortable journey which takes the traveller to a point for disembarking conveniently near to the final destination and a timely return service [14]. Considering public transport as main moving source group of elderly find it not easy. They mainly faced three difficulties as: 1) Facility and vehicle design. 2) Operation. 3) Cost [15].

In addition, within the accessibility problems there are several detail features that faced as obstacles for the elderly. Bad quality of the surfaces, the inaccessible approach of the communal toilets and the absence of railing in the ramp streets [13] are some of the examples. Some countries improved the design of steps, gaps, barriers and difficult changes in level which make accessibility easier for individuals in wheelchairs, the ambulant disabled and the elderly. Table 1 represents the summary of mode of transport choice for aged commuters in Melbourne. The table shows that elderly from Melbourne are more comfortable using private vehicle over public/active transport.

Table 1. Elderly mode choice in Melbourne (VISTA 2012 [16]).

Transport System	Transport Mode	Commuter (%)
Public Transport	Bus	1.5
	Tram	2.5
	Train	8.5
Active Transport	Walking	2.7
Private Vehicle	Vehicle Driver	79
	Vehicle Passenger	4
Other	Other	1.8

Majority of aged people stopped driving because of poor health condition. For health reason they step down from using public transport services as an option even if it is available. For example, older adults may be unable to walk to the bus stop, climb the stairs to a van, go to activities or pay for the transportation services. The financial problem is especially true for low-income older adults with limited financial resources. Particularly in crowded interchanges such as main train stations and central bus stations the older people find hard to organize travel. Lack of energy, confidence and anxiety about potential discomfort will exacerbate the impact of multiple minor disabilities. A typical trip using public transport will involve walking from home to the bus stop or station, identifying which service to take, boarding the bus or train, finding a seat, deciding where to alight, coping with interchanges as necessary and completing the journey to the final destination. For such a trip to be feasible the traveller must be confident that all the links in the chain will be manageable. The journey must be comprised of links that are accessible, coherent, compatible and continuous.

Melbourne, there is a long tradition of research into transport planning for aged people. This research provided information that until the early 1950s a majority of intra-urban travel was done by train, tram or bus [17]. But after the Suburban development outwardly radiating train and tram lines but this began to change. This leads a rise of car ownership levels in late 1950s [18]. As this continued through the 1960s and 1970s house building occurred further away from the fixed tram and train routes. With the development of Melbourne the extension of tram &train network was minimal [19]. Travel within and between the newly built housing areas required a car because tram and train services were beyond walking distance plus the low population densities made bus services uneconomical. In turn, the journey to work length for many Melburnians increased [20] as did the trip to other amenities including shops, health and welfare services. Social security is another obligation to use public transport for elderly specially women. This tends to favour car use. Poorly maintained bus stops (specially in winter) & not having accessible zebra crossing (in particularly for low vision people) [21] are another reason to avoid public transport for senior commuters. Table 2 represents the summary or average trip time comparison among different transport mode for elderly around greater Melbourne.

Table 2. Average trip time for elderly in Melbourne (VISTA 2012 [16]).

	Home sub Region		
Transport System	Melbourne -Inner, Mins	Melbourne- Middle, Mins	Melbourn e-Outer, Mins
Public Transport	44	71	66
Active Transport	16	14	13
Private Vehicle	17	18	19
Other	45	16	18

3 Applied strategies for aged friendly transport system

Promote mobility and accessibility in the transport system is a key element for the population with the reduced mobility conditions. The promotion is already developed and introduced to the existing transport system [22]. To improve the transport system for elderly we need to think some basic criteria or services. [23] Provides a list of qualities for aged friendly transport system. A number of studies have been conducted to provide this key service to maintain good quality of transport system.

Furthermore some strategic proposal has been made to attract elderly to use public transport. The studied proposal as explained as follows:

- According to [6] more flexible smart ticketing for public transport is one of the strategies. More discounts on off-peak rail travel will attract senior travel as well as helps the commercial revenue of transport system. A mixture of buses, community transport, taxis and lift-giving with good publicity and reliable services, would increase mobility amongst older people (particularly in rural areas). Also providing more benches, more public toilets and better street lighting would make walking easier and more attractive for elderly.
- Another study focuses on free bus pass to most common travel places for passenger aged over 60 [24].
- Provide STS (special transport service) for the people are unable to use public transport or car. STS often provides through public procurement of taxi services [25].
- Older people housing option should be more flexible which leads them easy access to daily necessity. New housing rules should be made to motivate future age-in-place situation. Older people need housing options which do not force them to depend on cars. This includes residential alternatives near secure shopping and medical choices to which they can walk or easily take public transit. New housing models are being developed. The elderly who choose

to stay their own place can get home delivery services for everyday goods [26].

- Another study focused on public transport stops improvement. For passengers with reduced vision, tactile markers led the way to the doorway of the bus and bus shelter. This will provide more mental safety to elderly [21].
- School & post office vehicles have limited use for a day. These can be used as a supportive vehicle for elderly. However unsuitability of insurance, administration, design, availability and routing arrangements can make difficulties to the contribution [15].
- In some countries a number of necessary improvements include space for a wheelchair with suitable safety provisions (including raised kerbs), a boarding device to enable wheelchair users to get on and off vehicles, a minimum number of priority seats on buses for disabled passengers, specification of the size and height of steps, handrails to assist disabled people, colour contrasting of features such as handrails and steps to help partially sighted people, easy to use bell pushes throughout a bus, audible and visual signals to stop a bus or to request a boarding device and equipment to display the route and destination [27]. Similar regulations need to apply for new rail vehicles and trams [28]. These policies and practices are being beneficiary not only to elderly but for those with luggage or children.
- The London concessionary travel scheme, funded by London local authorities, allows free off-peak travel for elderly and disabled residents on buses, the Underground, Docklands Light Railway and London rail services [14]. This represents the most advantageous scheme, in terms of cost to users and geographical extent of travel.

4 Proposed strategies

It is always challenging to deal with elderly persons' transport. One of the principal reasons that elderly have travel problems are related to their lack of physical resilience, which maybe either large or small. An accessible and comprehensive public transport system can provide a low cost, low emission alternative to the car. It is therefore important to understand trends in public transport use among the elderly. It helps to understand the long-term impacts of an ageing population on the transport network. The study identifies strategies to increase public transport accessibility for senior commuters. Proposed strategies to increase elderly public transport accessibility are:

- Shuttle bus route service: Analysing the travel time & destination a shuttle bus network will be introduced. From existing datasets mostly liveable elderly suburb, travel destination & travel time can be identified. The new shuttle bus service network depends on these three main variables. The route network of the shuttle service can use as elderly special service. It can be also modified according to needs of the elderly travel pattern.
- Existing public transport time table modify: As elderly mostly avoid peak hour travel so the suburban public transport time table can be changed accordingly. Service frequency can be adjusted when they travel most. This will drive elderly travel mode towards public transport. Specially for suburban bus service it can be adjusted.
- Ride share apps: Elderly persons are more comfortable with personal vehicle; ride share can be effective for day to day need. Through any ride share service a number of passengers can travel towards similar destination without waiting for public transport.

5 Conclusions

With active aging comes the ability and desire to remain living in one's own home and community safely, independently and comfortably for a longer period of time after retirement [16]. As people increasingly retire in the same neighbourhood where they lived most of their adult working life the growth of older adults is particularly rapid in the suburbs of metropolitan areas [29]. For the past two decades, this broader definition of 'ageing in place' has been adopted by researchers and policymakers to describe a number of seniors prefer to live. Research conducted in Australia and elsewhere in the world has consistently shown that most seniors prefer to live as long as possible in their homes [30]. This desire to 'age in their own place' long and active involvement seniors have had with the family home, their neighbours plus the surrounding local community. The option of moving into a retirement village is unattractive due to their high entry cost and ongoing charges. The nursing homes have low patient to staff ratios. Furthermore regular media reports are published of elderly patients being neglected. Hence living in own home provide flexibility & independence. If the public transport is easily accessible to elderly active aging can establish. Proposed strategies are straight forward to apply.

References

1. B. Aguiar, B & R. Macário, *'The need for an Elderly centred mobility policy'*, Transportation Research Procedia, **25**, 4355-4369 (2017)

2. S. Rosenbloom & J. Morris, *'Travel patterns of older Australians in an international context: Policy implications and options'*, Transportation Research Record, **1617**, 189-193 (1998)

3. Australian Bureau of Statistic, *Population projections: a tool for examining population ageing,*. Green, Robin, ed. Year book Australia 2008. Belconnen, ACT: Australian Bureau of Statistics, 2008 (Year book no.90), pp.188-191, ISSN: 0312-4746 (2008)

4. S. Kim & G.F. Ulfarsson, *Travel mode choice of the elderly: Effects of personal, household, neighborhood, and trip characteristics*, Transportation Research Record, **1894**, 117-126 (2004)

5. R. Alsnih & DA. Hensher, *'The mobility and accessibility expectations of seniors in an aging*

population', Transportation Research Part A: Policy and Practice, **37**, 10, 903-916 (2003)
6. R. Mackett, *'Improving accessibility for older people – Investing in a valuable asset'*, Journal of Transport & Health, **2**, 1, 5-13 (2015)
7. B. Engels & G-J. Liu, *'Ageing in Place: The Out-of-Home Travel Patterns of Seniors in Victoria and its Policy Implications'*, Urban Policy and Research, **31**, 2, 168-189 (2013)
8. JA. Davey, *'Older people and transport: coping without a car'*, Ageing and Society, **27**, 1, 49-65 (2007)
9. S. Rosenbloom & J. Morris, *'Travel patterns of older Australians in an international context: Policy implications and options'*, Transportation Research Record, **1617**, 189-193 (1998)
10. C. O'Fallon and C. Sullivan, *'Older people's travel patterns and transport sustainability in New Zealand cities'*. http://www.pinnacleresearch.co.nz/research/older_people.pdf
11. H. Li, R. Raeside, T. Chen & RW. McQuaid, *'Population ageing, gender and the transportation system'*, Research in Transportation Economics, **34**, 1, 39-47(2012)
12. S. Fobker & R. Grotz, *'Everyday Mobility of Elderly People in Different Urban Settings: The Example of the City of Bonn, Germany'*, Urban Studies, **43**, 1, 99-118 (2006)
13. D. Pramitasari & A. Sarwadi, *'A Study on Elderly's Going Out Activities and Environment Facilities'*, Procedia Environmental Sciences, **28**, 315-323(2015)
14. D. Metz, *'Transport policy for an ageing population'*, Transport Reviews, **23**, 4, 375-386(2003)
15. N. Ashford & WG. Bell, *'Transport for the elderly and the handicapped—an overview from the late 70'*, Transportation Planning and Technology, **5**, 2, 71-78 (1979)
16.Victorian Integrated Survey of Travel and Activity (VISTA). https://transport.vic.gov.au/data-and-research/vista
17. B. Engels & G-J. Liu, *'Ageing in Place: The Out-of-Home Travel Patterns of Seniors in Victoria and its Policy Implications'*, Urban Policy and Research, **31**, 2, 168-189 (2013)
18. G. Davison, S. Yelland & A. Davison, A, *Car wars: how the car won our hearts and conquered our cities, Journal of Australian studies*, **83**, 152-153, (2004)
19. P. Mess, *A very public solution: Transport in the dispersed city*, [Book Review], Melbourne University Press, (ISBN) 0522848672 :(ISBN) 0522848672
20. KEN. Johnson, & J. Black, *'An aid to understanding car ownership a study of households in four melbourne suburbs'*, International Journal of Transport Economics / Rivista internazionale di economia dei trasporti, **vol. 4**, 2, 147-167(1977)
21. J. Aarhaug & B. Elvebakk, *'The impact of Universally accessible public transport–a before and after study'*, Transport Policy, **44**, 143-150 (2015)
22. X. Karekla, T. Fujiyama & N. Tyler *'Evaluating accessibility enhancements to public transport including indirect as well as direct benefits'*, Research in Transportation Business & Management, **2**, 92-100 (2011)
23. R. Johnson, J. Shaw, J. Berding, M. Gather & M. Rebstock, *'European national government approaches to older people's transport system needs'*, Transport Policy, **59**, 17-27 (2017)
24. J. Barnes, A. Morris, R. Welsh, R, S. Summerskill, R. Marshall, D. Kendrick, P. Logan, A. Drummond, S. Conroy, B. Fildes & J. Bell, *'Injuries to older users of buses in the UK'*, Planning and Operations, **8**, 1, 25-38 (2016)
25. L. Hansson, & J. Holmgren, *'Reducing dependency on special transport services through public transport'*, Transportation Research Procedia, **25**, 2450-2460 (2017)
26. S. Rosenbloom, *'Sustainability and automobility among the elderly: An international assessment'*, Planning - Policy - Research - Practice, **28**, no. 4, 375-408 (2001)
27. S. Sharp & D. Macdonald, *'Regulating for Improved Accessibility to Buses'*, Towards Safety, Independence and Security. 9th International Conference on Mobility and Transport for Elderly and Disabled People, Warsaw , Poland (2001)
28. M. Deieammes & A. Frye, *'Accessible Railways in Europe: recommendations from the COST335 European research action'*, **1**, 68–74 (2001)
29. EE. Boschmann & SA. Brady, *'Travel behaviors, sustainable mobility, and transit-oriented developments: a travel counts analysis of older adults in the Denver, Colorado metropolitan area'*, Journal of Transport Geography, **33**, 1-11 (2013)
30. D. Olsberg & M. Winters, *Ageing in place: intergenerational and intrafamilial housing transfers and shifts in later life*, AHURI Final Report No. 88, Australian Housing and Urban Research Institute Limited,Melbourne,https://www.ahuri.edu.au/research/final-reports/88 (2005).

MATEC Web of Conferences **259**, 03007 (2019) https://doi.org/10.1051/matecconf/201925903007
ICTLE 2018

Ergonomic Evaluation and Redesign Manual Brick Work Station in The Village of Sail

Merry Siska[1], Eki Saputra[2] and Reski Mai Candra[3]

[1]Industrial Engineering Department, Universitas Islam Negeri Sultan Syarif Kasim Riau, HR Soebrantas Street 155, Pekanbaru, Indonesia
[2]Information System Department Universitas Islam Negeri Sultan Syarif Kasim Riau, HR Soebrantas Street 155, Pekanbaru, Indonesia
[3]Informatic Engineering Department, Universitas Islam Negeri Sultan Syarif Kasim Riau Riau, HR Soebrantas Street 155, Pekanbaru, Indonesia

Abstract. Working posture or work attitude is the working position is naturally formed by the body of workers due to the facilities used to interact with or work habits. Unsuitable working attitude can cause physical complaints such as pain in the muscles (Musculoskeletal Complain). Brick-making factories were encountered in the Village of Sail. This research is a field where once obtained the data needed and then do the processing of the data from the mapping process operations, risk evaluation work experienced by workers by using OCRA Checklist, calculate the value OCRA index, focus group discussions problems faced by workers, design repair work method proposed, calculates the data anthropometry workers, prototyping proposed workstation moulding bricks and evaluate the results of the optimization workstation moulding bricks. After all, factors rated the results of activities scored bricks have totalled a score of 23.19 for the body of the left and amounted to 25.13 to the right part of the body. This value is high because the number of ATA and RTA are very much different. If the number of recommended actions (RTA) is smaller than the number of actual action (ATA) then there is a risk assessment of musculoskeletal disorders.

1 Introduction

The number of companies in Indonesia currently lead the increasingly fierce competition, one of which is SMEs. The embryo of SMEs in Indonesia stems from activity *home industry* in the community, group of farmers, craftsmen, farmers groups, associations and others. Starting from people who want to become economically independent, with limited capital capability smaller groups can grow and develop into a Micro, Small and Medium Enterprises (SMEs) [1]. SME sector has a major contribution in employment, which absorbs more than 99.45% of the workforce and contribution to GDP is around 30% [2].

Until December 2012, Indonesia has 69,249 villages spread evenly across the archipelago. However, the growth of villages throughout Indonesia is not as evenly village existence. According to the Chairman of the Association of Disadvantaged District, Jayabaya Mulyadi, said as many as 32 thousand villages in Indonesia in the category of left behind. Indonesia Central Government continues to alleviate to catch up. † One way to improve the life and economy of the village is to promote small and medium enterprises (SMEs) in the village. In Indonesia, SMEs are the backbone of the Indonesian economy. The number of SMEs in Indonesia until 2011 to reach around 52 million. ‡ It accounts for 60% of GDP and holds 97% of the labor force. Although the quantity is very large and laborintensive, the share of national income is still about 57%. That figure is far less than its share of the conglomerates that are small but very dominating Indonesian economy [3].

MSDs are part of the occupational health problems in waste industries in developing countries [4-5]. The posture of a human body at work is influenced by several factors, including workstation layout (height of the workplace, the orientation of tools and work objects), hand tools design, work methods, and work habits, visual control and force exertion requirements and characteristics of the workers [5-6]. Poor and un-natural working posture have been associated with the onset of fatigue, body discomfort, pains and musculoskeletal disorders [5,7].

Work postures or work attitude is the working position is naturally formed by the body of workers due to the facilities used to interact with or work habits. Unsuitable working attitude can cause physical complaints such as pain in the muscles *(Musculoskeletal* Complain). *Musculoskeletal Disorders* (MSDs) is a disorder of skeletal muscle sections were perceived by someone from mild impairment to severe impairment. When someone receives a static load repeatedly and for a long time, will be able to cause a disruption in the form of damage to the joints, *ligaments* and *tendons* [8]. Thus we need to design a working posture and ergonomic workstations to provide comfort work to prevent complaints of occupational diseases and to improve productivity.

Brick-making factories were encountered in the Village of Sail. The Village of Sail in Pekanbaru city is the outer area of the municipal district of Pekanbaru. This area can be said to be still lagging behind areas - other areas. However, this area can also be said of small and medium industrial zones.

The worker will be working on two types of work or tasks to perform brick moulding process. The first task is to prepare the clay until ready to mould and moulding tasks into two is clay to be ready for drying wet bricks. The first task requires more force because the operators prepare the clay of the form of chunks enough weight to compact for easy moulding. While on assignment to two carriers that require precision moulded bricks have good results and solid. Because of his work process is still manual labour the rate specified by the operator. Figure 1-2 illustrates how operators work in the task of preparing the clay

Figure 1. Worker Bend Her Back to Get Clay on The Floor

Figure 2. Worker Mould The Clay Using Traditional Tools

Questionnaire PLIBEL is early screening tool in the form of a simple check sheet aimed at highlighting the risks of the musculoskeletal workplace. This method was chosen because this questionnaire targeting risk factors through a common question, simple, and can be used for any type of work [9]. The results of a questionnaire PLIBEL identification, risk exposure in the form of a percentage of musculoskeletal disorders in various areas of the body. These results will be used to support verbal interviews and can be seen in Table 1.

Table 1. The Results of PLIBEL's Questionnaire

Worker	*Musculoskeletal Risk Factor*				
	Neck, Shoulders, Back	Elbows, forear, hands	Feet	Knees and hips	lower back
1	34.60%	27.20%	16.20%	16%	14.20%
2	50%	54.50%	19.50%	13%	28.50%
3	30.70%	27.20%	18 70%	12.50%	14.20%
4	38.40%	36.30%	16.60%	14%	14.20%
Average	**38.42%**	**36.3%**	**17.75%**	**13.92%**	**17.77%**

Table 1. shows that the exposed body part most significant musculoskeletal disorders are the upper body / upper limb (neck, shoulders, upper back, elbows, hands, and arms). This is because of the manual process performed by the operator greater involvement of the upper body, especially the hands. The magnitude of this percentage is also caused by the operator answers to the questions concerning the working posture, tools, situations and work environments, as well as repetitive activities undertook more leads to the risk of exposure to the upper body musculoskeletal disorders.

2 Methods

One method for analyzing musculoskeletal disorders is OCRA method where this method can provide an assessment of the relative risks of musculoskeletal disorders are the number of risk factors are more and more detail than other methods. OCRA methods used to analyze the impact of workers to tasks that displays various risk factors work injury, especially in the upper body (repetition of work, strength, posture and movement that is not good, the lack of recovery period, and so forth). OCRA Index is a method that can generally be used to design or redesign both workstations and work there. While the OCRA checklist is the simplest way to apply is generally recommended as the initial examination workstation or work that displays repetitive activity[10].

The basic concept in OCRA is technical action, defined as a complex movement to complete a work task, which involves one or more joints of the upper limb. The OCRA index is calculated as a ratio of the foreseeable frequency (FF) of technical actions per minute and the corresponding number of recommended actions called the reference frequency (RF) of technical actions per minute. RF is the product of seven multipliers: posture multiplier (PoM), force multiplier (FoM), repetitiveness multiplier (ReM), recovery multiplier (RcM), duration multiplier (DuM), additional factors multiplier (AdM) and a constant of frequency (30 actions per minute) of technical actions per minute (CF) [11].

This method is used to direct and facilitate the troubleshooting process and analyze the results of processing through good research management so that the research conducted can be more qualified. Each stage in the research methodology is an important part that must be done properly and carefully.

The design tools in redesigning brick moulding station are designed based on anthropometric data of workers. In connection with this, the need for multiple data anthropometric measure that will be used to redesign. The use of anthropometric data associated with the user and the selection of subjects corresponding data

3 Results and Discussion

3.1.Comparison of Energy Consumption

Calculation workers heart rate data during the initial work is done to determine how much of the energy consumption of the job. Heart rate data moulding worker bricks obtained from the processed data collection to determine the energy consumption on the job, in the moments before and after the redesign. The comparison of the energy consumption of moulding workers brick before and after the redesign can be seen in Table 2.

Table 2. Comparison of Average Physiology Data

No.	Description	Before Redesign		After Redesign	
		Before Working	After Working	Before Working	After Working
1	Heartbeat (Pulse/min)	74.62	96.92	73.94	86 376
2	Oxygen Consumption (Liters/min	0.492	0.938	0.478	0.727
3	Energy consumption (kcal/min)	2.364	4.504	2.298	3.492

Table 2 shows the average energy consumption required by a moulding worker bricks before the design is equal to 2,364 kcal before work and after work amounted to 4,504 kcal. This means that the energy expended in the time before work amounted to 2,364 Kcal and the energy expended after work amounted to 4,504 Kcal workload that qualifies as*"light"* seen from heart rate, oxygen consumption, and energy consumption. While the condition of the drafting of the average energy consumption is issued upon the prior work amounted to 2,298 Kcal and after work amounting to 3,492 Kcal. This means that the energy expended before doing the work amounted to 2,298 Kcal and the energy expended after doing the work amounted to 3,492 Kcal workload that qualifies as*"light"* seen from heart rate, oxygen consumption, and energy consumption. Where workers can save energy as much as 1,012 Kcal/minute after using the tools and moulding facilities bricks after the redesign.

3.2 Comparison of Working Time

Data time existing work processes before and after redesign will be processed to determine the standard time working process. Test uniformity and adequacy of the data of standard time moulding the bricks before the redesign was observed from one worker for 30 times observation is at the upper control limit (UCL) and lower control limits (LCL) and test the adequacy of that has been done stating the samples taken have enough Because N'<N, namely 5.95 <900, then the data moulding time said enough bricks. Uniformity test data when moulding bricks after redesigning observed from one worker for 30 times observation is at the upper control limit (UCL) and lower control limits (LCL) and test the adequacy of the data that has been made stating the samples taken have been quite as N '< N is 7.42 <900, then the data moulding time said enough bricks.

After the test data uniformity and adequacy of the data has been completed, it can be processed further to determine the time of moulding raw brick, both before and after the redesign. [The comparison of the time standard brick moulding before and after the redesign can be seen in Table 3.

Table 3. Comparison of Average Time Moulding Bricks

Description	Before Redesign	Allowance	After Redesign	Allowance
Cycle Time	21.97 seconds	23.5%	16.41 seconds	20.5%
Normal time	24.13 seconds		18.93 seconds	
Standard Time	29.8 seconds		22.81 seconds	

3.3 Productivity Comparison

In determining the productivity of the condition before and after the design can be determined by output standard produced and labour time used by workers.

1. Before Redesign
   In the first working day for 420 minutes, the time standard brick moulding process time before redesigning amounted to 29.8 seconds and output the default is 845 units of bricks /day.
2. After Redesign
   Raw Time of brick moulding manufacturing process after the redesign amounted to 22.81 seconds and output the default is 1105 units of bricks /day.

Based on the data processing before the redesign, the standard time in a brick moulding process requires 29.8 seconds. Meanwhile, after the redesign, the standard time in the moulding becomes shorter brick is 22.81 seconds. Reduction of the standard time before and after the redesign affect the number of output standard produced is of 845 units of bricks/day produced before redesigning be 1105 units of bricks/day after the redesign. Labour productivity increased from 845 units of bricks/day before redesigning be 1105 units of bricks/day after a redesign or an increase in productivity by 30.76%.

The redesign results brick moulding workstations based participatory ergonomics can be seen in Figure 3-5.

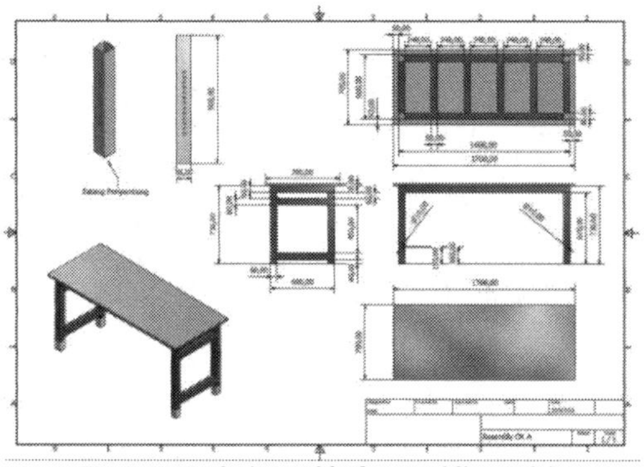

Figure 3. Redesign Table for Moulding Bricks

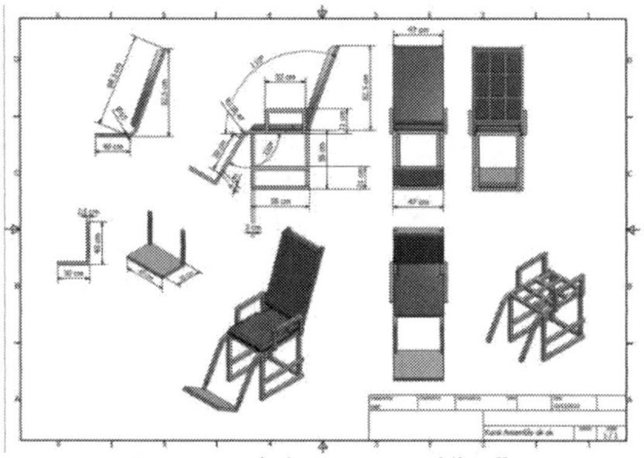

Figure 4. Redesign Rest Seat while Idle

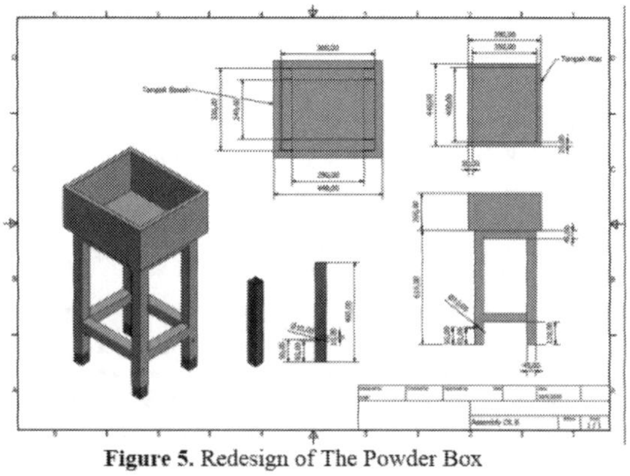

Figure 5. Redesign of The Powder Box

Recapitulation of musculoskeletal disorders risk classification criteria OCRA Checklist is shown in Table 4. Redesign table, seat and powder box in moulding bricks in real condition can be shown in Figure 6-8.

Table 4. Summary of musculoskeletal disorders risk classification based on criteria OCRA checklist bricks

Event type	Body Part	*OCRA Checklist*	Level	Risk Classification
Moulding Brick	Left	81.39	**Purple**	High Risk
	Right	178.29	**Purple**	High Risk

Figure 6. Redesign table for moulding bricks in real condition

Figure 7. Redesign rest seat while break time in real condition

Figure 8. Redesign of the powder box

4 Conclusion

The results of printing bricks have a score of 23.19 for left body and number 25.13 for the right body. This value is high because the number of ATA and RTA is very different. The high rate of repetition of each activity that occurs in brick printing multiplied by the number of production targets per day makes the number of actual technical actions that occur on the field very high.

The design table for printing bricks can also be set in height so that workers who have a waist height stand can use it according to the size of the waist height. It is intended that workers work with a comfortable work posture and can optimize movement in their work.

Redesigned wood powder tables can also be set in height so that workers who have different standing height variable can use it according to the size of their waist height. It is intended that workers work with a comfortable work posture and can optimize movement in their work. The percentile is 5th chosen so that the resting chair designed can be used by all workers who have different sizes. This is so that workers can use chairs to rest comfortably while fatigue at work.

References

1. N. W. Setyanto, R. Himawan, Zefry, E. Y. Arifianto, R. M. S. Puteri, and N. Kurnia, Perancangan Alat Pengering Mie Ramah Lingkungan, *Jurnal Rekayasa Mesin,* **3**, 411-420 (2012)
2. Y.R. Suci, Perkembangan UMKM (Usaha mikro kecil dan menengah) di Indonesia. *Cano Ekonomos* , **6**, 51-58 (2017)
3. Z. Hutabarat and M. Pandin, Absorptive Capacity of Business Incubator for SME's Rural Community Located in Indonesia's Village. *Procedia-Social and Behavioral Sciences*, **115**, 373-377 (2014)
4. E. Valero, A. Sivanathan, F. Bosché, and M. Abdel-Wahab, Musculoskeletal Disorders in Construction: A Review and a Novel System for Activity Tracking with Body Area Network. *Applied Ergonomics*, **54**, 120-130 (2016)
5. A. I. Musa, A. A. Yussouff, N. A. Raji, T.S. Ogedengbe, and R. O. Saheed, Ergonomics Investigation of Musculoskeletal Disorder Among the Workforce of Waste Management Industry in Nigeria. *TRANSACTIONS of the VŠB–Technical University of Ostrava, Safety Engineering Series*, **12**(2), 61-65 (2017)
6. D. B. Chaffin, G.B. Anderson, and D. J. Martin, *Occupational Biomechanics*, **3**rd ed John Wiley and Sons publisher, New York, 65-13 (1992)
7. O. Karhu, E. Kansi, and I. Akuorinka, Correcting Working Postures in Industry: A Practical Method for Analysis. *Applied Ergonomics*, **8**(4): 199-201. (1997)
8. HSE (Health and Safety Executive) Musculoskeletal Disorders in Great Britain (2014)
9. N. A. Stanton, Human factors and ergonomics methods. In *Handbook of human factors and ergonomics methods* , CRC Press, 27-38 (2004)
10. J. Malchaire, G. Gauthy, A. Piette, and F. Strambi, *A Classification of Methods for Assessing and/or Preventing the Risks of Musculoskeletal Disorders*. ETUI (2011).
11. D. Roman-Liu, A. Groborz, and T. Tokarski, Comparison of Risk Assessment Procedures Used in OCRA and ULRA Methods. *Ergonomics*, **56**(10), 1584-1598(2013)

Smart Logistics Framework: A Case Study of Phuket RO Water System

Warodom Werapun , Wisit Srimala, Kullawat Chaowanawatee and Thanakorn Karode

College of Computing, Prince of Songkla University, Phuket Campus, Thailand

Abstract. The product transportation is an important procedure in businesses, which consumes costs and times. Effective transportation management can reduce expenses significantly. New technologies such as Global Position System (GPS), maps, notification and network connection can be integrated all together in one small smart phone that provides accessing from anytime and anywhere. In this paper, we propose a smart logistics framework that combines different technologies in order to solve management and tracking issues in a real transportation environment.

1 Introduction

Nowadays a business of drinking water delivery has great potential for life that people would care more about their health. Drinking water demand has been significantly increased that widely attracts a large number of drinking water delivery companies. Even more companies try to join this businesses, they are still strongly competitive. A company that stays with traditional product delivery will face difficult management. For example, a manager needs to call to delivery staffs to know their locations, how it goes and manually checks the product stock periodically. All sale transactions are written in expensive sale slips. Moreover, transaction reports need to re-input all data from sale slips to a complex spreadsheet which leads human errors and can be easily lost.

All previously described issues require a smart system that offers the simple management to a manager. The system should provide features such as product stock checking and delivery staff route tracking. An effective logistics framework is the solution in order to manage the stock and location tracking for delivery staffs to a manager and show customer destinations to a delivery staff since one of the main costs is the product delivery. The products must be correctly shipped to customers on time in order to provide the greatest services and make business more valuable.

A smart system must be able to monitor location of delivery staffs to protect their cheating. Stock checking with notification can be used to effectively control products in warehouses. In this paper, the smart logistics framework has been proposed to introduce stock and tracking features by testing with Phuket RO Water company, Co., Ltd.

This paper is organized as follows. In Section 2 we present an overview of logistics and vehicle routing optimisation. In Section 3, we propose smart logistics framework and architecture based on the case study. In Section 4 we present our evaluation results and analysis. Finally, in Section 5 we summarize and conclude with open problems.

2 Literature review

2.1 Logistics and Milk Run

Logistics is the delivery of goods, information or resources from one place to another, such as from a seller to a buyer, from a supplier to a seller, from a supplier to a supplier. It can be divided into 3 types:

2.1.1. Direct shipment

Direct shipment is a direct delivery to a customer destination. It will not stop the increase or reduce products on the way, such as from a seller to a customer directly, this method will be faster, but it will cost more. A company must have a very good transport staff and the distance must not be too far.

2.1.2. Milk run

Milk-run is defined as "A route on which a truck either delivers product from a single supplier to multiple retailers or goes from multiple suppliers to a single buyer location" [1]. Milk-run comes from the concept that a milkman must bring a bottle to deliver to customers every morning when a delivery is to bring the old bottle back in order to prepare and deliver it again. It is the source of the milk transport method.

2.1.3 Transportation with cross docking

Transportation with cross docking is a delivery from a source to a distribution point or a warehouse and then from a warehouse to a warehouse nearby [2]. It allows products to be delivered to a customer faster but a staff has to check a warehouse. This may be done to change the type of conveyance, to sort material intended for different destinations, or to combine material from different origins into transport vehicles (or containers) with the same or similar destinations.

To create routes for delivering to customers easier, we need to group customers who stay closely to a zone, and then assign a delivery staff to that zone. However, if the zone is quite large or has complex constraints such as some customers have high priority to firstly receive products, some roads have been closed, there is a traffic jam, and an accident with road blocking, etc.; the routing algorithm with constraint checking must be added.

2.2 Vehicle Routing Optimisation

Minimizing path distance is a good way to reduce estimated time and fuel consumption in the product transportation procedure. Presently, the demands and supplies are bigger scale than the past and they are growing more in the future because the online shopping is easier and faster. Thus, the suppliers have to deliver many products to various locations each run.

Vehicle Routing Problem (VRP) is a problem given many vehicles, warehouse locations, customers' locations and try to find the optimal solution. The goal is to have all vehicle loaded at a warehouse, deliver items to customers and back to a warehouse. Moreover, other constraints, such as capacity of vehicle, fuel, delivery time, priority and more, can be taken into account.

2.2.1 Travelling Salesman Problem Heuristic

The vehicle routing problem can be composed of multiple travelling salesman problem (TSP). TSP is a well-known NP-hard combinatorial optimisation problem. Goal of the TSP is to find shortest tour for given locations. The direct method is to try all permutations of given locations and compute the path distance. However, the time complexity of this method is worst as it would take $O(n!)$. So this brute force method is impractical even for a problem instance with small amount of locations. Aside brute force algorithm, there are other better exact algorithms to find an optimal solution. For example, the Held-Karp dynamic programming algorithm is able to solve TSP in $O(n^2 2^n)$ [3].

The approximation algorithm using heuristic can find a solution faster than exact algorithm. The solution found by an approximation algorithm is not the global optimum but reasonably good enough for limited computation time. Moreover, they work well even on extremely huge (over ten-thousand locations) problem instance.

The nearest neighbour (NN) heuristic algorithm [4] is to repetitively pick a location and greedily find for nearest unvisited neighbour. After a whole repetition is done, the starting point can be changed to explore other possible solutions.

2.2.2 Swarm Intelligence

Swarm intelligence is a concept of optimisation algorithms that explore search space using many agents. Agents can interact or communicate to each other and adapt to search space to reach optimal solution. Most swarm optimisation has an ability to avoid local minima, which is not so good solution, to find global optimum.

Ant Colony algorithm is one of notable swarm optimisation algorithms. Its concept is sending out virtual ants (agents) to explore the map and leave pheromone trails inversely proportion to the tour length. So that the shortest tour, the most pheromone deposited path over time [5].

Other swarm optimisation also works on TSP. For example, Artificial Bee Colony (ABC) [6], Particle Swarm Optimisation (PSO) [7], Cuckoo Search (CS) [8].

2.2.3 Integer Linear Programming

Linear programming problem has a linear objective function and linear constraints. Given constraints are formed an intersection area called feasible region. The goal is to optimise the objective function having all variables within feasible region. The simplex algorithm and interior point method are known to solve linear programming efficiently.

The vehicle routing problem is resolvable using linear programming with integer constraints, namely Integer Linear Programming (ILP). A capacitated vehicle routing problem (CVRP) can be formulated [9] as

$$\text{minimize } \sum_{j=1}^{|R|} \gamma_j x_j$$

$$\text{subject to } \sum_{j=1}^{|R|} a_{ij} x_j = 1 \quad ; \quad \forall i \in V\backslash\{0\},$$

$$\sum_{j=1}^{|R|} x_j = K,$$

$$x_j \in \{0,1\} \qquad ; \quad j = 1, 2, \dots, |R|,$$

where R is a set of all feasible routes, each route R_j costs γ_j. The binary coefficient a_{ij} is 1 if and only if vertex j is covered in the route R_j. If the route R_j is selected in the solution, x_j will be 1 and will be 0 otherwise.

So the objective is to minimize the cost of chosen routes, having constraints that each vertex (customer) is visited by exactly one route and exactly K vehicles required. This problem formulation is similar to a set

https://doi.org/10.1051/matecconf/201925904001

partitioning problem. It can be relaxed to be a set covering model and solved using integer linear programming. Furthermore, this problem can be formulated using flow model then apply either column generation or branch-cut-and-price techniques to improve the performance [10].

Algorithm Comparison

The nearest neighbour heuristic algorithm is easy to implement and the local optima solution can be found over iterations. The solution is reasonably good for a limited computation time. Besides heuristic algorithms, swarm intelligence is capable of medium-large problem size and able to avoid local minima to search for better solutions. The simplex and interior-point methods, on the other hand, are optimisation algorithms for ILP formulated CVRP. It is able to solve even more constraints, such as time windows or priority customers, included in the problem instance.

3 Propose smart logistics framework

"Phuket RO Water System" is selected to be the delivery testbed as Milk Run delivery style for smart logistics framework implementation. Stock management and tracking System are the main components of this framework.

3.1 Architecture

Fig. 1 shows the smart logistics architecture. A delivery staff uses a mobile application, called Service Delivery, to deliver products to customers. It supports to pick up products from other warehouses. The application connects to the cloud server via 4G or WiFi and sends information for each transaction. A manager can check product stock, staffs, and customers in real-time. It will help to significantly increase business values.

3.2 Application Components

We design and implement mobile and web applications for service delivery to customers. The applications offer quickly accessing and reduce communication errors.

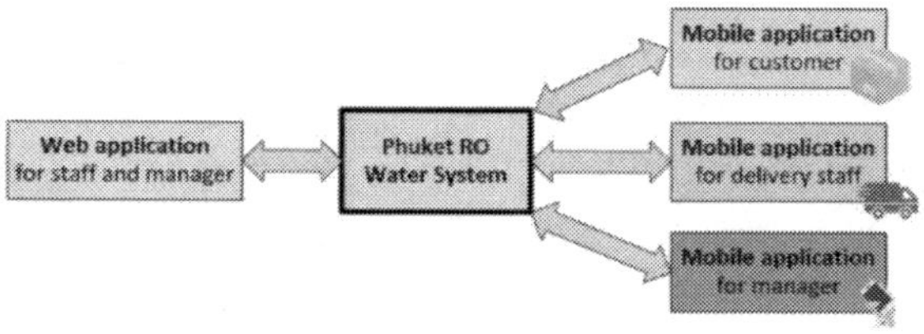

Figure 2. Smart logistics application components

The important features of smart logistics framework will be highlighted in order to present the different services of existing applications.

The framework, called Phuket RO Water system, consists of three mobile applications and one web application as summarized in Fig. 2.

3.2.1 Customer service mobile application

The customer service is developed for a customer who can order products online, check their order history, promotions, contact to the company, etc. One important problem of a customer is the product arrival time. With Geo-Fencing technology, a customer can setup application notification in order to alert after the delivery staff nearby in 500 meters approximately. Notification can be extended to alert via social network such as Facebook or Twitter, etc.

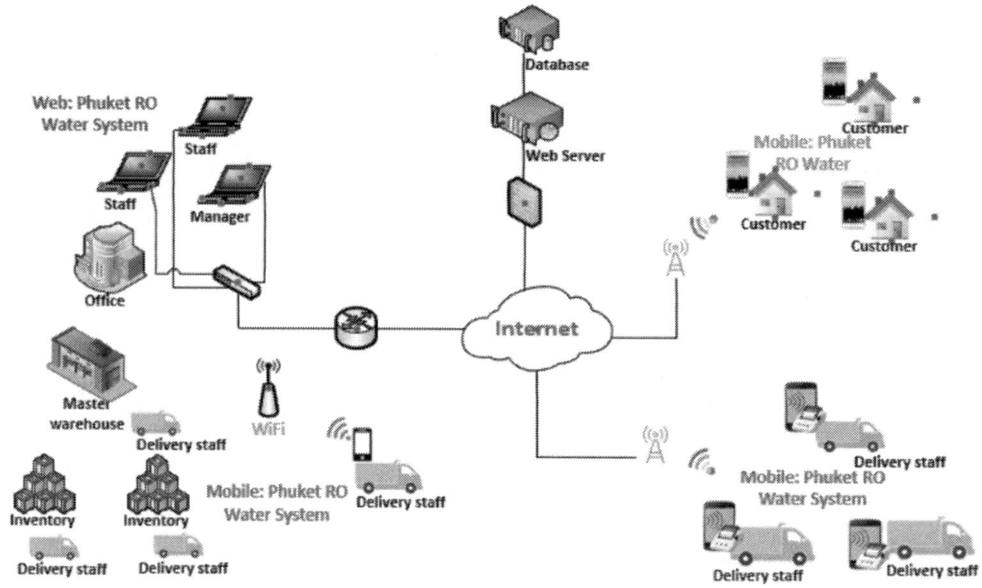

Figure 1. Smart logistics architecture

https://doi.org/10.1051/matecconf/201925904001

3.2.2 Service delivery mobile application

On the first step, a delivery staff takes products to a truck and asks for approval from a manager via a service delivery application. Next, he can drive to customers as illustrated in the map with route optimisation in Fig 3. The nearest neighbour algorithm is taken to apply with this application since it is the fastest and easiest way.

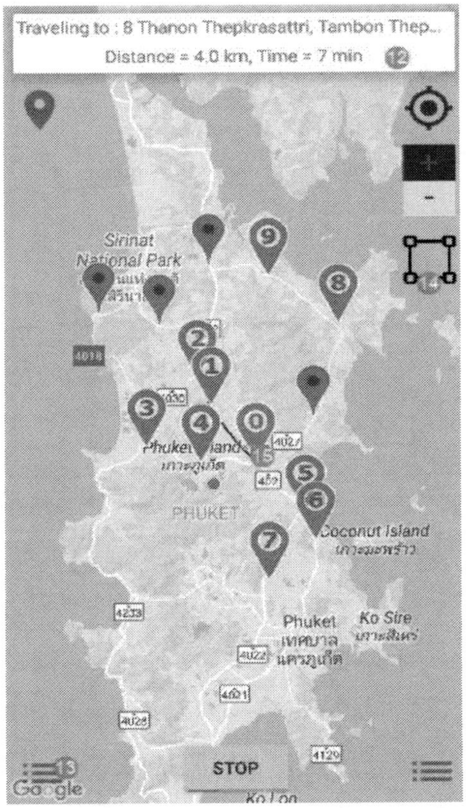

Figure 3. Route suggestion for product delivery

When a delivery staff arrives to a customer, he can deliver products, print slip using a small thermal printer connecting via Bluetooth. All transaction will be sent to the cloud server. Compare to the manual system, this will significantly reduce man-hours times from using hand writing slip and re-input information again in a complex spreadsheet that may cause human errors too. In case of no Internet signal, offline process can be done in an application, it will be synchronized again after reaching an Internet connection automatically.

3.2.3 Delivery manager mobile application

A manager can track all locations of delivery staffs. If they drive outside the customer routes, take a break for too much time, or accidents, etc. A manager can easily verify their cheating or performance and solve the problem rapidly from anywhere and anytime by his mobile phone.

3.2.4 Web application

An office staff can use a web application to fulfil some incomplete data from customers, reconcile data and access on some reports for stock monitoring. A manager can access a whole system to manage and assess a business plan.

The presented features may be existed, however, they are not completely integrated in one mobile phone by three mobile applications and one web application. For example, tracking service may be deployed by other tracking devices which are increasing the budget and most of them lag Geo-Fencing with notification features. Our proposed logistics framework make use of all integrated technologies in a mobile phone. Route optimization and zone of customers will help delivery staffs to service customers effectively.

4 Evaluation

After smart logistics framework has been developed, we test the system, collect transaction data in order to manage product stock and tracking effectively. The web and database servers are deployed on cloud servers from Digital Ocean [11] cloud platform with the cheapest droplet (5 USD/month). They run smoothly. NodeJS [12] and ReactJS [13] are chosen to be backend and frontend systems respectively. Responsive designed is used to support a mobile web view.

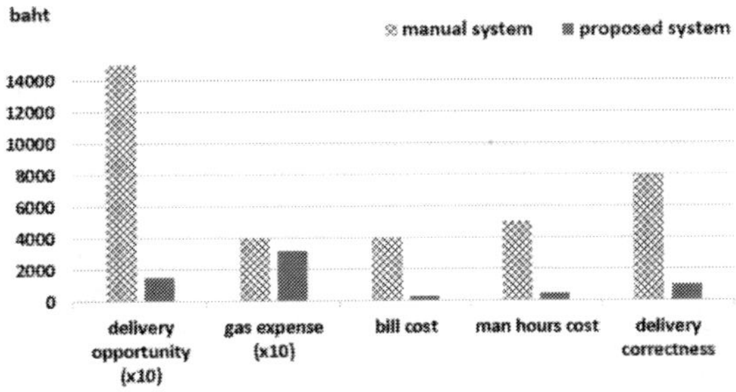

Figure 4. Monthly delivery expense analysis

Fig 4 illustrates the monthly delivery expenses. It reveals that our system can greatly reduce costs of transportation and management. The main impact is delivery opportunity cost since delivery staffs will deliver products to customers on time when customers required products. The explanation for all tasks is described in Table 1.

From table 1, these values are given by Phuket RO Water, Co., Ltd. which has used the smart logistics framework and collected data for several months with the previous manual system for many years. In addition, the proposed system is designed to support horizontal scale up by simply adding a number of cheap cloud servers in order to serve more customers with less increasing cost.

Table 1. Delivery cost analysis

Tasks	Manual system	Proposed system
Delivery opportunity, delivery staffs may arrive to a customer who may not need a product on that period	15% of product sales as 150,000 baht	With analytic and predictable system, the cost will be 10% of manual system, since less chance to deliver on wrong period
Gas expenses and truck maintenances	40,000 baht/month	20% decrease with routing optimisation
Bill slips	4,000 baht/month (0.5 - 1 baht/slip)	0.04 baht/slip for a thermal printing slip plus overhead of the first initial printer
Man-hours to input and analyse data to system	5,000 baht/month	Real-time information updating and less errors from human writing since it directly inputs from the mobile application
Delivery correctness, delivery staffs may forget some customers or cannot deliver products on time	8,000 baht/month	Less occurred with the proposed system since delivery staffs can check a customer list immediately.

5 Conclusion and future works

This research can be used to reduce several service expenses and open more opportunities to have more customers since the great services are provided. Our proposed mobile and web applications may be separately existing, however, the main contribution of this work is to serve seamless integration and implementation of the logistics application components that are deployed and used on a small smart phone. Logistics and shortest tour algorithms have been reviewed, we are going to test and implement more algorithms with more supported constraints to offer better services in the future. Moreover, other maps will be taken into account since Google is going to start having usage fees for maps APIs (Application Programming Interfaces).

Acknowledgements

This work is a part of Thailand Research Fund (Industry) research (No: RDG6050023). The authors would like to thank for the financial support. In addition, we wish to thank Kittipong Pulsawad for his information and ideas. Thanks to College of Computing, Prince of Songkla University, Phuket Campus.

References

1. C. Sunil, M. Peter, Supply Chain Management, 5th Edition, Prentice Hall, 400-425 (2013)
2. G. Mehmet, and H. B. James, J. of Business Logistics, **25**, 221 – 232 (2004)
3. M. Held, R.M. Karp, J. of SIAM, **10**, 196-210 (1962)
4. C.A.J. Hurkens, G.J. Woeginger, Operations Research Letters, **32**, 1-4 (2004)
5. M. Dorigo, L.M. Gambardella, IEEE Transactions on Evolutionary Computation, **1**, 53-66 (1997)
6. L.P. Wong, M.Y.H. Low, C.S. Chong, IEEE AICMS 08, 818-828 (2008)
7. X.H. Shi, Y.C. Liang, H.P. Lee, C. Lu, Q.X. Wang, Information Processing Letters, 103, 169-176 (2007)
8. A. Ouaarab, B. Ahiod, X.S. Yang, Neural Computing and Applications, 24, 1659-1669 (2014)
9. M.L. Balinski, R.E. Quandt, Operations Research, **12**, 300-304 (1964)
10. R. Fukasawa, H. Longo, J. Lysgaard, M.P. de Aragão, M. Reis, E. Uchoa, R.F. Werneck, Math. Program., 106, 491-511 (2006)
11. Digital Ocean, Simplicity at scale, [Online] Available : https://www.digitalocean.com
12. ReactJS, [Online] Available: https://nodejs.org
13. NodeJS, [Online] Available: https://reactjs.org

https://doi.org/10.1051/matecconf/201925904002

The effect of logistics service on firm financial performance in textile industry: evidence from Da Nang city, Vietnam

Hoang Van Hai[1], *Nguyen* Truong Son[2]

[1]The University of Da Nang, Campus in Kon Tum, Viet Nam
[2]The University of Economics - University Of Da Nang, Viet Nam

Abstract: The purpose of this study is to understand an effect of the logistics service on firm financial performance in textile industry of Da Nang city, Viet Nam. A questionnaire was developed to survey several critical factors in the logistics services. Factor analysis method was applied to find some major configurations for each influential factor, and multi-regression method was employed to analyze the influence of critical factor on the firm financial performance. The results showed that internal logistics, inbound logistics, outbound logistics, support activities, and cost of logistics have significantly positive relationship with the firm financial performance. Interestingly, the cost of logistics becomes the main factor affecting the firm financial performance in the textile industry.

1. Introduction

Logistical service has been known to be increasingly important to successful supply chain operations. Logistics service creates value by accommodating customers' delivery requirements in a cost effective manner [1]. Moreover, logistics is critical as a source of competitive advantage and as an importance factor improving the firm performance [2, 3, 4].

A vast variety of previous studies has been applied extensively in the supply chain area to examine how logistics service, as a firm resource, can affect the firm performance. Most of these studies focused on capabilities that are well-recognized supply chain imperatives ([5, 6, 7). However, there is a little attention to directly explore the effect of logistics service on the firm financial performance. Particularly, logistics service has positive influence on the financial performance for retail firms [8]. This result is supported by finding of Shang and Marlow, mentioning that there is a positive relationship between the logistics service and firm performance by using large manufacturing firm samples in Taiwan [9]. In addition, an evidence on the important role of logistics service has also been found in some developing countries, such as Mexico, India, China, and Viet Nam [10, 11].

On the other hand, textile-manufacturing firm recently plays an important role in some emerging markets, such as Viet Nam, and Indonesia. Firms from South East Asian countries, such as Viet Nam, have experienced low supply chain value. Thus, in order to have an effective and efficient flow of firm performance, these firms need to reduce their manufacture cost, especially for the textile firms. For this reason, it is of particular interest to study the current state of logistics activities, that are determined as the main factors of logistics service affecting the firm performance in the textile industry.

In this study, we focus on the textile industry in Da Nang city, one of the most developing regions in Vietnam. The logistics sector in the Da Nang city is fragmented into 130 players, but 70% of them are first party logistics provider (1PL) and second party logistics provider (2PL) services, and the total logistics costs reached USD 5 billion in the Da Nang City [12]. In Da Nang, most of logistics service firms are small and medium, as a result, their capacities are limited. In addition, a lack of professional training and experience, as well as labor of these firms make them to suffer strong competitiveness with other foreign firms in the same region.

Therefore, in order to improve the service quality, and to be more responsive to market requirements, this study focuses on identifying the critical factors of logistics service significantly affecting the firm performance. In particular, the main purpose of this study is to explore the factors of logistics service that impact the textile firm financial performance in Da Nang city,

Viet Nam. Textile is one of the key successful industry of Da Nang. Moreover, due to the globalization trend, Viet Nam government delivered several policies to promote Da Nang becoming one of the biggest logistics service center in Viet Nam.

This study is organised as follows. Section 2 presents a brief literature review about the study methods and previous results. Section 3 briefly outlines the data collection and methodology. In section 4, we provide a detailed discussion of the results. Finally, the last section discusses the implications of the results and concludes.

2. Literature Review and Hypotheses development

2.1. Firm performance

According to several previous standard studies, the firm financial performance can be divided into three types: operational, financial, and strategic performance [13]. The operational performance is more relative to improvement of the organizational activities, such as logistics cost reduction, on-time delivery, inventory turnover, and cycle time reduction. The firm financial performance is measured based on the relationship between total revenue and cost that can be proxy by profitability, return-on-investment, and return-on-sales [14, 15]. Finally, the strategic performance is the improvement of market goals, such as sales, market share, growth in sales and market share [16]. In this study, we focus on the firm financial performance as dependent variables to evaluate the role of logistics service factors.

2.2. The logistics service dimensions and the firm financial performance

In this part, we will provide the relationship between five potential critical factors and firm financial performance. The relationship between each logistics service factor and the firm financial performance is shown in Figure 1.

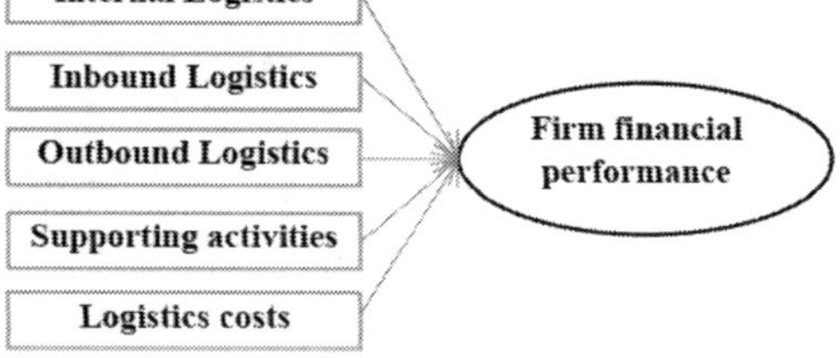

Fig. 1. An effect of logistic service factors on the firm financial performance.

2.3. An effect of the internal logistics

The internal logistics is identified as a task performance process considering the cooperation and effective information share among various departments, as well as effective internal decision-making [17]. The internal logistics system includes five main components: building design, equipment, systems, staffing, and process design [18]. Additionally, the internal logistics in each firm can be observed as cover tasks, such as internal transport, material handling, storage and packaging [19]. The internal logistics can enhance the efficiency of firm internal activities, improve the operational performance, such as on time delivery and cycle time reduction, to achieve a lower operational cost. Thus, the internal logistics has a positive effect on the firm performance [20]. Therefore, it is hypothesized that:

Hypothesis 1: There is a statistically significant positive relationship between the internal logistic quality and firm financial performance. In other words, high internal logistic quality can lead to high firm financial performance.

2.4. An effect of the inbound logistics

In this part, we demonstrate an effect of the inbound logistics quality effect on the firm financial performance. The inbound logistics is a movement process of all materials, collection, shipments, and inventory management, from providers to purchasing firms [21]. The purpose of inbound logistics services mostly focuses on cost saving, inventory optimization, and customer service [22]. A good transportation and firm system from supply to purchase, or from suppliers to factories, can improve the firm efficient activities [23]. Moreover, the role of warehousing management in the inbound logistics has been mentioned by several studies [24]. They mentioned that an efficiency of inbound logistics service systems could deliver optimal service at controlled costs [25]. Therefore, it is hypothesized that:

Hypothesis 2: There is a statistically significant positive relationship between the input logistics quality and firm financial performance. In other words, high input logistics quality can result in high firm financial performance.

2.5. An effect of the outbound logistics

According to the Council of Supply Chain Management Professionals, outbound logistics is the process related to movement and storage of products, from the final production line to final users [26]. The outbound logistics include the final step of the delivery process. The outbound logistics include four main indicators: inventory management system, distribution system, delivery systems to final customers, demand forecasting, and planning [27]. Therefore, it is hypothesized that:

Hypothesis 3: There is a statistically significant positive relationship between the outbound logistics quality and firm financial performance. In other words, high outbound logistics quality can lead to high firm financial performance.

2.6. An effect of the support activities

The role of support activities in logistics service has been mentioned by numerous standard studies. The support activities include infrastructure, human resource management, technology development, and procurement [28]. These activities are integrating functions that cut across the firm traditional functions, and directly affect the firm performance. The supporting activity system can enhance the customer trust, promote information exchange on the market needs, new product development, and production cost reduction, as well as increase the firm financial performance [29]. Therefore, it is hypothesized that:

Hypothesis 4: There is statistically significant positive relationship between the supporting logistics quality and firm financial performance. In other words, the high supporting logistics quality can also lead to high firm financial performance.

2.7. An effect of the logistics costs

In this section, we focus on the potential cost of logistics service that directly affect the firm financial performance. Whether outsourcing or self-service, businesses always have to spend on the logistics costs [28]. The inventory cost is a major element of total logistics cost [30]. Thus, the more reasonable the logistics cost, the higher the level of business efficiency and vice versa. At textile enterprise, the logistics costs can include transport costs, raw material inventory, inventory cost, cost of order processing and the information system, costly paperwork, delivery and payment. Therefore, it is hypothesized that:

Hypothesis 5: There is a statistically significant positive relationship between the cost of logistics quality and firm financial performance. In other words, the higher the cost logistics quality, the higher firm financial performance.

3. Methodology and data collection

A survey was conducted to collect data in this study. The sampling frame was primarily constructed from one of the best and biggest channel of enterprises' information in Vietnam, which aims at trade promotion and business connection (http://yellowpagesvn.com). Information of all firms are provided, such as company background, addresses, contact numbers, and email. The mailing list comprised of 100 firms in the textile industry of Da Nang. In this study, the questionnaires were directly sent to each firm in the sample list from June to October of 2017. The questionnaire, which was preceded by a cover letter on the letterhead of the authors' institutions, employed both fixed-alternative and opened-ended response questions.

The questionnaire were divided into two main parts: demographic questions, and respondent's attitude. For the demographic questions, we provided some information related to the respondent's business sector, their designation, and work experience. The respondent's attitude is measured using a five-point likert scale, ranging from 1, representing "strongly disagree", to 5, representing "strongly agree." In order to improve the liability of our samples, we conducted a pilot test within a small group of academics and logistics service companies to ensure the language clarity and validity of the measurement constructs.

Among the 90 obtained responses, 81 valid replies were used for further analysis. The valid response rate was 78.6%. As for years in business, 66.7% of the respondents started their works between five and 10 years prior to the survey time, 28.4% and 4.9% engaged in business between 11 and 15 years, as well as between 16 and 20 years prior to the survey time, respectively. A comparison of early (those responding to the first calling) and late (those responding to the second calling) respondents was carried out to test for non-response bias [31]. The t-test analyzing 27 resource items and 5 performance items indicated that there was no significant difference between the early and late respondents in all items. Moreover, a variable comparison, including firm size, firm status (ownership), and business duration, between the early and late respondents were conducted by using t-tests. The results revealed no significant differences. Therefore, non-response bias is not expected to be a problem.

The financial indicator variables are described and explained in the Table 1 below.

Table 1: Firm Financial Performance Indicator Variables

Variable	Explanation
ROA	- Arithmetic Average of the last three years of Return on Assets for firm. - (Net Income before Preferred Dividends + (Interest Expense on Debt-Interest Capitalized) / Last Year's Total Assets.
ROE	- Arithmetic Average of the last three years of Return on Equity Per-Share for firm. - Earnings Per Share / Last Year's Book Value Per Share.
ROIC	- Arithmetic average of the last three years of Return on Invested Capital for firm. - (Net Income before Preferred Dividends + Interest Expense on Debt – Interest Capitalized) / ((Last Year's Total Capital + Last Year's Short Term Debt & Current Portion of Long Term Debt)+(Current Year's Total Capital + Current Year's Short Term Debt & Current Portion of Long Term Debt)/2).
NI	- Net Income Growth; Represents the fiscal period income or loss reported by a company after subtracting expenses and losses from all revenues and gains for firm. - ((Net Income / Net Income)^(1/5))-1.

4. Analysis and Findings

4.1. Factor analysis

The measured scales were submitted to exploratory factor analysis. Varimax rotation was used as its common application, in addition, it can minimize the number of variables having high loadings on each factor, as well as can simplify the factor interpretation. Factors with eigenvalues greater than 1.0 are summarized in Table 2. The logistics service factors can be classified into five critical factors: Internal logistics, Inbound logistics, Outbound logistics, Support activities, and Logistics costs. The results are shown in Table 2.

Table 2: Factor analysis results for the five critical factors

Variables	Factor loading				
	1	2	3	4	5
Internal logistic					
Material inventory is suitable	0.796				
Cooperation activities among departments in the company are smooth	0.877				
Enterprises have sufficient resources on transport, human resources, warehouses, and yards to serve the production and business activities.	0.903				
Inbound logistic					
Material transportation process for production is favorable		0.822			
Ordering materials is easy		0.758			
Order time and material transportation ensure the production progress		0.844			
The company has good relationships with raw material suppliers		0.836			
Product defect (such as technical defect, packing defect, faulty quantity, missing product) occur rarely when dealing with raw material suppliers.		0.776			
Outbound logistics					
Packing process and product delivery are convenient			0.798		
Packing and product storage is well done			0.838		
On time delivery			0.881		
Good relationship between company and its business partners			0.829		
A little of mistakes (such as the number of products, product patterns) in delivery			0.630		
Supporting activities					
Information on clients and business partners is updated regularly				0.822	
Processing order process is efficient				0.869	
Handling complaint process is quick				0.824	
Procedure of goods delivery and payment is convenient and easy				0.689	
The company reaches customers' needs in terms of product specifications, delivery time, and location				0.810	
Finding professional logistic service providers is easy				0.821	
Logistics costs					
Transportation cost is suitable					0.815
Cost of raw material inventory is suitable					0.762
Cost of product inventory is suitable					0.829
Costs of order processing and information system are suitable					0.826
Costs of paperwork, delivery and payment are suitable					0.766
Costs of renting warehouse, checking products are suitable					0.768
Eigenvalue	5.094	4.323	3.619	2.199	1.813
Variance explained	21.226	18.013	15.080	9.164	7.555
Cumulative Variance explained	71.03%				
Cronbach's alpha for each dimension	0.724	0.839	0.886	0.868	0.888

Based on the factor analysis results, construct validity of this study could be confirmed. The reliability analysis was also conducted. It can be seen from the Table 2 that the smallest Cronbach's alpha value is 0.724. This overall reliability is 0.84. It implies that the sampling results are reliable.

In the next, we provide the correlations between the logistics service factors and the firm financial performance. In this study, FFP denotes the firm financial performance, LNB represents the internal logistic, LIN is the inbound logistic, LOU indicates the inbound logistic, LSE is the supporting activities, and LCO represents the logistic cost. The results are shown in Table 3.

Table 3: Correlation analysis results

Variables	FFP	LNB	LIN	LOU	LSE	LCO
FFP	1					
LNB	0.380**	1				
LIN	0.267*	0.006	1			
LOU	0.257*	0.205	0.206	1		
LSE	0.557**	0.107	0.280*	0.030	1	
LCO	0.613**	0.268*	-0.105	-0.036	0.257*	1

*: $p < 0.05$ **: $p < 0.01$

The correlation matrix provides initial evidence on our hypotheses: these logistics factors are positively associated with the firm financial performance. In other words, a high correlation was found between the FFP and LSE, LCO. The others were not highly correlated.

4.2. Multiple regression analysis

In order to understand an effect of the logistics service factors on the firm financial performance, multiple regression analysis was utilized. Based on the factor analysis results, we defined the dependent variables as the firm financial formance, such as the ROA, ROE, ROIC, and NI. The independent variables include the internal logistics, inbound logistics, outbound logistics, support activities, and logistics costs. Moreover, we also used other variables, such as the number of employees, capital size, and company age, as control variables in the multiple regression analysis. The regression model can be written as below:

Firm financial performance = α + $\beta1$* Internal logistics + $\beta2$* Inbound logistics + $\beta3$ * Outbound Logistics + $\beta4$ * Support activities + $\beta5$ * Logistics costs.

To increase the liability of our results, prior to running the multiple regression model, we examined satisfaction of the regression assumptions of homoscedasticity, linearity, normality, independence of residuals, and the absence of multi-collinearity. Based on plot of residuals versus the predicted values, an achievement of the homoscedasticity and residual independence for our sample data could be confirmed. The other assumptions were also satisfied. In particular, we conducted the normality of residuals by using the Anderson – Darling test. We found that the test statistic A square was 0.43 (p-value > 0.1), suggesting that the normality assumptions were satisfied. For checking the multi-collinearity, we utilized the Variance Inflation Factor (VIF), which are shown in Table 4. The results showed that all VIF values are less than 10, indicating that the multi-collinearity effect in the multiple regression could not occur. Moreover, the evidence of autocorrelation test for residuals are also provided. The main results of the multiple regression are shown in Table 4:

Table 4: Multiple regression results

Dependent variables: Firm financial performance					
	Model 1		Model 2		
Predictors	Coefficient β	*t-statistic*	Coefficient β	*t-statistic*	VIF
Control variables					
Company age	0.004	0.372	0.002	0.214	1.172
Number of employee	0.021	0.521	0.017	0.921	1.054
Size	0.009	1.820	0.005	1.712	1.212
Critical factors					
Internal logistics			0.167	2.346	1.135
Inbound logistics			0.180	2.493	1.173
Outbound logistics			0.193	2.770	1.099
Support activities			0.354	4.860	1.195
Logistics costs			0.504	6.931	1.190
R^2	0.041		0.667		
Adj. R^2	0.032		0.645		
F	0.504		6.117***		
Durbin-Watson value	2.001			2.042	

From the model 1 of Table 4, it can be clearly seen that the size variable is positive (0.009) and statistically significant within *t*-statistic is 1.820, which means that the large firms tend to be have outperformance compared to the smaller firms. The others two control variables are positive and insignificant. When moving to model 2 of Table 4, the size effect also remains higher effect level. Importantly, all the five critical logistics service factors are statistically significant positive. In fact, the coefficient of the internal logistics is 0.167 within *t*-statistic 2.346, the coefficient of the inbound logistics is 0.180 within *t*-statistic 2.493, that of the outbound logistics is 0.193 within *t*-statistic 2.770, that of support activities is 0.354 within *t*-statistic 4.860, and that of logistics cost is 0.504 within *t*-statistic 6.931.

The results from the model 2 suggested that the larger firms could also contribute to enhancement of the logistics service and firm financial performance. Interestingly, the acceptability of our hypothesis was found, indicating that the five critical logistic factors positively affect the firm financial performance. Among these five factors, the logistics costs became a main factor successfully contributing to the firm financial performance. This result is consistent with Juuso Töyli, reporting that when firms get more attention on logistics service, they might gain competitive advantage relatively easy by focusing more on the logistics performance, at least in a short-term [19].

5. Conclusion

This study examined the effect of logistics service on the firm financial performance in the textile industry sample in Da Nang city, Viet Nam. The model suggested that a five-dimension conceptualization of the logistics service factor can affect the firm financial performance that use data from the logistics service customer to assess the critical elements. The results indicated that the five main factors, such as the internal logistics, inbound logistics, outbound logistics, support activities, and cost of logistics, were statistically significant positive relation to the firm financial performance. In addition, among the five factors of logistics service, the logistics cost became the main factor affecting the firm financial performance in the textile industry. In particular, the total costs are higher than average. An explanation could be because the textile firms have to utilize the outsourcing logistics services, thus, in case they conduct the services by themselves, the outsourcing costs could be higher. Based on the results from this study, we suggest that the firms can improve their information systems by training to enhance their employee professional skills. They should also apply technical innovation to their logistics service for capacity improvement.

This study contributes to the existing literature by exploring the composition of the logistics service quality and investigating its impact on the firm financial performance in the textile sector. In consideration the values of this study, some points need to be considered: firstly, this study concerns only the textile industry, secondly, the results can be affected by the firm location. Despite these limitations, this study provides some valuable insights from theoretical and practical perspectives. As such, this finding is unique for the textile sector because it introduced and empirically validated the measurement of logistics service quality in the textile industry.

Finally, future studies should assess a direct relationship between the logistics service and firm performance for large data sample and different kind of industries in Viet Nam, as well as include more factors, such as external and situational factors affecting the logistics service. In addition, future studies should also investigate an impact of relationship length on the interrelationships identified in the model.

Reference

1. Bowersox, DJ, Closs, DJ & Cooper, MB. *Supply Chain Logistics Management*, (2002)
2. Bowersox, DJ & Closs, DJ. *Logistics Management: the Integrated Supply Chain Process*, (1996)
3. Gadde, L & Hulthén, K. IMM **38**, 633 (2009)
4. Hsiao, H, Kemp, R, van der Vorst, J & Omta, S. IJPE, **124**, 75 (2010)
5. Cho, J, Ozment, J & Sink, H., IJPDLM **38**, 336 (2008)
6. Lai, F, Li, D, Wang, Q & Zhao, X. JSCM **44**, 22 (2008)
7. Nir Kshetri. IJIM **39**, 80 (2018)
8. Schramm-Klein & Morschett. IRRDCR **16**, 277 (2006)
9. Shang & Marlow, Transportation Research Part E **41**, 217 (2005)
10. Arroyo, P, Gaytan, J & Boer, L. IJOPM **26,** 639 (2006)
11. Lieb, R. IJPDLM **38**, 495 (2008)
12. Sabine Limbourga, Ho Thi Quynh Giang & Mario Coolsc. Procedia Engineering, **142**, 124 (2016)
13. Fabbe-Costes, N. & Jahre, M. IJLM **19,** 130 (2008)
14. Chenhall, R. H., & Langfield-Smith, K. EMJ **25,** 266 (2007)
15. Morgan, N. A., JAMS **40**, 102 (2012)
16. Woojung Chang, Alexander E. Ellinger, Kyoungmi (Kate) Kim, & George R. Franke., EOM **34**, 282 (2016)
17. Wong, C. Y., Boon-Itt, S., & Wong, C. W. Y., JOM **29**, 604 (2011)
18. Baker, P., IJLR&A **9**, 207 (2006)
19. Juuso Töyli, Lotta Häkkinen, Lauri Ojala, Tapio Naula. IJPDLM, **38**, 57 (2008)
20. Lordache, V., Cormos, A.C., Nemtanu, F.C. & Lopez, V.R.T., VJES **3**, 85 (2012)
21. Shyam, M. IJSSH **1**, 1 (2012)
22. Bowersox D. J., Closs D. J. & Cooper M. B. *Supply Chain Logistics Management.* (2010)
23. Coyle, J.J., Bardi, E.J. & Langley Jr., C.J… *The Management of Business Logistics: A Supply Chain Perspective* (2003)
24. Jea-Il Sohn, Su-Han Woo & Taek-Won Kim. IJLM, **28,** 680 (2017)
25. Isaac Elking, John-Patrick Paraskevas, Curtis Grimm, Thomas Corsi & Adams Steven. JSCM. **53**, 22 (2017)
26. CSCMP (Council of Supply Chain Management Professionals), (2013)
27. Liberatore MJ, Miller T. BEJ **7** (2016)
28. Christopher, M., *Logistics and Supply chain Management*, (1998)
29. Haemoon Oh & Kawon Kim. IJCHM **29**, 2 (2017)
30. Jouni Juntunen, Mari Juntunen & Jari Juga. IJLRA **18**, 442 (2015)
31. Armstrong, J.S., Overton, T.S.,. JMR **14**, 396 (1977)

A Hybrid Meta heuristic Algorithm for the Balanced Line Production under Uncertainty

Pasura Aungkulanon[1] , *Pongchanun* Luangpaiboon[2], and *Roberto* Montemanni[3]

[1]Faculty of Industrial Technology, Phranakhon Rajabhat University, Bangkok, 10220, THAILAND
[2]Industrial Statistics and Operational Research Unit (ISO-RU), Faculty of Engineering, Thammasat University, Pathumthani, 12120, THAILAND
[3]Dalle Molle Institute for Artificial Intelligence (IDSIA), University of Applied Sciences and Arts of Southern Switzerland (SUPSI), Galleria 2, Manno 6928, SWITZERLAND

Abstract. This study proposes a hybrid Golden Ball Algorithm for solving a balanced line production for a garment firm in Thailand. At present, production lines are those in which the timing of the job movement between stations is coordinated in such a way that all of the jobs are indexed simultaneously via some heuristic sequencing or dispatching rules. This research studies the balanced line production problem with some stochastic patterns, and develops a Golden Ball Algorithm or GBA and its variants to solve the problem. To assess the effectiveness of the proposed hybrid algorithm, a computational study is conducted for both deterministic and stochastic patterns of the problem. The comparisons are made for two different levels of processing times and due date. It can be concluded that the variant HGBA2 of the algorithm by adjusting answers of the successor function on both custom training and successor phases, is slightly more effective than the other hybrid approaches in terms of quality of solutions under uncertainty.

1 Introduction

Generally, the job shop deals with a work location in which a number of general purposed work stations exist and are used to perform a variety of jobs. There are some special characteristics when compared to traditional machine shop [1]. The factors describing the job shop consist of an arrival pattern (static or dynamic), number of machines or work stations, work sequence (fixed or random sequence) and performance evaluation criterion (make span, average time of jobs in shop, lateness, average number of jobs in shop, utilization of machines or workers). The job shop can be described via a Gantt chart. It is simple to display graphically results and schedule evaluating results. However, it is not suitable for complex situations.

For the problem with N Jobs and M Machines, the number of possible schedules is extremely large (N!)M. Heuristics can be practically solve this problems based on sequencing or dispatching rules [2-4]. The conventional ones include a random pick with equal probability (RANDOM), first come first serve with earliest release date (FCFS), shortest processing time (SPT), earliest due date (EDD), critical ratio of processing time and time until due (CR), slack time (ST) and slack time per remaining operation (ST/O).

Recently, production lines have become the primary mechanism of productive industrial systems. Although there are various problems relative to the line production in the reality, the literature on this issue has been focusing on a few specific problems [5-6]. Many new production lines have been focusing on the Just In Time (JIT) production principle. In this article, the Golden Ball Algorithm has been presented, an effective metaheuristic approach, to seek optimal solutions for the balanced line production problem. Moreover, there is an introduction of a robust optimisation approaches to balanced line production with the input data uncertainty of task times and due date changes. The objective of this research is to apply new hybrid methods based on the Golden Ball Algorithm for the job arrangement in order to balance workload in three production sections. Workload has impact on overtime cost, due date and employees' opinion. The remainder of this article is organised as follows. The background of the problem definition and the proposed algorithm with its hybridisations are reviewed in Sections 2 and 3, respectively. The problem and computational results are described in Section 4. Finally, present areas for future research are concluded in Section 5.

2 Balanced line production

A firm for this study is one of the small and medium enterprises or SMEs garment manufacturer of cleanroom apparels. It develops, produces and distributes a wide range of other cleanroom components. Production is done in a main assembly building and completed in 4-5 assembly stations that vary from apparel to apparel. The steps of the cleanroom apparel production can be summarized as follows. The production of the apparel starts at the cutting and pattern department. The cut and patterned parts that require sawing are taken to the sawing sub-assembly shop. After the completion of sawing the parts are stored in an after sawing area. Then these parts are taken to the final assembly line according to the scheduling. When the assembly is done in the production department then the final part is packed and stored in the warehouse for shipping. Lean and just In Time (JIT) principles drive the firm as an assembling technology.

The company is currently developing a new cleanroom apparel line production. The current status of this project is that a design of the line production is in the final phase and the process engineers are working on the possible assembling methods under uncertainty on some important information. This is the background why the company now asks for a thorough analysis of the cleanroom apparel line production and an assembly line with shortest and balanced line production system. Balancing an production line is a strategy in which tasks are distributed evenly to each assembly line so that each line has the same amount of the work. The significant thing is to balance the workload of the operator at every line or stations, reducing the operator idle time over the takt time which means the decrease of unused idle station capacity. The objective is then to balance line production which gives various advantages in reducing wastes, idle workers, and operator changing, faulty product and stocks.

3 Hybrid golden ball algorithm (HGBA)

Golden ball algorithm (GBA), introduced by Osaba, Diaz & Onieva [7], is a method first used in solving the close-open mixed vehicle routing problem. The GBA algorithmic procedures are based on the principle of football competition as shown in Figure 1[8]. This method is similar to the league football competition. The league consists of a number of football clubs whose members are football players. Once a football team of a specified number of members is formed, the playing season starts. Every team will be matched up together in a form of league. A team with good football players will have a good chance to win as team strength depends on the quality of team members. Before each competition, every football player will have a chance to practice enhancing their capability while each player can move to other teams through trading. Details of the Golden ball algorithm presented in each step are as follows:

Step 1: Build a football team from a standard prototype problem. In this step, n football teams will be built randomly, and each team will create m football players.

Step 2: Calculate the objective function which will be the value of the team's power.

Step 3: The competition is to develop an answer or Competition Phase dividing into two stages.

The first step involves the answer adjustment process using the principle of a self-practicing athlete in order to increase their individual capability, or the custom training function. The second step is to build a new answer using the principle of practicing the team's plan where football players are generated, practice the training plan among team members to enhance the overall team's capability using the successor function created from the answer's neighbourhood. This new answer will be compared with the previous one. If it is better, it will be kept. If it is not better, then the existing answer will be used for further development in the next round.

Step 4: Transfer period is referred to as the trading period of football players. When the competing season ends, a football team with higher scores will have a chance to exchange high performance football players with a team with lower scores to increase the team's strength. There are two cases, i.e. exchanging between teams and generating answers by randomly inputting new values which is similar to trading football players from other sources.

Step 5: Repeat the competition. After finishing each match, the team with better answers will be a winner with three points and the losing team will get 0 point. For the case of a tie score, each team will get 1 point. In each season, the competition will end only if all teams compete together. Competition will continue until completing the specified number or the termination criterion.

In this paper, three methods were generated to modify answers. The first and the second methods were represented the answer improvements in the custom training function. The third were used to adjust answers of the successor function.

Method 1: Randomize two jobs from all jobs and randomly swap positions.

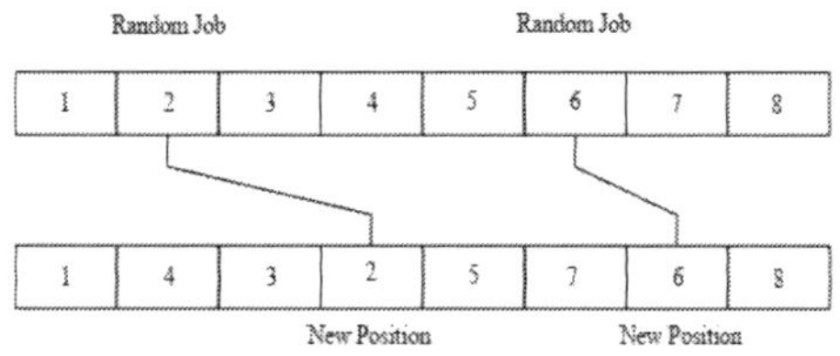

Method 2: Randomize four jobs from all jobs and randomly swap positions.

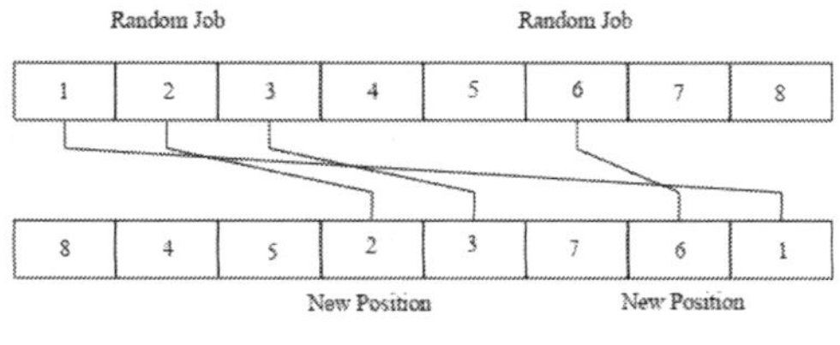

Method 3: Randomize six jobs from all jobs and randomly swap positions.

https://doi.org/10.1051/matecconf/201925904003

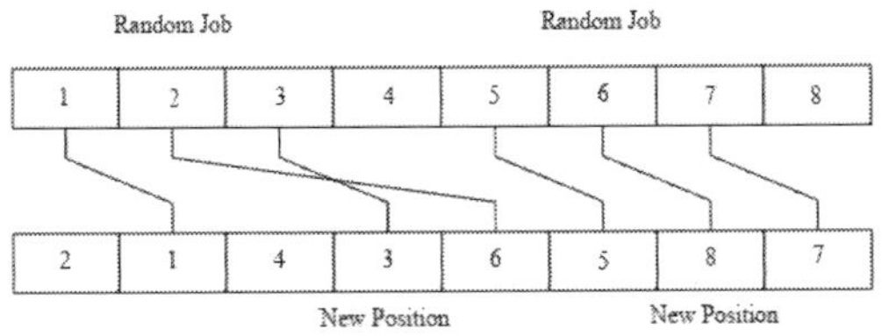

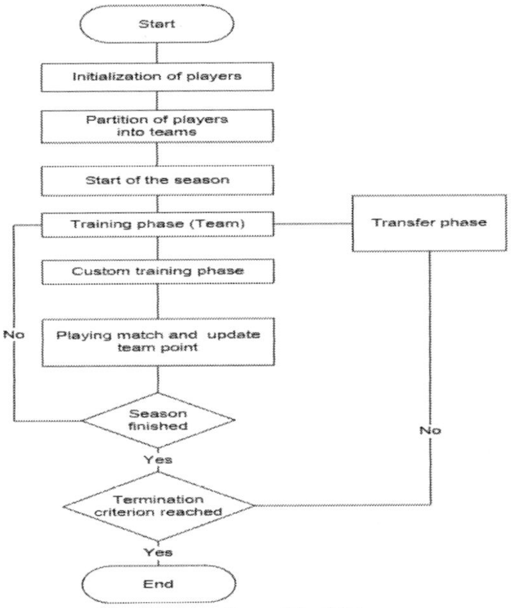

Fig. 1. Process flow of the GBA

Hybrid method type 1 (HGBA1) was calculated by the first and third methods for a custom training and successor phase. Hybrid method type 2 (HGBA2) was analysed by the second and third methods for a custom training and successor phase. Hybrid method type 3 (HGBA3) was applied by the third method on both custom training and successor phases.

4 Computational results and analyses

The case study of balanced load schedule problem is based on real data from a cleanroom apparel company in Thailand. We test our methods on the orders of a three-month timeframe. The company's quantity demand is currently over its production capacity. In the SMEs case study, the main processes are cutting and pattern process, sawing sub-assembly, final assembly and packaging. There are eight regular working hours per day and three overtime hours per days. Based on the information, Table 1 shows the ordered items for processing processes. It contains 30 jobs including order arrival, with different amount of quantities, set up time, production processes, cycle times, and due date. The average cycle time is 50 units per day. The operation units as described above can be categorized as three sets of OP1, OP2 and OP3. The production problems were missing due date time data, high over time cost and unbalanced load in three operation units. Thus, the objective of this research was to balance load in three operation units [9]. If balance load of all operation units are equal, overtime cost will decrease and can make product already in due date time.

In this work, the computational procedure described above was implemented as a computer simulation program in a Visual C#2008 computer program. A Desktop computer with CPU AMD FX6300 Series was used for computational experiments. A comparison of the new method procedures of HGBA1, HGBA2 and HGBA3 results are obtained in this section. GBA algorithm has its own influential parameters that affect its performance in terms of solution quality and process time. In each algorithm, the parameter setting values were based on previous literatures. For all optimization problems presented in this paper, the parameter levels were applied throughout. For parameter setting, the initial population was set at 48 players, 8 teams and 6 players per team.

All algorithms were applied 1000 evaluation points. There were fifteen trial runs in each problem. The performance of the different algorithms was compared based on criteria which comprise of balancing work load in each operation and minimizing an average of the total difference values. Table 2 shows the work load of each operation units of new three hybrid methods compared with original heuristic scheduling methods such as the first come first serve (FCFS), the longest processing time (LPT), the shortest processing time (SPT) and early due date (EDD). Sum of difference value (SDV) was calculated from an absolute difference in time among OP1, OP2 and OP3. The SDV value of unbalanced workload of HGBA3 was lower than any other methods.

In real business industries, the due date and production processing time are uncertain so we created two patterns of noise uncertainty in this study. The first pattern, production process time was randomly ranged between [-1, 1] via the uniform distribution add to average process time. The second pattern, the due date could be changed by customer. The random value ranged between [-5, 5] via the uniform distribution which was designed for adjusting the due date before balancing workload in each operation unit. From tables 3 and 4, performance of HGBA2 was better than other methods in minimizing SDV when the process was uncertain.

TABLE 1. Data from company

Clean room Apparel								
No of job	Start	S	M	L	XL	Total	Lead time	Due date
1	1/3/2017	1	110	153		264	17	1/20/2017
2	1/4/2017		192	100	19	311	11	1/15/2017
3	1/6/2017	81	50	26	63	220	13	1/19/2017
4	1/7/2017		58	48	12	118	18	1/25/2017
5	1/8/2017		40		25	65	21	1/29/2017
6	1/9/2017	3	1			4	18	1/27/2017
7	1/10/2017	5	110	25	35	175	14	1/24/2017
8	1/14/2017		105	20	63	188	19	2/2/2017
9	1/16/2017		15	15		30	20	2/5/2017
10	1/20/2017		1	9	10	20	16	2/5/2017
11	1/21/2017		66	2	2	70	21	2/11/2017
12	1/22/2017	5		40	20	65	12	2/3/2017
13	1/23/2017		55	100	40	195	10	2/2/2017
14	1/25/2017		148		30	178	16	2/10/2017
15	1/27/2017		115			115	12	2/8/2017
16	1/29/2017	174				174	20	2/18/2017
17	1/30/2017	12	15	15	4	46	20	2/19/2017
18	1/31/2017			108	47	155	15	2/15/2017
19	2/3/2017				38	38	16	2/19/2017
20	2/4/2017	320				320	17	2/21/2017
21	2/5/2017			170		170	20	2/25/2017
22	2/7/2017		112	91	4	207	19	2/26/2017
23	2/8/2017		100			100	19	2/27/2017
24	2/9/2017	2	24	29	13	68	18	2/27/2017
25	2/11/2017	40	21	9	5	75	15	2/26/2017
26	2/12/2017	8	6	6	1	21	17	2/29/2017
27	2/13/2017	2	5	71	15	93	15	2/28/2017
28	2/14/2017		130	20	67	217	15	2/29/2017
29	2/15/2017			98	33	131	13	2/28/2017
30	2/17/2017		14			14	19	3/7/2017

TABLE 2. Summary workload (day) for each operation unit from the nominal demand

Operation Unit	OP1	OP2	OP3	SDV
HGBA1	24.42	26.54	25.98	4.24
HGBA2	25.7	25.72	25.52	0.4
HGBA3	25.72	25.64	25.58	0.28
FCFS	26.84	24.34	25.76	5
SPT	25.28	25.82	25.84	1.12
LPT	22.22	27.52	27.2	10.6
EDD	24.24	27.52	25.18	6.56

TABLE 3. Summary workload (day) for each operation unit from the first uncertain pattern (Adjust process time)

Operation Unit	SPT	HGBA1	HGBA2	HGBA3
OP1	24.47	25.377	26.785	25.049
OP2	26.366	27.568	26.366	26.505
OP3	29.124	27.014	26.809	28.405
SDV	9.31	4.38	0.89	6.71

TABLE 4. Summary workload (day) for each operation unit from the second uncertain pattern (adjust due date time)

Operation Unit	SPT	HGBA1	HGBA2	HGBA3
OP1	32.46	25.18	25.52	25.18
OP2	23.42	25.88	25.86	25.88
OP3	21.06	25.88	25.56	25.88
SDV	22.80	1.40	0.68	1.40

5 Conclusions

This study proposed the algorithmic procedures of the GBA to simulate the parameter of number of players, number of teams and number of players per team. Various hybridisations applied to assess performance and created new possible solution methods. The hybrid techniques apply a random location swap of jobs for balancing workload. The new hybrid method called HGBA2 is a compromise tools for this target. The hybrid technology is a new way to increase the performance of original techniques to find better solution. To confirm the performance of our new method, it should apply to solve other combinatorial problems such as quadratic assignment problem and travelling sales man problem. Also, various types of production lines can be considered, such as mixed-model lines, parallel stations, U-shaped lines, etc. The algorithm parameters should be optimised by means of the design and analyses of experiments for finding suitable parameter levels when solving complex problems. Future studies should determine whether other proposed hybrid metaheuristic algorithms can be used for solving this problem as well as enhancing the speed of convergence. In addition, other rules of the custom training function seem challenging during searching the optimum.

Acknowledgment

The first and second authors wish to thank the Faculty of Industrial Technology, Phranakhon Rajabhat University and the Faculty of Engineering, Thammasat University, respectively, for the financial support.

References

1. C. Bierwirth, OR Spektrum, **17** (1995)
2. J. Blazewicz, W. Domschke, and E.Pesch, Eur. J. Oper. Res. **93**,1 (1996)
3. M. Dell'Amico, and M.Trubian, Ann. Oper. Res., **41** (1993)
4. U. Dorndorf, and E. Pesch, Comput. Oper. Res., **22** (1995)
5. L. Li, Y.L. Qian, K. Du, and Y.M. Yang, Int. J. Prod. Res., **54**,3 (2016)
6. W.C. Chiang, T.L. Urban, and X. Xu, Int. J. Prod. Res., **50**,1 (2012)
7. E. Osaba, F. Diaz, and E. Onieva, Appl. Intell., **41**,1 (2014)
8. E. Osaba, X.S. Yang, F. Diaz, P. Lopez-Garcia, and R. Carballedo, Eng. Appl. Artif. Intell., **48** (2016)
9. I. Kucukkoc, and D.Z. Zhang, Int. J. Prod. Res., **55**, 20 (2017)

MATEC Web of Conferences **259**, 04004 (2019) https://doi.org/10.1051/matecconf/201925904004
ICTLE 2018

Inventory lot sizing and supplier selection for multiple products, multiple suppliers, multiple periods with storage space using lingo program

Muoyleang Sambatt[1], Chirawat Woarawichai[2], Thanakorn Naenna[3]

[1&3]*Department of Industrial Engineering, Mahidol University, Nakhonpathom 73170, Thailand*
[2]*Department of Industrial Engineering, Rajamangala University of Technology Lanna Tak, 63000, Thailand*

Abstract. This study focuses on inventory lot sizing for supplier selection of multiple products; suppliers and periods with storage space constraint. The objective is to develop mathematical model. Lingo software program is used as solution tool. The proposed problem is found as mix integer linear programming (MILP) which beneficial for decision making of choosing optimal supplier and product in right period of time with minimum overall inventory cost.

1 Introduction

Lot sizing and supplier selection has become interesting topic in term of business practice as well as in academic study. Defining lot size and supplier selection strategy consist grave impact in purchasing as well as production and inventory cost. Lot sizing and supplier selection are two impressive decision makings procedure to minimize cost while maximize quality and service level. To this problem, the buyer could not decide separately so that it is required to combine them together by concerning on one main goal is to minimize cost as well as maximize quality and service level.

Meanwhile, there are more studying investigating on not only classical model, until now there is extension and broad research done focusing on several various aspects of lot sizing considering on supplier selection. There is study on periodic review concerning to find number of ordering quantity with fixed review time while some study on continuous review lot sizing with fixed quantity of ordering. Some studying on single period single product inventory lot sizing with capacitated resource, some are curious about multiple periods, multiple products lot sizing with deterministic or stochastic demand and so on with different aspect.

All of above description is the crucial factor lead this study to focus on inventory lot sizing for selecting supplier with multiple products, multiple suppliers, multiple periods considering on storage space in periodic review. This study expects to develop mathematical formulation of proposed problem and obtain a model for a decision making to choose optimal supplier providing right product at right time.

2 Literature Review

Good inventory management can be done based on good production planning as well as replenishment controlling. Math model is required to develop following proposed problem including linear programming, non-linear programming, mixed integer programming, geometric programming, gradient-based nonlinear programming and dynamic programming. It was illustrated that when problem become NP-hard, it is required to use heuristic tool to solve those problems [2].

2.1 Lot sizing

Lot sizing is the most crucial issue and one of most complicated problem in production planning. Lot sizing defines time and amount of product to produce to minimize overall cost. Lot size objective is to define quantity to order to reduce replenishment time. Lot-sizing is an impressive factor of production or order and inventory planning [1, 3, 5]. Lot sizing planning and scheduling of production is really required for many kinds of production environment. There are much attention focusing on integration of lot sizing in practice in real activity [14]. Lot sizing is the crucial determinant in supply chain management to select optimal supplier to obtain the best product, lot size, and right time to order during a finite planning horizon [13].

2.2 Inventory model classification

There are several criteria to define classification of inventory model such as finite or infinite characteristic of planning horizon, single or multiple number of items, fixed or variable order quantity which found as periodic or continuous review, zero or non-zero lead time, capacitated or non-capacitated capacity, deterministic, static or dynamic and stochastic demand, single or multiple echelon stocking and finally consider to allow or not allow unsatisfied demand condition as illustrating in figure 1 [11]. There is literature review of inventory lot sizing problem. It was indicated that inventory planning has been a complicated problem. The study has been review focusing on four classifications such as single-level unconstrained resources, single-level constrained resources, multiple-level constrained resources, and multiple-level unconstrained resources. This review illustrated both real practice as well as academic study. It was found that capacity factor and fact of plant environment such as scrap, demand uncertainty or unspecific data are the root for research in lot sizing field [9].

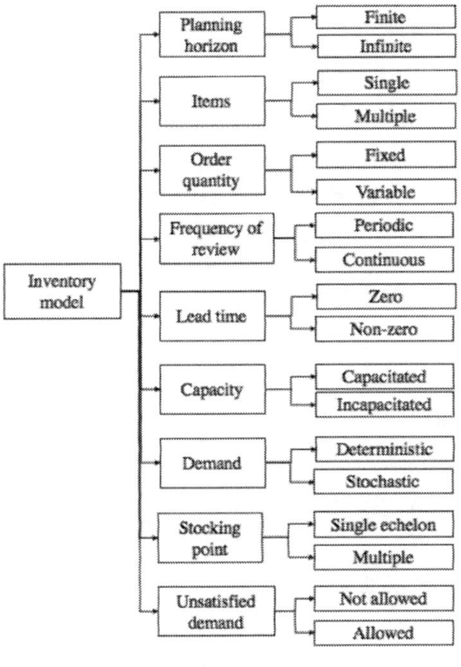

Fig. 1. Inventory model classification.

2.3 Inventory lot sizing and supplier selection

Supplier selection is an important indicator and significant complexity in organization [8]. Supplier selection is found as a crucial concerning in lot sizing [16]. Meanwhile, lot sizing and supplier selection are two impressive decision- making procedures which aims to define size, is really main thing to concern to minimize total cost [11]. A problem with storage space constrained buyer procures a single product in multiple periods from multiple suppliers was studied [6]. Lot sizing is a kind of tactical decision, the crucial determinant in supply chain management. It is really important to select good supplier to obtain the best product, lot size, and right time to order during a finite planning horizon [5, 12, 14]. There are many study focus on supplier selection and procurement lot sizing. The study is to select the best suppliers and determine optimum quantity from selected supplier focusing both tangible and intangible procedure as well as time. Supplier selection is found as multiple criteria decision making [17]. There are plenty of study pay more attention on lot sizing along with supplier selection. The crucial objective is to select which supplier and how much to make ordering from that choice of supplier. Defining lot size as well as supplier selection strategy consists grave impact in term of purchasing as well as production and inventory cost. Problem of lot sizing and scheduling is investigated [13]. Furthermore, there is study on multiple period lot sizing with supplier selection problem. The purpose is to decide what products to order in what quantities with which suppliers in which periods. An enumerative search algorithm and a heuristic was used to solve the problem; however, it is not process fast enough for real case of complicated problem [4]. There is study on lot sizing and supplier selection of multiple products and multiple suppliers which consist two multiple objective models concerning on shortage [12]. The study of multiple suppliers, multiple periods with quantity discount was investigated. The proposed problem focuses on finite planning horizon, fixed ordering quantity, independent and known demand with lead time. The proposed problem is formulated as mix integer programming and genetic algorithm was used as solution tool [2]. Lot sizing with supplier selection problem with multi-echelon, multi products defective supply chain network with stochastic demand was studied [7].

Right supplier selection could provide more benefit by eliminate purchasing cost as well as high competency in term of good corporation as known that supplier selection is the most pivotal indicator in purchasing management. The research was done focusing on decision making on appropriate selection of suppliers, and on assigning order quantities to these suppliers, in case of multiple sourcing, with multiple criteria and with suppliers' capacity constraints. The objective of this research is to minimize summation of purchasing cost, ordering costs and storage costs considering on some specific constraints such as limitations on buyer's budget, quality and service. The proposed model is formulated as a mixed integer non-linear programming model in order to solve multiple sourcing problem relating to total cost of logistic, net price, storage, transportation and ordering cost. GINO algorithm was used to solve proposed problem [19].

There is study on problem of multiple products, multiple period with deterministic zero lead time with known planning horizon demand and focus on storage space and budget constraints. This aims to optimize lot sizing for each supplier as well as reduce total inventory cost considering on join purchasing cost, transaction cost for suppliers and holding cost for left inventory. The proposed problem is developed as mix integer linear programming and LINGO 12.0 software program was used as solution tool [5].

2.4 Lingo program

Lingo program is a recommended optimization tool which illustrating mathematical model and see process of model by lingo solver. The model indicates computational time and solution method. There are four various kinds of solver such as direct, a linear, a nonlinear and a branch and bound solver. Direct solver help lingo defines all fixed variables and constraint from model and seeking whether model is linear or nonlinear. If there is nonlinear constraint or variables, so model is found as nonlinear programming. Lingo roles to build and solve mathematical model easier and effectively. It is capable to solve several kinds of model while it is required computational time and input data [15].

3 Methodology

3.1 Math model development

This study investigates on inventory lot sizing with supplier selection of multiple products, multiple suppliers, multiple period with storage floor space constraint in term of periodic review as seen in figure 2. Based on this, it is required to considered on ordering quantity X_{ijt} which is vary depending on different period while reviewing time is fixed. This is done to define quantity of which product to order from which supplier in which right time period.

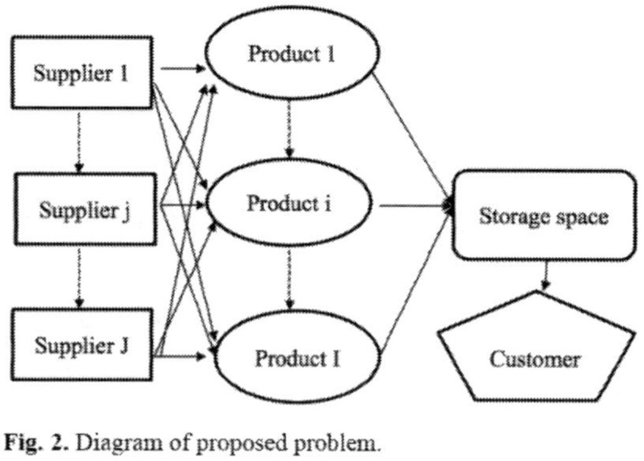

Fig. 2. Diagram of proposed problem.

Assumption

Random demand
Shortage is not allowed
Limited storage floor space
Inventory of item i at beginning of planning period is 0

Indices

i= 1 2 3…I index of products
j= 1 2 3…J index of suppliers
t= 1 2 3…T index of periods

Parameters

D_{it}: demand of product i in period t
P_{ij}: purchase price of product i from supplier j
H_i: holding price of product i
O_j: transaction price from supplier j
W: total available storage space
f_i: storage space for the ith product
I_{it}: Total inventory of each period

Decision variable

X_{ijt}: ordering quantity of item i from supplier j with in time t.
Y_{jt}: equal to 1 if make ordering for supplier j at time t, equal to 0 otherwise

Objective function

The objective is to minimize total inventory cost as following description:

Minimum total cost= Transaction cost + Purchasing cost+ Holding cost

$$\text{Min} = \sum_{j}^{J}\sum_{t}^{T} O_j Y_{jt} + \sum_{t}^{T}\sum_{j}^{J}\sum_{i}^{I} P_{ij} X_{ijt} + \sum_{t}^{T}\sum_{i}^{I} H_i I_{it}\,;\ \forall i,j,t \quad (1)$$

Subject to:

$$I_{it} = I_{it-1} + \sum_{t}^{T}\sum_{j}^{J}\sum_{i}^{I} X_{ijt} - \sum_{t}^{T}\sum_{i}^{I} D_{it};\ \forall i,j,t \quad (2)$$

$$\sum_{t}^{T}\sum_{i}^{I} D_{it} \sum_{t}^{T}\sum_{j}^{J} Y_{jt} - \sum_{t}^{T}\sum_{j}^{J}\sum_{i}^{I} X_{ijt} \geq 0;\ \forall i,j,t \quad (3)$$

$$\sum_{t}^{T}\sum_{i}^{I} f_i I_{it} \leq W;\ \forall i,j,t \quad (4)$$

$$Y_{jt} = 0 \text{ or } 1;\ \forall j,t \quad (5)$$

Equation (1) indicating objective function which composed of transaction cost, purchasing cost and holding cost. Transaction cost is summation of transaction price O_j multiply with binary number Y_{jt}, purchasing cost is summation of purchasing price P_{ij} multiply with ordering quantity X_{ijt} [4]. Holding cost is summation of holding price H_i multiply with total inventory I_{it} [4,10,18]; (2) Inventory constraint is summation of inventory of each previous period with ordering number minus demand. (3) Shortage is not allowed [12], (4) Storage floor space required to be lower than total space W [5]. (5) binary number Y_{jt} which equal to 1 if make purchasing from supplier j at period t and 0, otherwise [12].

3.2 Computational experiment

In this section, a commercially optimization package LINGO 12.0 is used for solving proposed problem. Experiments are conducted on a personal computer with described feature as table 1.

Moreover, data is generating in order to see the performance of proposed problem. The transaction price is generated in interval [1; 3] mean random uniform distribution between 1 to 3. Holding price is from interval [1; 5], purchasing price is from interval [5;15]; storage space for each product is from interval [0.1; 0.3], total available storage space is 1800 square feet, demands are from interval [30; 6500], inventory for item i at the beginning of a planning period is assumed to be 0. The interval of randomization of each parameter is shown in Table 2.

Table 1: Hardware and software

Feature	Units
MacBook Air	13 Inch, 2017
Processor	1.8 GHz Intel Core i5
Memory	8GB 1600Mhz DDR3
Graphic Intel HD Graphics	6000 1536 MB
Lingo Software Program	12.0

Table 2: Interval value of data

Parameter	Interval of random value
D_{it}	[30, 6500]
H_i	[1,5]
O_j	[1,3]
P_{ij}	[5,15]
f_i	[0.1, 0.3]
I_{i0}	0
W	1800

4 Result and discussion

The values of each parameter are generated randomly. Numerical examples with problem size of 3 products, 4 suppliers, 6 periods are presented to examine the performance of Lingo 12.0 as indicating in figure 3. This example, lingo output provides for selection product, supplier in each period such as following:

Product 1:
X_{131}=542, X_{132}=5077, X_{133}=5077, X_{134}=2889, X_{135}=6257, X_{136}=2156

Product 2:
X_{231}=4948, X_{232}=1147, X_{233}=31, X_{234}=3808, X_{235}=4836, X_{236}=5994

Product 3:
X_{311}=1241, X_{312}=361, X_{313}=4364, X_{314}=2365, X_{315}=2130, X_{316}=4706

Furthermore, problem is extended to be more complicated. The interval of randomization of each parameter is shown in Table 3. This problem is extended by consider on multiple products i from 3 items until 20 items, multiple supplier j from 4 to 22 with three different periods such as 6 periods, 12 periods and 24 periods.

As the proposed problem is extended to be greater and more complicated, lingo program could solve, and found that its mathematical model as mix integer linear programming (MILP) and provide global optimal solution which is right criteria to select optimal supplier, product with right period of time to reduce total inventory cost. Based on this result, lingo program software is effective to use as tool for optimize this problem and it is reach to proposed objective to this study.

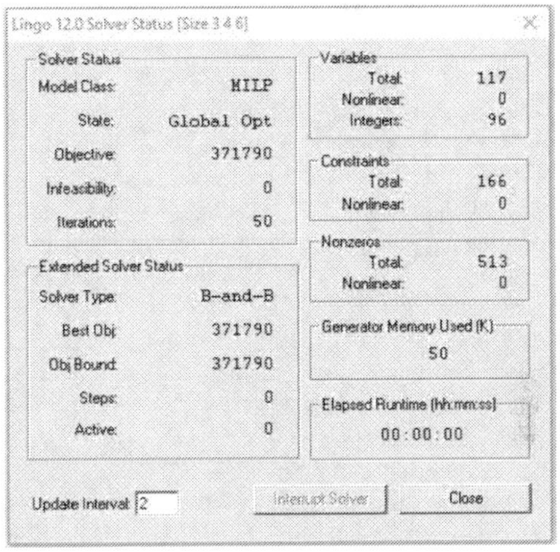

Fig. 3. Lingo output of problem size 3x3x6

Table 3: Problem size and output of lingo

Problem size	Total inventory cost
3x4x6	371790
3x4x12	760584
3x4x24	1464030
6x8x6	586880
6x8x12	1142240
6x8x24	2282020
10x12x6	992208
10x12x12	1987310
10x12x24	3983870
16x18x6	1579020
16x18x12	3162200
16x18x24	6592040
20x22x6	1878730
20x22x12	3761870
20x22x24	7790270

5 Conclusion

This study focuses on inventory lot sizing for supplier selection of multiple products, multiple periods of supplier selection with storage space constraint which is done in periodic review. The proposed problem is formulated as mix integer linear program (MILP) which is optimized in Lingo 12.0 software program. The result obtains a mathematical model which is able to defines optimal number of product to order from right supplier, right period to reach the objective of minimizing overall inventory cost.

References

1. A. Atamtürk, & S. Küçükyavuz. An O (n2) algorithm for lot sizing with inventory bounds and fixed costs. Oper. Res. Lett. **36**, 297–299 (2008)
2. A.H.I. Lee, H.Y. Kang, C.M. Lai, & W.Y. Hong. An integrated model for lot sizing with supplier selection and quantity discounts. Appl. Math. Model. **37**, 4733–4746 (2013)
3. B. Karimi, S. M.T. Fatemi Ghomi , & J.M. Wilson. The capacitated lot sizing problem: A review of models and algorithms. Omega. **31**, 365–378 (2003)
4. C. Basnet, & J. M. Y. Leung. Inventory lot-sizing with supplier selection. Comput. Oper. Res. **32**, 1–14 (2005)
5. C. Woarawichai, T. Kullpattaranirun, & V. Rungreunganun. Inventory Lot-Sizing Problem with Supplier Selection under Storage Space and Budget Constraints. Int. J. Comput. Sci. **8**, 250–256 (2011)
6. D. Choudhary, & R. Shankar. A goal programming model for joint decision making of inventory lot-size, supplier selection and carrier selection. Comput. Ind. Eng. **71**, 1–9 (2014)
7. E. Senyigit, & I. Soylemez. The Analysis of Heuristics for Lot Sizing with Supplier Selection Problem. Procedia-Soc. Behav. Sci. **62**, 672–676 (2012)
8. F. Çebi, & I. Otay. A two-stage fuzzy approach for supplier evaluation and order allocation problem with quantity discounts and lead time. Inf. Sci. **339**, 143–157 (2016)
9. H. C. Bahl, L. P. Ritzman, and J. N. D. Gupta. OR Practice Determining Lot Sizes and Resource Requirements: A Review. Oper. Res. **35**, 329–345 (1987)
10. H.K. Alfares & R. Turnadi. Lot sizing and supplier selection with multiple items, multiple periods, quantity discounts, and backordering. Comput. Ind. Eng. **116**, 59–71 (2018)
11. H. Ullah and S. Parveen. A Literature Review on Inventory Lot Sizing. Glob. J. Res. Eng. **10**, 21–36 (2010)
12. J. Rezaei, & M. Davoodi. Multi-objective models for lot-sizing with supplier selection. Int. J. Prod. Econ. **130**, 77–86 (2011)
13. L.E. Cárdenas-Barrón, J.L. González-Velarde, & G. Treviño-Garza. A new approach to solve the multi-product multi-period inventory lot sizing with supplier selection problem. *Comput. Oper. Res.* **64**, 225–232 (2015).
14. L. Guimarães, D. Klabjan, & B. Almada-Lobo . Modeling lotsizing and scheduling problems with sequence dependent setups. Eur. J. Oper. Res. **239**, 644–662 (2014)
15. LINDO Systems Inc. Lingo the modeling language and Optimizer. Chigago (2017)
16. M.M. Mazdeh, M. Emadikhiav, & I. Parsa. A heuristic to solve the dynamic lot sizing problem with supplier selection and quantity discounts. Comput. Ind. Eng. **85**, 33–43 (2015)
17. O. Ustun, & E.A. Demirtas. An integrated multi-objective decision-making process for multi-period lot-sizing with supplier selection. Omega. **36**, 509–521 (2008a)
18. R. Uthayakumar, & S. Priyan. Pharmaceutical supply chain and inventory management strategies: Optimization for a pharmaceutical company and a hospital, Oper. Res. Heal. Care. **2**, 52–64 (2013)
19. S.H. Ghodsypour, & C. O'Brien. The total cost of logistics in supplier selection, under conditions of multiple sourcing, multiple criteria and capacity constraint, Int. J. Prod. Econ. **73**, 15–27 (2001)

A two-stage policy for improving the efficiency of the traffic network by providing value-added ATIS service

Jingjing Liang, Xiaoning Zhang and Zhe Liang

School of Economics and Management, Tongji University, 200092 Shanghai, China

Abstract. Based on the pay-per-use charge in mobile Apps, value-added advanced traveler information systems (VATIS) service can be provided in traffic networks. From the perspective of the traffic manager, we propose a two-stage policy for improving the efficiency of traffic networks by providing VATIS service. At the first stage, the traffic manager selects several ATIS service providers, then offers seed fund to each of them to encourage VATIS service provision. At the second stage, the traffic manager announces a given target of the total market penetration of VATIS service and encourages service providers to achieve it together. If they could complete the task, compensation will be paid. To evaluate the efficiency of this policy, four scenarios are formulated by four bi-level models, and the total travel time is set as the evaluation indicator. Numerical experiments show that, the suggested policy can reduce total travel time dramatically if the market penetration target has been appropriately selected.

1 Introduction

In previous years, people used to make their travel decisions with the aid of vehicle navigation systems, which are the predecessor of Advanced Traveler Information System (ATIS). At that time, ATIS service providers can sell navigators and charge for map updating services to gain profit. However, due to the severe competition in the ATIS market, nowadays almost all ATIS service providers (e.g., Goggle Map or Amap) offer the basic route guidance service for free. In this case, they begin to seek for profit by providing the value-added ATIS (VATIS) service based on the evidence that people are willing to pay to the save travel time [1-5]. As an actual example, Amap offers several kinds of value-added services to travelers, such as taxi service, hotel reservation, and online coupons obtaining, which are bringing Amap a tidy profit. Consequently, ATIS service providers are dedicated to maximize their market scales in this new era. That can also contribute to enhancing the efficiency of the traffic networks [6]. But the expensive operating cost for collecting traffic information is the major limitation for ATIS service providers [7], especially when the information quality is required to be as accurate as possible. Thus, policies for supporting ATIS service providers, mainly by finance support, need to be issued.

In this study, we suggest a two-stage policy from the perspective of the traffic manager. At the first stage, the traffic manager selects several ATIS service providers, and offers seed funds to each of them. Considering that ATIS services of different service providers are coessential products, service providers will cooperate with each other after severe competition, then act as a community to maximize their profits. Then the policy should move to the second stage. The traffic manager announces a given target of the total market penetration of VATIS services and asks service providers to achieve it together. If they could complete the target, reasonable compensations will be paid to them.

If the traffic manager compensates service providers directly during the period of market competition and asks them to achieve fixed targets of the market penetration of VATIS services, it might be a tough work to decide which values of fixed targets should be set for each service providers, and how much compensates should be given. By dividing this policy into two stages, the policy can make full use of the market competition in the first stage, because the traffic manager can get a clear understanding of each service provider, so that both compensations and the fixed target of the total market penetration of VATIS services can be made more wisely and efficiently. Besides, compensations will be given to the community of service providers, this approach can avoid the trouble of setting sub-targets for every selected service provider.

We illustrate the evolution of traffic state of the traffic network where the two-stage policy implemented, by four cases, namely the basic case, the competition case, the cooperation case, and the restricted case. The basic case describes the initial state of the traffic network where only the basic ATIS service is provided, and it will be treated as the comparative standard of other three cases. The competition case and the cooperation case are used to depict market behaviour of VATIS service providers in the first stage. The restricted case presents the state of the traffic network in the second stage. Then, four models are proposed to formulate these cases, respectively. At last, the efficiency of the suggested policy is tested by the total travel time based on a traffic survey which is carried out in Hongkou district of Shanghai, China.

The study is structured as follows. In section 2, four models are developed for demonstrating the states of the traffic network after implemented the two-stage policy. Numerical analyses are given in Section 3 based on a traffic survey. Section 4 concludes this study.

2 Modeling

2.1 Basic case: the initial state of the traffic network

Consider a traffic network $\mathbf{G}=(\mathbf{N},\mathbf{A})$ which has $\mathbf{N}$ nodes and $\mathbf{A}$ links, its origin-destination (OD) matrices are given and fixed, and d^{rs} (the set $\mathbf{d}$) denotes the total travel demand from origin $r\in\mathbf{R}$ to destination $s\in\mathbf{S}$. We assume that the basic ATIS service is provided for free, so there is no need to consider about their profit. Denote the basic ATIS service quality as θ_0, then it can be used as the parameter of the Logit-based traffic assignment in the network. Therefore, the link flow pattern can be formulated by the following mathematical program:

[Model-1]
$$\min FB(\mathbf{f})=\sum_{a\in\mathbf{A}}\int_0^{v_a}t_a(x)dx+\sum_{r\in\mathbf{R}}\sum_{s\in\mathbf{S}}\sum_{j\in\mathbf{J}_{rs}}f_j^{rs}\frac{1}{\theta_0}\ln f_j^{rs}, \quad (1)$$

subject to

$$\sum_{j\in\mathbf{J}_{rs}}f_j^{rs}=d^{rs}\quad\forall r\in\mathbf{R},s\in\mathbf{S}, \quad (2)$$

$$v_a=\sum_{r\in\mathbf{R}}\sum_{s\in\mathbf{S}}\sum_{j\in\mathbf{J}_{rs}}f_j^{rs}\delta_{aj}^{rs}\quad\forall a\in\mathbf{A}, \quad (3)$$

$$f_j^{rs}\geq 0\quad\forall r\in\mathbf{R},s\in\mathbf{S},j\in\mathbf{J}_{rs}. \quad (4)$$

In Model-1, j (the path set $\mathbf{J}_{rs}$) is the index of path in OD pair rs, f_j^{rs} is the flow of path j in OD pair rs. v_a (the link flow set $\mathbf{v}$) represents the flow in link a. $t_a(v_a)$ is the link travel time function of link a. δ_{aj}^{rs} equals 1 if path j in OD pair rs uses the link a, and δ_{aj}^{rs} equals 0 otherwise.

Solving Model-1, the initial equilibrium state of the traffic network can be obtained and defined as

$$\mathbf{f}^0:=\{\mathbf{f}\in\arg\min FB(\mathbf{f})\}. \quad (5)$$

2.2 Models for evaluating the efficiency of the two-stage policy

As we have mentioned in the introduction, the state of the traffic network after implemented the two-stage policy can be described by four cases: the basic case, the competition case, the cooperation case, and the restricted case. In this subsection, they are modeled in mathematical forms based on the bi-level models. In the lower-levels of these models, stochastic equilibrium theories will be applied to describe the equilibrium state of the traffic network. At the same time, the upper-levels of these models are formulated by profit functions of VATIS service providers. We will present the common lower-level model at first, and then give the upper-level models of three bi-level models one after another.

2.2.1 The mixed stochastic equilibrium state of the traffic network

Suppose that there has M VATIS service providers in the traffic network, and the set of VATIS service providers is denoted by $\mathbf{M}=\{m:1,2,3...M\}$. Let $m=0$ be the initial class of travelers who only use the basic ATIS service for free, and $m\in M$ be the class of travelers who use the service of VATIS service provider m and pay for the service. Therefore, the whole set of travelers in the traffic network can be expressed as $\mathbf{M}_0=\mathbf{M}\bigcup\{0\}=\{m:0,1,2,3...M\}$ where m is the index of both VATIS service providers and the classes of travelers. d_m^{rs} denotes the demand of the m th class of travelers. Under assumptions of homogeneous travelers, principles of stochastic user equilibrium (SUE) [8], the mixed stochastic equilibrium model can be formulated as below:

[Model-L]
$$\min_{\mathbf{f},\mathbf{d}}F(\mathbf{f},\mathbf{d},\boldsymbol{\theta},\boldsymbol{\tau})=\sum_{a\in\mathbf{A}}\int_0^{v_a}t_a(x)dx+\sum_{r\in\mathbf{R}}\sum_{s\in\mathbf{S}}\sum_{m\in\mathbf{M}_0}\sum_{j\in\mathbf{J}_{rs}}f_{mj}^{rs}\left(\tau_m^{rs}+\frac{1}{\theta_m}\ln f_{mj}^{rs}\right), \quad (6)$$

subject to

$$\sum_{j\in\mathbf{J}_{rs}}f_{mj}^{rs}=d_m^{rs}\quad\forall r\in\mathbf{R},s\in\mathbf{S},m\in\mathbf{M}_0, \quad (7)$$

$$\sum_{m\in\mathbf{M}_0}d_m^{rs}=d^{rs}\quad\forall r\in\mathbf{R},s\in\mathbf{S}, \quad (8)$$

$$f_{mj}^{rs}\geq 0\quad\forall r\in\mathbf{R},s\in\mathbf{S},m\in\mathbf{M}_0,j\in\mathbf{J}_{rs}, \quad (9)$$

$$v_a=\sum_{r\in\mathbf{R}}\sum_{s\in\mathbf{S}}\sum_{m\in\mathbf{M}_0}\sum_{j\in\mathbf{J}_{rs}}f_{mj}^{rs}\delta_{aj}^{rs}\quad\forall a\in\mathbf{A}, \quad (10)$$

$$\tilde{c}_{mj}^{rs}=\eta\sum_{a\in\mathbf{A}}t_a(v_a)\delta_{aj}^{rs}+\tau_m^{rs}\quad\forall r\in\mathbf{R},s\in\mathbf{S},m\in\mathbf{M}_0,j\in\mathbf{J}_{rs} \quad (11)$$

$$P_{mj}^{rs}=\frac{\exp\left(-\theta_m\tilde{c}_{mj}^{rs}\right)}{\sum_{k\in\mathbf{J}_{rs}}\exp\left(-\theta_m\tilde{c}_{mk}^{rs}\right)}\quad\forall r\in\mathbf{R},s\in\mathbf{S},m\in\mathbf{M}_0,j\in\mathbf{J}_{rs}, \quad (12)$$

$$c_m^{rs}=\sum_{j\in\mathbf{J}_{rs}}P_{mj}^{rs}\tilde{c}_{mj}^{rs}\quad\forall r\in\mathbf{R},s\in\mathbf{S},m\in\mathbf{M}_0, \quad (13)$$

$$Q_m^{rs}=\frac{\exp\left(-\varphi c_m^{rs}\right)}{\sum_{k\in\mathbf{M}}\exp\left(-\varphi c_k^{rs}\right)+\exp\left(\alpha-\varphi c_0^{rs}\right)}\quad\forall r\in\mathbf{R},s\in\mathbf{S},m\in\mathbf{M} \quad (14)$$

$$Q_0^{rs}=1-\sum_{m\in\mathbf{M}}Q_m^{rs}\quad\forall r\in\mathbf{R},s\in\mathbf{S}, \quad (15)$$

$$d_m^{rs}=d^{rs}Q_m^{rs}\quad\forall r\in\mathbf{R},s\in\mathbf{S},m\in\mathbf{M}_0. \quad (16)$$

In (6)-(16), variables and parameters are defined as below. f_{mj}^{rs} (the path flow set $\mathbf{f}$) denotes the flow of path j in the m th class of travelers in OD pair rs. τ_m^{rs} is the service price of the m th VATIS service provider in OD pair rs, and $\tau_0^{rs}=0$ for all OD pair rs. θ_0 is the service quality of the basic ATIS service, θ_m is the service quality of the m th VATIS service provider, and there has $\theta_m>\theta_0$ for all $m\in\mathbf{M}$. $\bar{c}_{mj}^{rs}$ is the travel cost of a traveler of the m th class when he chooses path j in OD pair rs. η is a positive parameter for transferring travel times into monetary value. P_{mj}^{rs} is the probability of a traveler of the m th class for choosing path j in OD pair rs. c_m^{rs} is the average travel cost of travelers of the m th class. Q_m^{rs} is the probability of a traveler for choosing the service of the m th VATIS service provider in OD pair rs, and Q_0^{rs} is the probability of a traveler for choosing the basic ATIS service in OD pair rs. The choosing process (14)-(15) is formulated in the similar way to Yang [9] has did, φ and α are parameters in it.

2.2.2 Competition case in the first stage

At the first stage of the suggested two-stage policy, VATIS service providers will compete by altering their service strategies, including alter service prices and service qualities of themselves to maximize their profits. Denote Ω_m as the profit of the m th VATIS provider, its pricing strategy as $\mathbf{x_m}=(\theta_m,\tau_m^{rs})$, and the group pricing strategy as $\mathbf{z}=(\theta_0,\tau_0^{rs},\theta_1,\tau_1^{rs},...,\theta_M,\tau_M^{rs})$. The bi-level model for describing competition scenario can be formulated as below:

[Model-2]
$$\max_{\mathbf{x_m}}\ \Omega_m(\mathbf{d},\mathbf{x_m}) =\sum_{r\in\mathbf{R}}\sum_{s\in\mathbf{S}}\tau_m^{rs}d_m^{rs}-\sum_{i\in I}\phi_{im}\theta_m^{i}\quad \forall m\in\mathbf{M}, \tag{17}$$

subject to

$$\mathbf{d}\in\arg\min_{\mathbf{f},\mathbf{d}}\mathbf{F}(\mathbf{f},\mathbf{d},\mathbf{z}), \tag{18}$$

$$\theta_m\ge 0,\quad \tau_m^{rs}\ge 0\quad \forall r\in\mathbf{R},s\in\mathbf{S},m\in\mathbf{M}. \tag{19}$$

In equation (17), the first term is the total revenue of the m th VATIS service provider, the second term is the total operation cost of the m th VATIS service provider, ϕ_{im} is the cost coefficient of the cost related to the information quality and $\mathbf{I}=\{i:i\in R^{+}\setminus[0,1]\}$. We construct the operation cost function in the exponential form under the consideration of arguments of related literatures on ATIS service pricing [10-12].

It should be noted that the service quality and the service price are two sets of decision variables in the upper-level model of Model-2. If we simplify the competition case into a process of sequential decision making, the competition among VATIS service providers would turn out to be a Nash game [13]. The reaction method in game theory can be applied in analyzing the optimal solution of Model-2. Obtain the first order partial derivatives of Ω_m with respect to θ_m and τ_m^{rs}, and let them equal to 0, we can obtain the optimal decision of the m th VATIS service provider:

$$\begin{cases}\dfrac{\partial\Omega_m(\mathbf{d},\mathbf{x_m})}{\partial\theta_m}=0 & \forall m\in\mathbf{M}\\ \dfrac{\partial\Omega_m(\mathbf{d},\mathbf{x_m})}{\partial\tau_m^{rs}}=0 & \forall r\in\mathbf{R},s\in\mathbf{S},m\in\mathbf{M}\end{cases} \tag{20}$$

Define set $\mathbf{K}=\{k:k\in\mathbf{M}\setminus\{m\}\}$, and denote R^1 and R^2 as the implicit reaction functions of θ_m and τ_m^{rs}, respectively. The corresponding reaction functions can be presented as

$$\theta_m=R^1_{k\in\mathbf{K}}(\theta_k,\tau_k^{rs}),\tau_m^{rs}=R^2_{k\in\mathbf{K}}(\theta_k,\tau_k^{rs})\ \forall r\in\mathbf{R},s\in\mathbf{S},m\in\mathbf{M} \tag{21}$$

The optimal solution should meet

$$\begin{cases}\theta_m^{*}=R^1_{k\in\mathbf{K}}(\theta_k^{*},\tau_k^{rs*}) & \forall m\in\mathbf{M}\\ \tau_m^{rs*}=R^2_{k\in\mathbf{K}}(\theta_k^{*},\tau_k^{rs*}) & \forall r\in\mathbf{R},s\in\mathbf{S},m\in\mathbf{M}\end{cases}. \tag{22}$$

Solving Model-2, we have its optimal solution:

$$\mathbf{f}^{*}:=\left\{\mathbf{f}\in\arg\max_{\mathbf{x_m}}\Omega_m(\mathbf{d},\mathbf{z}),\forall m\in\mathbf{M}\right\}. \tag{23}$$

2.2.3 Cooperation case in the first stage

In most cases, service providers can gain more profit by cooperation, so in the first stage, VATIS service providers will choose cooperation after the drastic market competition. We develop another bi-level model called Model-3 for this case. Denote G as the total profit of M VATIS service providers and $\mathbf{X}:=\mathbf{z}\setminus\{(\theta_0,\tau_0^{rs})\}$, then the cooperation scenario can be formulated as below:

[Model-3]
$$\max_{\mathbf{X}}G(\mathbf{d},\mathbf{X})=\sum_{m\in\mathbf{M}}\Omega_m =\sum_{m\in\mathbf{M}}\sum_{r\in\mathbf{R}}\sum_{s\in\mathbf{S}}\tau_m^{rs}d_m^{rs}-\sum_{m\in\mathbf{M}}\sum_{i\in I}\phi_{im}\theta_m^{i}, \tag{24}$$

subject to (18) -(19).

Denote the optimal solution of Model-3 as $(\bar{\mathbf{d}},\bar{\mathbf{X}})=\arg\max_{\mathbf{X}}G(\mathbf{d},\mathbf{X})$, where $\bar{\mathbf{d}}=(\bar{d}_1^{rs},\bar{d}_2^{rs},...,\bar{d}_M^{rs})$, $\bar{\mathbf{X}}=(\bar{\theta}_1,\bar{\tau}_1^{rs},\bar{\theta}_2,\bar{\tau}_2^{rs},...,\bar{\theta}_M,\bar{\tau}_M^{rs})$, and denote the stable solution of Model-2 as $(\mathbf{d}^{*},\mathbf{x}^{*}_{\mathbf{m}})$ for all $m\in M$, we can know that, only if the follow constraint was met, the cooperation would happen:

$$G(\bar{\mathbf{d}},\bar{\mathbf{X}})>\sum_{m\in\mathbf{M}}\Omega_m^{*}(\mathbf{d}^{*},\mathbf{x}^{*}_{\mathbf{m}}). \tag{25}$$

Solving Model-3, we have its optimal solution:

$$\bar{\mathbf{f}}:=\left\{\mathbf{f}\in\arg\max_{\mathbf{X}}G(\mathbf{d},\mathbf{X})\right\}. \tag{26}$$

2.2.4 Restricted case in the second stage

Cooperation among VATIS service providers may lead to the increasing of the total travel time and other loses of social welfare issues while they pursuing higher profit. The second stage of this policy is aimed at dealing with this problem by restricted measures. The specific approach is, the traffic manager promises to compensate

VATIS service providers, but a given and fixed target of the total market penetration of VATIS service should be achieved. Denote the current total profit of M VATIS services as GR, and denote Q_{fixed} as the fixed target, then the restricted scenario can be formulated as

[Model-4]

$$\begin{aligned}\max_{\mathbf{X}} GR(\mathbf{d},\mathbf{X}) = & \sum_{m\in\mathbf{M}}\sum_{r\in\mathbf{R}}\sum_{s\in\mathbf{S}}\left(\tau_m^{rs} d_m^{rs} - \sum_{i\in I}\phi_{im}\theta_m^i\right) \\ & - N\cdot\max\left\{0, Q_{fixed} - \sum_{m\in\mathbf{M}}\sum_{r\in\mathbf{R}}\sum_{s\in\mathbf{S}} w^{rs} Q_m^{rs}\right\}\end{aligned} \quad (27)$$

subject to (18) - (19). In (27), the first term is the summation of profit functions of all VATIS service providers, and the second term is a penalty (opportunity cost) for not achieving the fixed market penetration target Q_{fixed} of VATIS service in the given highway network. N is a very large number, and $w^{rs} = \dfrac{d^{rs}}{\sum_{r\in\mathbf{R}}\sum_{s\in\mathbf{S}} d^{rs}}, \forall r\in\mathbf{R}, s\in\mathbf{S}$.

Consider a fixed market penetration target Q_{fixed} of VATIS services in the traffic work, denote the optimal solution set of VATIS service providers as $(\hat{\mathbf{d}},\hat{\mathbf{X}}) = \arg\max_{\mathbf{X}} GR(\mathbf{d},\mathbf{X})$. Denote the total compensation as β, then the lower bound of β will be determined by

$$\beta \ge \max\left\{0, G(\bar{\mathbf{d}},\bar{\mathbf{X}}) - GR(\hat{\mathbf{d}},\hat{\mathbf{X}})\right\}. \quad (28)$$

Solving Model-4, we obtain its solution:

$$\hat{\mathbf{f}} := \left\{\mathbf{f} \in \arg\max_{\mathbf{X}} GR(\mathbf{d},\mathbf{X})\right\}. \quad (29)$$

2.3 The evaluation indicator and solving method

The total travel time has been extensively used as the indicator in evaluating the efficiency of traffic networks both in studies and practices, it can be calculated by

[Indicator]

$$T(\mathbf{f}) = \sum_{r\in\mathbf{R}}\sum_{s\in\mathbf{S}}\sum_{j\in\mathbf{J}_{rs}} f_j^{rs}\left[\sum_{a\in\mathbf{A}} t_a(v_a)\delta_{aj}^{rs}\right], \quad (30)$$

where

$$v_a = \sum_{r\in\mathbf{R}}\sum_{s\in\mathbf{S}}\sum_{j\in\mathbf{J}_{rs}} f_j^{rs}\delta_{aj}^{rs} \quad \forall a\in\mathbf{A}. \quad (31)$$

If the implement of the suggested two-stage policy could improve the efficiency of the traffic network, $T(\mathbf{f}^*)$, $T(\bar{\mathbf{f}})$ and $T(\hat{\mathbf{f}})$ will all be smaller than $T(\mathbf{f}^0)$. Numerical experiments are also presented to demonstrate this argument.

To solve these models, the method of successive averages (MSA) and particle swarm optimization (PSO) are tested and used in this study. MSA is used to solve the lower-levels of bi-level models and Model-1. PSO is used in solving the upper-level models of all the bi-level models. Besides, the detail steps of MSA and its extensions are elaborated in Yang [9]. PSO also is a well-developed method which has been provided in many kinds of mathematical programming software [14]. Thusly, we can use these methods directly.

3 Numerical experiments

3.1 Data employed and parameter settings

The traffic survey was carried out during the evening rush hours (from 3:30 pm-7:30 pm) from August 9th to August 19th, 2016 (only weekdays) in Hongkou district of Shanghai, China. As shown in Figure 1, Figure 2 and Table 1, 30 main roads of Hongkou district are inspected. Figure 1 is the spatial distribution map of Hongkou district, and Figure 2 is the structure sketch of the road network in Hongkou district.

Figure 1. Map of Hongkou

Table 1. Survey results of Hongkou district

Node A	Node B	AB flows	BA flows	C_a	t_a^0
1	4	2490	2638	2069	3
2	5	3764	3454	2652	3
3	4	3876	3170	2621	1
4	5	--	--	4118	2
4	11	--	--	1034	1
5	6	3706	3704	1211	1
5	8	2446	2624	796	1
7	5	--	--	849	1
7	8	--	--	647	1
8	9	1776	1926	636	1
8	13	2680	2582	796	1
10	11	--	--	1248	1
11	12	3856	3904	1622	1
12	13	2650	2924	1273	1
12	7	2958	2584	1273	1
12	16	1582	1966	1379	1
13	14	1748	2120	2334	2
13	17	1980	2286	1008	1
15	16	--	--	1061	1
16	18	1265	1375	1591	2
17	19	2997	2991	1662	1
17	14	3391	3101	1248	1
18	21	2896	2612	624	1

18	19	2825	2690	1662	1
19	22	552	639	451	1
20	21	3548	3958	998	1
21	22	3076	3110	1036	1
21	24	--	--	1186	1
22	23	3010	2990	1747	1
22	25	--	--	1167	1

Unit: persons/hour is the unit of C_a, minutes is the unit of t_a^0.

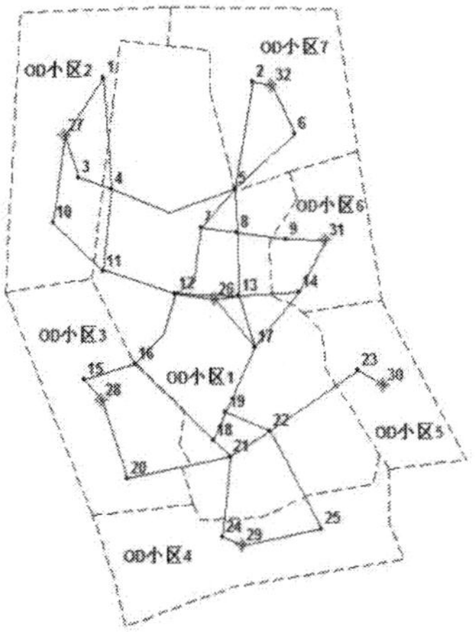

Figure 2. Structure of Hongkou

Figure 2 is created in TransCAD and intersections are labeled as node 1 to node 25. At the same time, 26 to 32 are centroids of traffic zones corresponding to OD zone 1 to OD zone 7. Note that 26 to 32 are virtual nodes in this map but they are necessary for OD estimation, so we assume that every link connects with these centroids have the capacity of 10000 persons/hour and free flow travel time of 0 minute. The SUE method has been used in OD estimation for 20 iterations, and the default convergence value is 0.01. Two parameters of BPR function are set as 0.15 and 4, respectively. Finally, we have obtained the matrix of OD travel demands of Hongkou district, as shown in Table 2.

Table 2. Results of OD estimation in TransCAD (unit: persons/hour)

OD zone	1	2	3	4	5	6	7
1	0	790	67	19	21	3316	2778
2	531	0	161	40	38	1012	2252
3	140	311	0	1321	1538	487	967
4	97	148	3684	0	651	197	428
5	69	111	2610	398	0	132	273
6	1413	1385	245	49	45	0	3166
7	4469	2812	1020	243	158	2369	0

To simplify the computation process of our numerical experiments, we only select two OD pairs from the network in which freeways are included. Two selected OD pairs are $OD(1,7)$ and $OD(2,7)$. Their travel demands are $OD(1,7)=2779$ (persons/hour) and $OD(2,7)=2253$ (persons/hour), which have been estimated in Table 2.

In our numerical experiments, other parameters are set as below. θ_0 is 0.1, BPR function is $t_a = t_a^0\left[1+0.15\left(\frac{v_a}{C_a}\right)^4\right]$, $\alpha=1.75$, $\varphi=0.35$, $\eta=1$. There are two VATIS service providers in this numerical network, so $M=2$. their parameters of profit functions: $\phi_{11}=400$, $\phi_{12}=1000$, $\phi_{21}=500$, $\phi_{22}=1200$ and $i=2$. The boundaries of decision variables are $\underline{\overline{\mathbf{X}}}=\left\{\begin{matrix}\mathbf{X_m}=(\theta_m,\tau_m^{rs}):0.1\le\theta_m\le1.0,0\le\tau_m^{rs}\le10,\\ \forall m\in\{1,2\},r\in\{1,2\},s\in\{7\}\end{matrix}\right\}$.

3.2 Optimal solutions of the four cases

Table 3 depicts optimal solutions of the four cases. Look through each stage of the suggested policy, we can find that, the result of the restricted scenario is the best which has the lowest total travel cost. It indicates that setting up the market penetration target of VATIS services is an effective way to decrease the total travel time if VATIS service providers are motivated to achieve the target.

Table 3. Optimal solutions of four cases

Cases	$\left(\theta_1,\theta_2,\tau_1^{17},\tau_2^{17},\tau_1^{27},\tau_2^{27},\Omega\setminus G\setminus GR,T(f)\right)$
Basic	$(0.1,0.1,0,0,0,0,\text{—},84865)$
Competition	$(0.56,0.52,4.13,4.05,4.39,4.32,7671,58652)$
Cooperation	$(0.57,0.53,6.35,6.28,6.73,6.70,8851,64953)$
Restricted	$(0.53,0.48,1.48,1.15,4.10,3.93,5728,57047)$

Unit: θ_i has no unit, yuan/per time is the unit of τ_i^{rs}, yuan is the unit of $\Omega\setminus G\setminus GR$ and $T(f)$.

3.3 Find the suitable fixed market penetration target of VATIS service

To find the most suitable fixed market penetration target of VATIS services in the traffic network, an analysis on the fixed market penetration target of VATIS services has been conducted. The fixed market penetration target Q_{fixed} changes from 0.1 to 0.9 and each step is 0.05. The setting of other parameters is the same as that of Section 3.2 apart from Q_{fixed}.

Note that Q_{fixed} is no more than a fixed target for VATIS service providers, so if Q_{fixed} was too high, service providers might not be able to attain it. Thus, we need to check the realized market penetration $Q_{actually}$ to find a suitable value of Q_{fixed}. Figure 3 illustrates the values of $Q_{actually}$ in the numerical network under varied levels of Q_{fixed}. We can see that $Q_{actually}$ will be lower than Q_{fixed}, if Q_{fixed} was set too high. Reasons are twofold. On the one hand, if Q_{fixed} was easy to achieve, so the traffic manager should choose Q_{fixed} properly and reasonably.

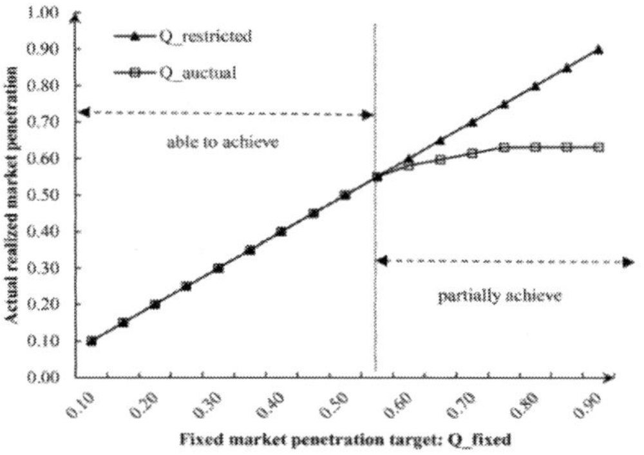

Figure 3. The differences between fixed market penetration target and realized market

Besides, $Q_{actually}$ has a tendency to increase to a stable value ($Q_{actually} = 0.63$ when $Q_{fixed} \geq 0.80$, as shown in Figure 3), that is the reasonable value of Q_{fixed} for pushing VATIS service providers spare no effort in decreasing the total travel time. At the same time, the saving time will be $\Delta T' = 18404$ (minutes) (28.33\% improvement), and the total saving time will be $\Delta T = 38316$ (minutes) (45.15\% improvement), when compared with the total travel time in the basic scenario. Moreover, the total compensation needs to be no less than $\Delta\beta = G(\bar{\mathbf{d}}, \bar{\mathbf{z}}) - GR(\hat{\mathbf{d}}, \hat{\mathbf{z}}) = 10811$ (yuan).

In summary, we find that there has a stable value of the fixed market penetration Q_{fixed} , in our experiments, it equals 0.80, where the biggest effectiveness of reducing the total travel time of the network can be attained. At the same time, a total compensation $\beta \geq \Delta\beta$ (in our experiments, β should be larger than 10811 yuan) also needs to be guaranteed, with the aim of motivating VATIS service providers to do their best in improving the efficiency of the traffic network.

4 Conclusion

In this study, a two-stage policy aiming at improving the efficiency of the traffic network is proposed based on the feasibility of charging value-added ATIS services. Models of describing each case in this policy are formulated and the evaluation indicators are pointed out. Numerical results show that the suggested policy can decrease the total travel time substantially.

The proposed two-stage policy can be very helpful to the traffic manager in regulating the behavior of ATIS service providers, especially when they are competing for ATIS market shares. However, this study also has limitations. It only has discussed about competition and cooperation among ATIS service providers by considering pricing strategy and the information quality they provided. Other potential factors, such as the influence of word-of-mouth, preferences of travelers, also need to be discussed in the future study.

Further extensions of this study can be investigated as below. Constitutes of the profit function of VATIS service providers would be an important issue in analyzing the effectiveness of this policy. Investigating its constituent part can enhance the veracity and rationality of the suggested policies. Besides, the evaluation criteria of target completion in the second stage of the suggested policy are interesting extension, hierarchic criteria might be the proper way of the evaluation of the suggested policy. Moreover, the suitable total compensation paid by the traffic manager needs to be further discussed.

5 Acknowledgement

This study is substantially supported by the National Natural Science Foundation Council of China (#71531011) and a grant of Shanghai Shuguang Program(13SG23).

References

1. D. Brownstone, A. Ghosh, T.F. Golob, C. Kazimi, DV. Amelsfort, Drivers' willingness-to-pay to reduce travel time: evidence from the San Diego I-15 congestion pricing project. TRANSPORT RES A-POL. **37(4)**:373-87. (2003)
2. R.C. Jou, K.H. Chen. A study of freeway drivers' demand for real-time traffic information along main freeways and alternative routes. TRANSPORT RES C-EMER. **31(31)**:62-72. (2013)
3. Z. Li, D.A. Hensher, J.M. Rose. Willingness to pay for travel time reliability in passenger transport: A review and some new empirical evidence. TRANSPORT RES E-LOG. **46(3)**:384-403. (2010)
4. J. Ma, F. Zhou, C. Lee. Providing personalized system optimum traveler information in a congested traffic network with mixed users. *Journal of Intelligent Transportation Systems Technology Planning & Operations*. (2016)
5. S. Nakayama. Effect of providing traffic information estimated by a stochastic network equilibrium model with stochastic demand. TRANSPORT RES C-EMER. **70**:238-51. (2016)
6. S.C. Welburn, K. Arnold, I. Maudlin, G.W. Gooday. Designing next generation multimodal traveler information systems to support sustainability-oriented decisions. ENVIRON MODELL SOFTW. **56(56)**:83-93. (2014)
7. R. Zhang, E.T. Verhoef. A monopolistic market for advanced traveller information systems and road use efficiency. TRANSPORT RES A-POL. **40(5)**:424-43. (2006)
8. Y. Sheffi. Urban Transportation Networks: Equilibrium Analysis with Matheatical Programming Methods: *Prentice-Hall*. (1984)
9. H. Yang. Multiple equilibrium behaviors and advanced traveler information systems with endogenous market penetration. TRANSPORT RES B-METH. **32(3)**:205-18. (1998)
10. C. Shapiro, H.R. Varian. Information rules: a strategic guide to the network economy: *Harvard Business School Press*; 62–7 p. (2008)

11. K.L. Hong, W.Y. Szeto. A methodology for sustainable traveler information services. TRANSPORT RES B-METH. **36(2)**:113-30. (2002)
12. H. Yang, X. Zhang. Modeling competitive transit and road traffic information services with heterogeneous endogenous demand, *Transportation Research Record Journal of the Transportation Research Board*, **1783(1)**:7-18. 7-18 p. (2001)
13. N. Hu, J. Wang, D. Mu. Research on the cooperation and competition strategies of port based on hinterland traffic condition. *International Conference on Logistics, Informatics and Service Sciences*; (2017)
14. Y. Da, G. Xiurun. An improved PSO-based ANN with simulated annealing technique. NEUROCOMPUTING. **63**:527-33. (2005)

MATEC Web of Conferences **259**, 05002 (2019) https://doi.org/10.1051/matecconf/201925905002
ICTLE 2018

Public preference towards the integration of krl and transjakarta tariff

Andyka Kusuma , Tri Tjahjono and Alviana Dwi Syaputri

Civil Engineering Department, Universitas Indonesia, Depok, Indonesia

Abstract. Transportation costs in Jakarta reach 30% of the total income of the community every month while the standard that has been set by World Bank is only 10%. This paper discusses the result of willingness-to-pay study conducted at Sudirman railway station in the downtown of Jakarta with the purpose of reducing transportation cost by implementing the tariff integration system of KRL and Transjakarta. The data was collected by the methods of stated preference survey with commuters as the respondent.The survey was conducted by asking the respondents some questions including respondents' socio-economic characteristic, travel distance, and travel choice exercise. The existing modes that will be compared are motorcycle and bus (kopaja). Motorcycle is the mode whom commuters mostly use with 62% while kopaja is only 38%. The data was analyzed by using discrete choice model and utility function. Based on the analysis, parameters that influence commuters to shift from existing modes to transjakarta are saving time, saving fare, transportation expenditure, and education level. Moreover, the study found that the cheaper the integration tariff offered, the greater the probability of choosing transjakarta services and the faster the travel time of Transjakarta, the greater the probability of choosing transjakarta. Furthermore, each comparison mode has different tariff preferences on the tariff integration system of Transjakarta and KRL.

1 Introduction

Jakarta as the capital city of Indonesia with more than 10 million populations. This situation creates high demand of transportation in Jakarta. The demand is not only generated by people in Jakarta but also from the sub-urban area around Jakarta which called BODETABEK. According to *Badan Pusat Statistik* in 2014 there are 1.38 million commuters from BODETABEK come to Jakarta each day.

Considering the high demand of transportation, Jakarta is now developing and improving the public transportation system. The revamping was started in 2004 where Bus Rapid Transit (Transjakarta) operated and followed with the revitalization of electric railway (KRL) IN 2012. Unfortunately, however some revitalization has been conducted by the government, 70% of commuters still use the private vehicle.

This high mobility of people in Jakarta using private vehicle or public transportation has an impact to the transportation costs they spend everyday. Based on Indonesian Transportation Society (MTI), transportation costs in Jakarta reach 30% of total community income every month. It is three times higher than the standard that has been set by World Bank which is only 10%. It caused by the lack of integration in every aspect of public transportation system. This problem should be addressed by implementing the system that allow people to cap their transportation costs a day.

Based on this problem, this research aims to observe public preference about how much they willing to pay in a day to use public transportation, in the case of the integration of KRL and transjakarta tariffs.

2 Literature review

2.1 Tariff Integration

Various study found in observing about public transportation integration. There are three components in transit integration, physical integration, service and operational integration, and payment integration (Setijowarno, 2016). Physical integration can be defined as modal shifting infrastructure that allows passengers to move intra or to other modes easily. Service and operational integration is a well-informed schedule of arrival and departure of public transportation. This integration also reducing passengers' waiting time. While payment integration is the concept of smartcard use which allows an effective and efficient service network. Moreover, one of the most important aspect in public transportation integration is the tariff integration.

Tariff integration can be interpreted as the same pricing scheme from different public transport operators, for example some bus companies apply the same rate for a single trip or some train companies apply the same rate for a certain distance trip. In the wider integration, this same tariff scheme applies even to more than one operator service which means that fares for travel within a certain distance will be the same whether one uses the services of one operator or mixes them with the services of different operators. In developed countries the concept of tariff integration is applied by combining fixed tariffs in which they sell valid tickets for an unlimited number of modes including subway, tram and bus with different operators within a specified time period, for instance in one day (Takahashi, 2016).

Summarizing many case studies relate with tariff integration particularly in developed countries, (Abrate et al, 2009) estimated that the tariff integration system in Italy could increase the number of passengers' trip by 2.2% in the short trip and 12% in the long trip. (Matas, 2004) estimated that the introduction of the tariff integration system in Madrid in 1987 led to a growth in bus of 3.4% and underground patronage of 5.3% in short-run while bus grew 7% and underground patronage 15% in long-run. (Taylor and Carter,1998) stated in Maryland the tariff integration system replaced a five-zone fare structure with a flat fare. The 1% annual decline in transit ridership shifted to 4% growth per annum. While in New York the tariff integration system which was introduced in 1994 and 1999 increased weekday ridership by more than 12% on the subway and by more than 40% on the bus. (Shiftan & Sharaby, 2012) concluded from the survey and model that fare is the significant factor in attracting passenger.

2.2 Jakarta case study

Jakarta is the most populated city in Indonesia and has high demand of transportation. Based on Department of Transportation, in 2010 the total trips in Jakarta reach 21.9 million trips per day where 15.3 trips are by private vehicles. This condition creates bad traffic particularly in rush hours. Revitalization of public transportation system has been done in order to facilitate the citizen to shift from private to public transportation.

The first modern public transportation in Jakarta was Bus Rapid Transit (BRT) called Transjakarta. This BRT was the first BRT in Southeast Asia. Transjakarta is a BRT system with the longest track in the world with a total length of 208 km in 13 corridors. Transjakarta mostly operated from 5 a.m to 10 p.m but now some corridors operate 24 hours.

Transjakarta is one of the favorite public transportation in Jakarta. Based on *Badan Pusat Statistik* (BPS), in 2011 Transjakarta reach the peak of its passenger number 114,769,431 but it declined to 111,260, 869 passengers in 2012. The increase started again in 2013 with 112,522,638 passengers, unfortunately in 2015 the decline was higher than before because it can only reach 102,950,384 passengers. While in 2016 the passengers increase again for 20%.

Since 2013, the ticketing system for Transjakarta started to use the electronic ticket system or e-ticket. The type of smartcard that can be used in Transjakarta is issued by several banks such as Bank Rakyat Indonesie (BRIZZI), Bank Central Asia (Flazz), Bank Negara Indonesia (TapCash), Bank Mandiri (e-money), Bank DKI (JakCard), dan Bank Mega (Mega Card).

Tariff for Transjakarta is divided into two types, the tariff from 5 a.m to 7 a.m is Rp 2000,- while the the tariff for 7 a.m to 5 a.m is Rp 3500,-.This tariff is already subsidized by the government of Jakarta. The other favorite public transportation in Jakarta is the electric rail called KRL Commuter Line. KRL has been operating since 1925 and now KRL's routes serve to the sub-urban area of Jakarta such as Bogor, Tangerang, Depok, Bekasi, and Serpong. Based on PT. Kereta Commuter Indonesia (KCI), till August 2017 KRL has covered 72 stations throughout Jabodetabek, Banten, and Cikarang with the length of route 418.5 Km.

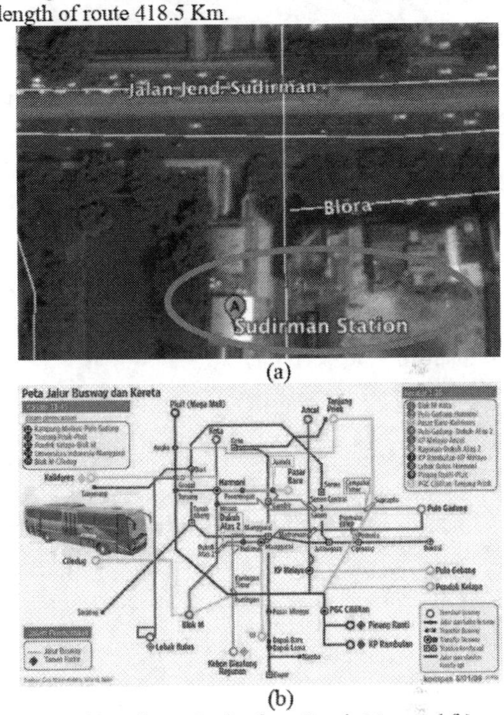

Figure 1. (a) Sudirman Station from Google Maps and (b) Route Maps of KRL and Transjakarta

Table 1. Number of KRL's passengers per year

Year	Percentage
2011	121.1 Million
2012	134.1 Million
2013	158.5 Million
2014	208.5 Million
2015	257.5 Million
2016	280.6 Million

2.3 Discrete choice model

Discrete choice model is a model where the function is used to find the probability of an individual to determine a choice of some alternatives based on socio-economic characteristics and the relative attractiveness of an options (Ortuzar W., 2011). The discrete choice applications in transportation research field are to find the individual transport mode choice for a particular condition. This model will represent the decision making of an individual to choose the mode through a function called Utility Function. The utility function is specified as follows:

$$U_i = \beta_0 + \beta_1 X_1 + \beta_2 X_2 \dots . \beta_n X_n \quad (1)$$

where U_j is utility of i-choice, $X_i \dots X_n$ is the product variables observed, β_0 is the alternative specific constant (ASC) and β_1-β_n is the coefficient of variables. The choice in this study is binary where the respondents face two alternatives. In this case, the binomial logit model is used to compare between two modes of transportation. The binomial logit can be written as follows:

$$P = \frac{1}{1+e^{-(Z_2-Z_1)}} \quad (2)$$

where the P is the probability of individu to choose the specific transportation mode while Z_1 and Z_2 are considered as the utility function of each alternatives mode.

3 Methodology

The following analysis of the public's preference on tariff integration of KRL and Transjakarta is based on various data sources. The data is divided into two, the primary data and secondary data.

3.1. Primary data

The primary data is based on stated-preference survey conducted in Sudirman Station. Stated preference survey aims to validate public's preference to shift from the existing mode to Transjakarta. This survey will show the potential additional demand of Transjakarta and the preference tariff people willing to pay.

In this survey, the respondent is the commuter that travel the main trip use KRL and use motorcycle or conventional bus on their last mile trip. The survey conducted in Sudirman Station where the people in weekdays mostly ended their main trip to work.

3.2 Secondary data

The secondary data is the additional data that was obtained through some books, journal, or paper. This data will be used to support the primary data collection. The secondary data for this research is as follows:

1. The distance from Sudirman Station to some sub-urban area within a certain radius.
2. Transportation cost per week.
3. Travel time for KRL, Transjakarta, motorcycle, and the conventional bus within a certain radius.

3.3 The model

The model is based on stated-preference survey, asking the respondents how much they willing to pay if the tariff of KRL and Transjakarta are integrated and would they shift to use Transjakarta. There are two alternatives asked to the respondent and their response should be yes or no.

The model is specified as a binomial logit model and based on discrete choice model. The utility function in this study using combination among four variables. There are two measured components and two socio-economic variables. The measured components are the different between existing fare of transjakarta and other modes (saving fare) and the travel time different between transjakarta and other modes (saving time). The other socio-economic variables are expenditure of transportation and education. In general, the utility function of this study can be written as follows:

$$U_{(TJ\text{-}AM)} = \beta_0 + \beta_1(\text{SAVINGFARE}) + \beta_2(\text{SAVINGTIME}) + \beta_3(\text{EXPTRANS}) + \beta_4(\text{EDU}) \quad (3)$$

where $U_{(TJ\text{-}AM)}$ is utility of transjakarta-choice, $X_i \dots X_n$ is the product variables observed, β_0 is the alternative specific constant (ASC) and β_1-β_n is the coefficient of variables.

4 Data and analysis

The first step to analyze the data is by presenting the survey data. The data of stated-preference survey is showed as follows :

4.1. Proportion Respondent

Total respondents that have been surveyed is 92 respondents with the following details :

Table 2. The Proportion of respondents' mode

Mode	Percentage
Motorcycle	62%
Kopaja	38%

Table 3. The proportion of respondents' gender

Gender	Percentage
Female	53%
Male	47%

Table 4. The distribution of respondents' age

Age Group	Percentage
15-20 years old	14%
21-29 years old	53%
30-39 years old	20%
40-49 years old	9%
>50 years old	4%

Table 5. The Proportion of respondents' occupation

Occupation	Percentage
Student	27%
Employee	48%
Teacher/Lecturer	3%
Others	22%

Table 6. The distribution of education level

Education Level	Percentage
High School	40%
Diploma	30%
Bachelor	11%
Master	19%

Table 7. The distribution of respondents' transportation expenditure per week

Expenditure of Transportation	Percentage
<Rp 100.000	20%
Rp 100.001 – Rp 250.000	49%
Rp 250.001 – Rp 500.000	21%
Rp 500.001 – Rp 1.000.000	7%
>Rp 1.000.001	3%

Table 8. The distribution of respondents' expenditure

Expenditure	Percentage
<Rp 500.000	3%
Rp 500.001 – Rp 1.000.000	6%
Rp 1.000.001 – Rp 5.000.000	56%
Rp 5.000.001 – Rp 10.000.000	20%
>Rp 10.000.001	15%

Table 9. The distribution of respondents' main trip distance

Main Trip Distance (KRL's Trip)	Percentage
50 Km	19%
40 Km	6%
35 – 30 Km	19%
25 Km	56%

4.2. Model and analysis

R programming is used as a tool to build a modal choice model (binomial logit) between Transjakarta – motorcycle and Transjakarta – kopaja. After the data was run by R programming, the model obtained as follows:

Table 10. Summary of models

Transjakarta - Motorcycle							
No	Variables	Koef	Std. Error	T Value	Pr (>)	Sig	Log Likelihood
1.	ASC	-0.5578	4.910e-01	-1.136	0.25593		-266.8975
2.	Savingfare	-6.353e-05	2.439e-05	-2.605	0.00920	**	
3.	Exp-trans	2.028e-01	8.00e-02	2.535	0.01123	*	
4.	Education	-1.486e-01	4.576e-02	-3.247	0.00116	**	
5.	savingtime	-3.065e-03	1.392e-02	0.220	0.82581		
Transjakarta - Kopaja							
No	Variables	Koef	Std.Error	T value	Pr (>)	Sign	Log Likelihood
1.	ASC	1.769	3.692e-01	4.791	1.66e-06	**	-99.63882
2.	Savingfare	-6.895e-05	2.202e-05	-3.131	0.00174	**	
3.	Exp-trans	-9.626e-01	1.579e-01	-6.095	1.09e-09	***	
4.	Education	5.136e-01	1.173e-02	4.378	1.20e-05	***	
5.	Savingtime	-1.973e-02	1.737e-02	-1.136	0.25609		

Based on the model as shown in table 3, the utility function can be arranged as follows:

Table 11. The utility function of transjakarta

No	Mode	Utility Function
1.	Motorcycle	U = –0.5578 – 0.00006353 (*Savingfare*) + 0.2018 (*Exptrans*) –0.1486 (*Education*) – 0.003065 (*Savingtime*)
2.	Kopaja	U = 1.769 – 0.00006895 (*Savingfare*) – 0.9626 (*Exptrans*) + 0.5136 (*Education*) – 0.01973 (*Savingtime*)

In the model result, there is a positive and negative sign that can be interpreted as follows:

1. The ASC (alternative specific constant) from motorcycle has a negative value. It means that there is a negative chance on the possibility of people to shift from motorcycle to transjakarta without any other influences. While the bus model has a positive value which means that there is such a positive chance that people will shift from bus to transjakarta without any other influences.
2. The variable of saving fare showed both motorcycle and bus have a negative value. It can be interpreted that if the tariff difference between Transjakarta and motorcycle has a negative value or Transjakarta's tariff is cheaper than motorcycle automatically the utility value will be higher so that the probability of shifting to Transjakarta is bigger.
3. The variable of saving time is negative sign which means that if the transjakarta travel time is faster than the existing mode (the saving time is negative) then the probability to shift is increasing. While if the existing modes is faster then it reduces the probability to shift.

While the other socio-economic variables like expenditure of transportation and education showed an opposite result. The result can be influenced by the data distribution.

4.3. The sensitivity of model

Sensitivity of model towards tariff change represents how influential the tariff changes to the probability of mode shifting from bus or motorcycle to Transjakarta. The sensitivity of model is represented in graphic so that the probability changes can be observed along with the gradually tariff changes.

The value on horizontal axis represented the T_{TJ}-T_{Bus}. The negative value showed the condition that using Transjakarta is cheaper than the bus while the positive value pointed the condition when using Transjakarta is more expensive than kopaja.

The graphic showed the comparison of tariff changes towards probability people to shift from each mode, motorcycle and bus. The slope of the lines on the graph tend to negative values, it is stated that the higher the saving fare, the smaller the probability of choosing transjakarta. As seen on the graphic, the slope of bus line is steeper than the motorcycle. It indicated that a little change in the fares in model will cause significant changed in the probability of modal choice.

Graphic 1. Model Sensitivity Graphic towards Tariff Changes for Both Motorcycle and Kopaja

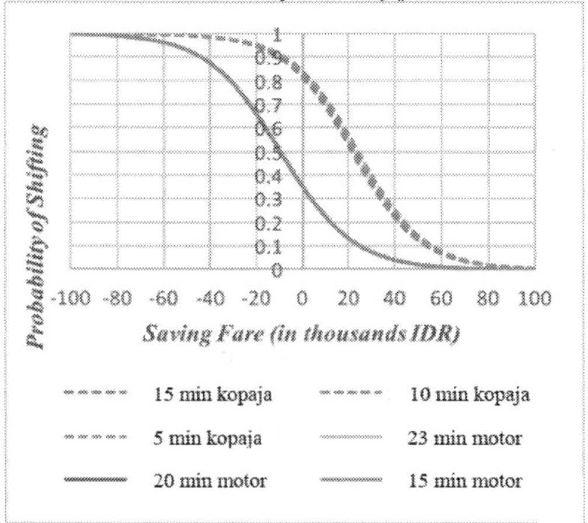

4.4. Willingness-to-Pay

Willingness to Pay (WTP) value is a tariff that is willing to be paid by users in the KRL and Transjakarta tariff integration system. The value of WTP can be obtained when probability of each individual in choosing each mode is 50% while the difference of utility values between HST and other modes equal zero (0).

Table 12. Summary of willingness to pay

Comparison Mode	Saving Time (Ttransjakarta – Tother modes)	Savingfare at the 0.5 probability
Kopaja (Bus)	15	Rp 21.364,-
Motorcycle	15	(-) Rp 9.503,-

From the table above we can see that the kopaja users are willing to pay Rp 21.364,- higher than the existing fares to the system of tariff integration. The number seems like impossible but there are some reasons that might create the result. Firstly, the (+) sign of ASC represent the desire of bus users that is more likely to use transjakarta. On other hand, the existing tariff between transjakarta and the kopaja is not too much different but based on stated preference survey, the respondents stated that the far walking distance from the railway station to the bus stop is the main excuse so they do not use transjakarta.

While for the motorcycle, the users are willing to pay Rp 9.503,- cheaper to shift to transjakarta. The respondents describe that based on safety and comfort, transjakarta is superior than motorcycle but from the efficiency they choose motorcycle. Thus, they only willing to pay Rp 9.503 cheaper to shift to transjakarta.
At the end, revealed survey is also conducted to see the exact number respondents are willing to pay for the tariff integration system. This following table shows the result:

Table 13. The summary of revealed preference survey

Modes	*Desirable Willingness to Pay*				
	Average	*75th percentile*	*Median*	*STDev*	*Max Tariff*
Kopaja	Rp 6,250,-	Rp 7,500,-	Rp 6,000,-	Rp 2,098,-	Rp 10,000,-
Motor cycle	Rp 6,403,-	Rp 7,375,-	Rp 6,000,-	Rp 2,022,-	Rp 15,000,-
All	**Rp 6,555,-**	Rp 7,500,-	Rp 6,000,-	Rp 2,242,-	Rp 15,000,-

According to the table, the respondents are willing to pay Rp 6,555,- to use the services of KRL and transjakarta as an integrated public transportation system.

4.4. Potential impacts of tariff integration

Jakarta as the capital city of Indonesia now is having a huge development and revamping of public transportation. The development starts from the infrastructure construction to the system improvement. The governor of Jakarta is now having a program called 'oktrip'. It is a program that allow people to pay a certain amount of money for using public transportation.

This study can be a reference for the governor to decide the capping system. The number of Rp 6.555,- might be representative as the tariff for integrated public transporatation system in Jakarta. Soon, when the policy is implemented, it is expected to create a lot of people to shift to use transjakarta and KRL. In the future, this research could be continued to analyze the additional demand of transjakarta due to tariff integration system.

5 Conclusion and future research

This paper investigated the public preference of tariff integration for KRL and Transjakarta locating in the downtown of Jakarta, Sudirman Station as a case study. The research showed that people tend to shift to transjakarta would be influenced by some variables, such as saving fare, saving time, expenditure of transportation, and education. This study focuses to the measured variables such as saving fare and saving time. The result shows that the more negative saving fare or the cheaper transjakarta tariff would be increasing the probability to shift to transjakarta while the more negative saving time or the faster transjakarta also increasing the probability to shift to transjakarta.

The research also clearly showed that there is different tariff preference between motorcycle and kopaja. Based on stated preference survey, kopaja tends to shift to transjakarta with higher fares than motorcycle. Kopaja users are willing to pay Rp 21,364,- higher to the system of tariff integration while motorcycle users only willing to shift if the tariff integration is Rp 9,503,- cheaper than the existing tariff. This different result is caused by different existing condition of each modes where kopaja tariff and travel time are not much different with transjakarta's while motorcycle tariff is way higher with faster travel time. At the end of the survey, respondents also asked their exact preference tariff for this integration system. This revealed survey produced that averagely the respondents are willing to pay Rp 6,555,- for this KRL and Transjakarta tariff integration system.

The outcome of this study particularly has great potential impact on the program created by Jakarta government. Despite reducing the transportation cost, this study provide the exact number of how much the people of Jakarta are willing to pay to the tariff integration system. This result is expected to be the reference of Jakarta government in order to decide the public policy related with the famous program called 'oktrip'.

The result of this study suggested that tariff integration system should be implemented as the first step of public transportation improvement. Further research is required to examine the additional demand due to the implementation of tariff integration system.

Acknowledgement

This study is funded by Universitas Indonesia under the PITTA scheme No: 2376/UN.R3.1/HKP.05.00/2018. This fund is very useful for developing research both for the authors and the institution.

References

1. Fitrianingsih, & Paramitarani. (2004). *Kesediaan dan Kemampuan Penumpang Kereta Api "Pandawangi".* Universitas Katolik Soegijapranata, Semarang.
2. Leland Blank, Anthony Tarquin. (2012). *The McGraw-Hill Companies.* New York: Raghothaman Srinivasan.
3. Lembaga Teknologi Fakultas Teknik. (2017). *Laporan Akhir Studi Travel Diaries Penumpang LRT Jabodetabek* . Jakarta, DKI Jakarta: Universitas Indonesia.
4. Lembaga Teknologi Fakultas Teknik Universitas Indonesia. (2017). *Laporan Akhir Studi Willingness to Pay Penumpang LRT Jabodetabek.* Jakarta, DKI Jakarta: Universitas Indonesia.
5. Manning, C. (2007). *Logistics Regression (with R).* Stanford: Stanford University.
6. McFadden, D. (1997). Measuring Willingness-To-Pay For Transportation Improvement .
7. Ortuzar, J. d., & Willumsem, L. G. (2011). *Modelling Transport.* WILEY.
8. Prabhakaran, S. (2016). Retrieved November 23, 2017, from r-statistics.co: http://r-statistics.co/Logistic-Regression-With-R.html
9. Sharaby, N., & Shiftan, Y. (2012). The Impact of Fare Integration on Travel Behavior and Transit Ridership . *Elsevier* , 63-70.
10. Badan Pusat Statistik. (2016). *Statistik Transportasi DKI Jakarta* . Jakarta, DKI Jakarta: Badan Pusat Statistik DKI Jakarta.
11. Breidert, C. (2005). *Estimation of Willingness-to-Pay.* Theory, *measurement, and application.* Vienna University of Economics and Business , Vienna.
12. Sihombing, D. A., & Surbakti, M. S. (2010). *Analisa Pemilihan Moda Kereta Api dan Bus (Studi Kasus: Medan - Pematang Siantar).* Departemen Teknik Sipil. Medan: Universitas Sumatera Utara.
13. Tamin, R. O., H., K., Munandar, A., & Setiadji, B. (1999). Studi Evaluasi Tarif Angkutan Umum dan Analisa Ability to Pay (ATP) dan Willingness to Pay (WTP) di DKI Jakarta. *Jurnal Transportasi* , *1*, 122-135.
14. Torres-Reyna, O. (2014). *Logit, Probit, and* Multinomial *Models in R.* Princeton: Princeton University.

Analysis of Transit Oriented Development Potential on Light Rail Transit Palembang, Simpang Polda Station Area

Fadhil Muhammad Fajri[1] and Jachrizal Sumabrata[2]

[1]Civil Engineering Departement, Universitas Indonesia, Indonesia
[2]Civil Engineering Departement, Universitas Indonesia, Indonesia

Abstract. The government of South Sumatera Province, decided to build Light Rail Transit (LRT) in Palembang City with the route from Sultan Mahmud Badaruddin II Airport to Jakabaring area to address the health, social and economic impacts from traffic congestion. However, will the development of LRT in Palembang city be enough to prevent total traffic congestion in 2019 and utilized by the people of Palembang City who have been accustomed to using private vehicles? With this rationale, the Transit Oriented Development (TOD) theory was adopted in this study as the answer to those questions. This study focused on analyzing the potential application of the TOD concept on the Simpang Polda Station Area by using several 'TOD-ness' measurement indicators from some previous research and TOD Standard by ITDP [1]. Using 8 new potential indicators, the assessment of the potential application of the TOD concept on the Palembang City LRT Simpang Polda showed that this transit area has enough potential for the implementation of the TOD concept by applying several supporting development plans.

1 Introduction

Palembang is the capital of South Sumatera province, Indonesia. In urban areas, such as Palembang itself has problems, because urban space is increasingly widespread randomly and not evenly to the area around the city causing urban sprawl, and will experience vehicle traffic in 2019. To anticipate the congestion, the government of South Sumatera Province build Light Rail Transit (LRT) line from Sultan Mahmud Badaruddin (SMB) II Airport to Jakabaring area. Transit Oriented Development (TOD) theory adopted to analyze whether the development of LRT in Palembang City is enough to prevent the total congestion in 2019 and will be utilized by the people who have been accustomed to using personal vehicles and existing public transportation, and more importantly, wether TOD Concept can be applied in Palembang City.

Author decided to analyze the potential of TOD application in one of Palembang LRT station areas, that was Simpang Polda Station area, in hope that this research could be utilized to develop the area around the station to be transit-oriented and support LRT Palembang to become a sustainable public transportation system.

This document is strucutured in five main section:

Section 1 Introduction, which introduces the themes, states the main objective of the paper and displays paper's structure;

Section 2 Literature Review, where the state of problem and TOD definition are collected;

Section 3 Methodology, propose a methodoly to define TOD potential of the station area;

Section 4 Result & Discussion, introduces a preliminary characterization of the station area and analysis of TOD potential of the station area;

Section 5 Conclusions & Future Steps, encompass a summing up of the points and a statement of opinion or decision reached, included the tasks to be done in the future to improve TOD potential of the station area.

2 Literature Review

2.1 Palembang City

Based on PP. 23 in 1988 [2], the area of Palembang City was 400.61 km^2 and administratively divided into 16 sub-districts and 107 urban villages, where according to the Palembang City Regulation 2007, had been expanding from 14 districts and 103 urban villages. With the population of 1,602,071 persons, Palembang City was classified as Metropolitan city (Statistic of Palembang Municipality, 2017) [3].

Land use in Palembang City was dominated by residential and commercial area. While the perimeter of the city was dominated by residential areas, the downtown area was dominated by commercial areas, although there were also some residential areas in some parts of the central city of Palembang.

Transportation nodes in Palembang City was already quite a lot, there are 13 bus terminals connecting areas within the city of Palembang, Kertapati railway station connecting Palembang City - Lubuk Linggau, and Palembang City - Tanjung Karang (Lampung), river crossing port in the Palembang urban area, the Boom Baru Sea Port in Ilir Barat II connecting Palembang City - Mentok City (Bangka), Palembang City - Batam and Palembang City - Jakarta, and the Sultan Mahmud Badaruddin II International Airport. These transportation nodes can be seen on Figure 1.

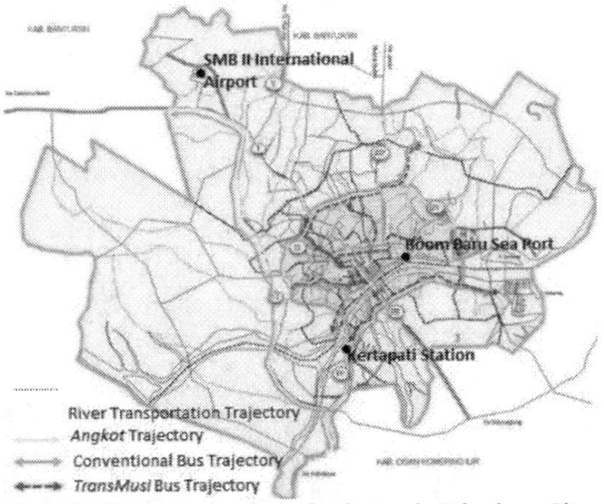

Fig. 1. Public transport nodes and trajectory in Palembang City

2.2 Light Rail Transit (LRT) Palembang

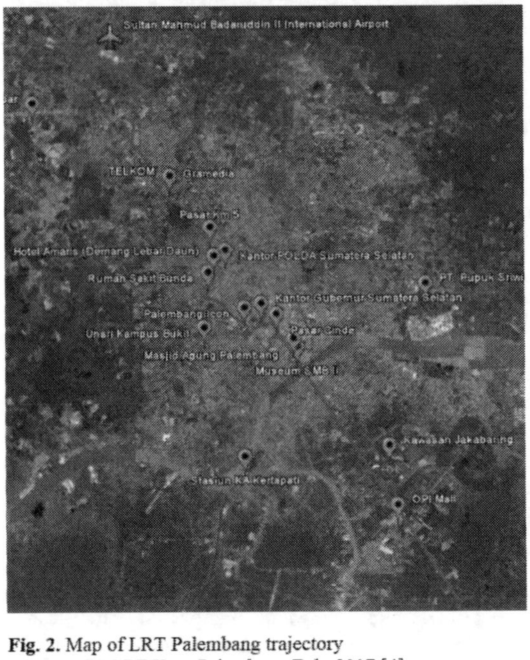

Fig. 2. Map of LRT Palembang trajectory
Source : FS LRT Kota Palembang Dsk, 2017 [4]

Light Rail Transit (LRT) Palembang was a light rail mass transit system connected Sultan Mahmud Badaruddin II International Airport with Jakabaring Sports Complex. In order to welcome the 2018 Asian Games in Palembang City, the planned monorail development that was originally to be implemented was later canceled due to the difficulty of finding investors who could complete the project on time and the project was considered less profitable. The monorail was later replaced by a more effective LRT. This 7.2 trillion rupiah project was required to be financed by the Central Government through the State Budget and the assignment of construction was on PT. Waskita Karya.

This project started on January 1st, 2015, and would be done at most on July 2018 (Indonesia Republic, 2015) [5]. In Figure 2 can be seen there is 13 stations, 9 electricity substations, and 1 Depot on the Palembang LRT line, with the stations order as follows:

1. Sultan Mahmud Badaruddin II Airport Station
2. Asrama Haji Station
3. Talang Buruk Station
4. RSUP Station
5. Simpang Polda Station
6. Demang Lebar Daun Station
7. Palembang Icon Mall Station
8. Dinas Perhubungan Station
9. Pasar Cinde Station
10. Ampera Station
11. Poltabes Station
12. Jakabaring Station
13. OPI Mall Station
14. Light Rail Transit Depot

Total length of LRT Palembang was 23.40 km as can be seen in Figure 1. The train formation used is 3 cars, i.e. 2 units of motor cars and 1 unit trailer car. The capacity of the train that was 8 persons/m^2 and accommodate 736 passengers (PT. Tribula Magenta, PT. Waskita Karya, 2017)[4].

2.3 Transit Oriented Development (TOD)

Peter Calthorpe, a student of the environmental sustainability movement, developed TOD to address the ecology of communities. Peter Calthorpe codified the concept of Transit-Oriented Development (TOD) in the late 1980's and, while others had promoted similar concepts and contributed to the design, TOD became a fixture of modern planning when Calthorpe published "The New American Metropolis" in 1993 (Caltron, 2007) [6]. TOD has been defined generally as "a mixed-use community that encourages people to live near transit services and to decrease their dependence on driving (Still, 2002) [7]." Calthorpe saw it as a neo-traditional guide to sustainable community design. Beyond its definition of built form, it was also a community design theory that promised to address a myriad of social issues.

3 Methodology

3.1 Data Collection Method

The data collection in this research was conducted by conducting field survey to get the data of land use, building density, road condition (for pedestrians and motor vehicles), and transportation system in Simpang Polda Station area of LRT Palembang City within 800 m radius with Simpang Polda Station as its buffer point. In addition, other data such as population density, frequency and capacity of LRT Kota Palembang, and the prediction of the number of LRT users of Palembang City was obtained from the parties related to the Palembang LRT development project, namely the Local Government of Palembang, *BAPPEDA* South Sumatra Province, Statistics Palembang, PT. Waskita Karya Tbk, Geography Laboratory of Universitas Indonesia, *RTRW Kota Palembang*, and Google Earth. Data collection in this study was conducted in April 2018.

3.2 Analysis Method

The data obtained were then analyzed by TOD measurement method by Sulistyaningrum, S. (2017) [8] and TOD Evaluation by Galelo, Ana *et al.* (2014) [9], i.e. by quantifying the data of population density and building density in the area of the station used as the data collection location, namely the Simpang Polda Station area. In addition, to know the "TOD-ness" of the area, was also used TOD Standard v2.1 issued by Institute for Transportation & Development Policy (ITDP) (2014) [1] as the measurement index.

Based on the assessment system of TOD Standard [1], TOD measurement method by Sulistyaningrum, S. (2018) [8], and TOD Index [9], the indicators and proposed TOD assessment system on this paper was walkways amenities, ratio of pedestrian intersections to motor vehicle intersections, total road bed area used for motor vehicle travel and on-street parking as percentage of total developed area, building density, percentage of residential building on developed area (mix-use), population density, transit system service, and the number of interchanges to other modes of transportation.

4 Results & Discussion

4.1 Walkways Amenities

The convenience of pedestrians in accessing stations in the Simpang Polda Station area could be assessed from pedestrian facilities on every street, such as sidewalks and sharing roads, zebra-cross, and pedestrian bridges that are in good condition. However, only 250 meters pedestrian walkways on both lane at Demang Lebar Daun Street and 250 meters at Sudirman Street had width of more than 2 meters, while other sidewalks along the arterial road this area was 2 meters or less with poor conditions. Meanwhile, on the local roads were applied share roads (without walkways), but in terms of road safety was still less safe for pedestrians so that pedestrians needed to walk on the main road that were at risk of accidents and disrupt the smooth flow of traffic on the road.

In addition, the road crossing facility in Simpang Polda Station area was rather inadequate, because in that area there were only 2 (two) pedestrian bridges located close to Simpang Polda Station and zebra-cross at Simpang Polda junction. The distance from one pedestrian bridge to the other one was 585 meters, where the minimum crossing standard was less than 500 meters. Meanwhile, the distance from zebra-cross to the nearby pedestrian bridge was 290 meters. Beside those three crossing facilities, no other crossing facilities were found on arterial roads in the area. In addition, the crossings still did not support wheelchair users to cross roads, like pedestrian bridges that have no rail for wheelchairs. As for the crossings on the small roads were not provided special facilities so pedestrians could cross anywhere. This gave minus points for this indicator because of the risk of accidents and the inhibition of road traffic caused by the system.

Based on these analysis, to support the implementation of the TOD concept in this region and refine the scores on this indicator, it was expected for the government to immediately improve pedestrian facilities so that pedestrian facilities meet the safety, liveliness, and comfort principles and enliven the pedestrians in the region. Therefore, 8 points was given on this indicator.

4.2 Ratio of Pedestrian Intersections to Motor Vehicle intersections

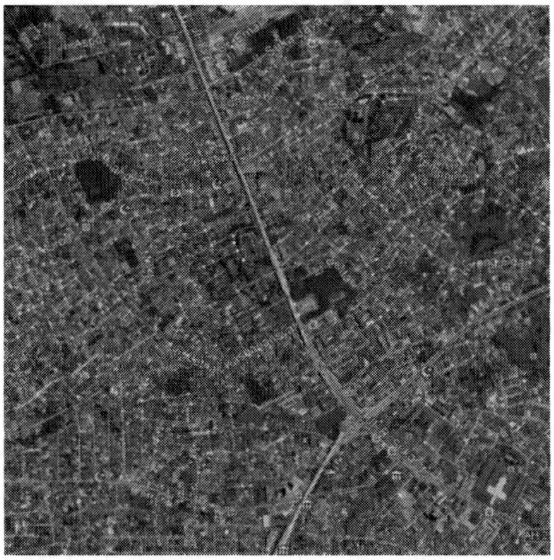

Fig. 2. Pedestrian intersections (yellow) and motor vehicle intersections (red) in the Simpang Polda Station area

In Figure 3 can be seen total pedestrian intersections in the area were 122 intersections, while total motor vehicles intersections were 132 intersections. Thus the ratio of pedestrian intersections to motor vehicles intersections was 0.924: 1. Since the pedestrian intersection ratio was less than 1, it means the area was still on the side of the vehicle compared to the pedestrians. Based on this value, to support the implementation of the TOD concept in this region and refine the value on this indicator, it was expected for immediate alteration the designation of the existing road to be made more pro-pedestrians. Therefore, 4 points was given on this indicator.

4.3 Total Road Bed Area Used for Motor Vehicles and On-Street Parking Area

In the this area, the length of the road that has a width of 28 meters of road was 2,537.18 meters, the length of the road that has a width of 9 meters of road was 4,558.12

meters long, the length of the road that has a width of 5 meters road was 13,256.56 meters, and a total parking area of 71,343 m^2. Thus, the total area of roads in the area was 249,689.92 m^2, i.e. 16.46% of the total developed area. Since the area of road bed was above 15%, to support the implementation of the TOD concept in this region and improve the value of this indicator, it was necessary to do the widening of the walkways or sideways and the addition of the canopy on the side of the arterial and shoulder roads on local roads so pedestrians can feel more comfortable and secure. Therefore, 8 points were given on this indicator.

4.4 Building Density

Fig. 3. Building area on Simpang Polda Station area

The building area in this region was 1,266,846.7 m^2, while the total area was 2,011,428.57 m^2 (Figure 4). Thus, the percentage of developed area in this area was 62.98%. On the basis of the assessment of analyzing the potential of the assessed area as a TOD-based area, this area could be said to have potential for application of the TOD concept since the percentage of developed area in this area was lower 70%. Based on the analysis, to support the implementation of the TOD concept in this region and improve the value of this indicator, vertical development (densification) needed to be enforced, such as apartment and other hi-rise building construction. Therefore, 12 points were given on this indicator.

4.5 Percentage of Residential Building on Developed Area (Mix-Use)

Mix-use here was referred to the percentage of building area designated for residence in the total developed area in this area. Number of resident in this region reached 70% of the total developed area in this area. Based on the analysis, 6 points were given to this indicator. Based on the analysis, to support the implementation of the TOD concept in this region and refine the value of this indicator, the land in this area needed to be utilized for development that encourages the integration of land use, such as supporting development in the commercial and retail sectors of the region. Therefore, 6 points were given to this indicator.

4.6 Population Density

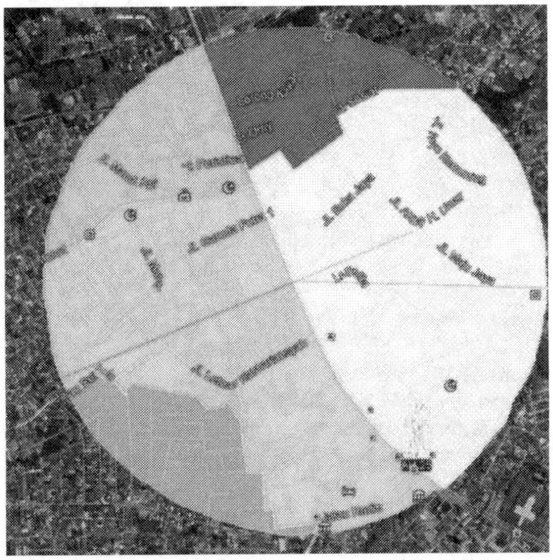

Fig. 4. Area distribution by sub-district (red: Sukarami; yellow: Kemuning; green: Alang-Alang Lebar; blue: Ilir Timur I; orange: Ilir Barat I)

The population of Palembang City in 2016 was 1,602,071 persons. Population density in Palembang City in 2016 reached 3,999.08 inhabitants/km^2. Population density in 16 sub-districts was quite diverse with the highest population density located in Ilir Timur I sub-district with population density of 11,137 inhabitants/km^2 and the lowest was in Gandus sub-district of 916 inhabitants/km^2.

In the Simpang Polda Station area with an 800 meters buffer, the sub-districts included in this area were Sukarami, Kemuning, Alang-alang Lebar, Ilir Timur I and Ilir Barat I sub-district (Figure 5). With a total area of 2,011,428,57 m^2, this area consisted of Sukarami sub-district area of 195.160 m^2, Kemuning sub-district of 770,436 m^2, Alang-Alang sub-district of 209,681 m^2, Ilir Timur I of 650,456 m^2, and Ilir Barat I of 185,696 m^2. By factoring the population with the area of each sub-district in the area, it was estimated that the total population of the region was 17,186 persons, which is the population density of this area was 8,544 inhabitants/km^2. By comparison the TOD area around the train station in Florida, which was about 17,000 inhabitants/km^2 (Florida Departement of Transportation, 2009) [10], to support the application of the TOD concept to the region and to refine the value of this indicator, as mentioned in point 4.6, vertical development needed to be developed to diversify the area and integrated the development of the non-residential sector in the region as mentioned in point D so that all daily activities could be carried out within this area and draw the masses to condense this region. Therefore, 6 points was given on this indicator.

4.7 Transit System Service

LRT will operate for 18 hours every day with peak-hour scheduled for 3 hours i.e. 06.00-09.00 am and 04.00-07.00 pm. The frequency of service at off-peak hour is every 30 minutes and at peak-hour morning and night is every 15 minutes. LRT travel time from SMB II to OPI Mall is about 37 minutes where most runs 30 minutes. At peak-hour, time headway is 15 minutes with average waiting time of 7.5 minutes. In the off-peak period, headway can be reduced to 30 minutes. So the travel time of one way is 37 minutes with a 10 minute layover which is used as the basic assumption to calculate the need of the facility. The travel time used on alternating trips is 94 minutes.

Based on feasibility study conducted by PT. Waskita Karya and PT. Trisula Magenta (2017) [4] potential demand for travel along the LRT corridor of Palembang City reaches 637,881 passengers/day. Based on these data, the analysis of mode selection predicts LRT users of Palembang City in 2017 was estimated at 74,288 passengers/day. While at Simpang Polda Station, passengers of LRT of Palembang city predicted 8,786 passengers/day with 7,406 passengers/day heading toward Jakabaring and 1,380 passengers/day heading toward SMB II airport and 8,392 passengers/day drop off the station. The number of passengers LRT Palembang in this region as much as 2,853 trips per day with the average number of trips per-day population of this area about 1.43 trips per day.

The LRT operating pattern assuming 10% of total travel would occur at peak-hour and there was no difference between passenger volumes both directions along the corridor, then peak hour demand would be split evenly to 725 passengers per-hour per-direction (pphpd). For off-peak state, the passenger demand would be about 480 pphpd. Considering that passenger demand at peak-hour was less than 800 passengers and off-peak condition was less than 500 passengers, the LRT system could be optimized using 2-3 carriages for all fleet with capacity of 8 passenger standing/m^2.

Based on the analysis, to support the implementation of TOD concept in this region and improve the value of this indicator, it was necessary to improve the operational performance of LRT of Palembang City to increase the amenities and interest of the community in using LRT Palembang City accompanied by densification of Simpang Polda Station area. Therefore, 7 points were given in this indicator.

4.8 Number of Interchanges to Other Modes of Transportation

Other transportation modes in Simpang Polda Station i.e.:

1. TransMusi coridor I (Alang-alang Lebar – Ampera) on JPO KM 5 A, Muhammadiyah, and Gloria bus stops;
2. TransMusi coridor II (Terminal Sako – PIM) on Rumah Kemuning A bus stop, Bayangan RRI Terminal;
3. TransMusi coridor V (Bandara SMB II – Bukit Siguntang) Karya Sejati, Rimba Kemuning bus stops;
4. City transport *(angkot)* on Ampera – Perumnas line, could be stopped along the road; and
5. Taxi and online-based transportation.

However, TransMusi and *angkot* was only passing through the city artery road (Sudirman Street, Demang Lebar Daun Street, and Basuki Rachmat Street) and not through the local roads. Based on this analysis, to support the implementation of the TOD concept in this region and refine the value of this indicator, there was a need of a perfect integration between each type of public transport that was active in this area so that the public transport range could be more widespread and avoid the use of the same line by the transport different public. Therefore, 6 points were given on this indicator.

4.9 TOD Potential Assessment of Simpang Polda Station Area

Based on the analysis that had been done, the result of TOD potential indicator assessment of Simpang Polda Station area can be seen on Table 1.

Table 1. TOD potential indicators assessment of Simpang Polda Station area

Indicators	Max. Point	Point
Walkways amenities	15	8
Ratio of pedestrian intersections to motor vehicle intersections	10	4
Total road bed area used for motor vehicle travel and on-street parking as percentage of total developed area	10	8
Building density	15	12
Percentage of residential building on developed area (mix-use)	15	6
Population density	15	6
Transit system service	10	7
The number of interchanges to other modes of transportation	10	6
Total	**100**	**57**

As the total point of TOD potential of Simpang Polda Station area is more than 55, it can be said that, overall, the Simpang Polda Station area had enough potential for the implementation of TOD concept

5 Conclusion & Future Steps

With the aim of directing the people of Palembang City to move from the habit of using personal vehicles to become using public transportation and the establishment of a sustainable city, the authors tried to analyze the potential of implementing the TOD concept in the Palembang LRT Simpang Polda Station within 800 meters radius. This study used the TOD measurement method by Sulistyaningrum, S. (2017) [9] and TOD Evaluation by Galelo, Ana et al. (2014) [8], and the TOD Standard issued by ITDP [1] to form 8 indicators to measure the potential implementation of the TOD concept in the Palembang LRT Regional Police Station intersection.

Based on the analysis that had been done, the result of TOD potential indicator assessment of Simpang Polda Station area showed that, overall, the Simpang Polda Station area had enough potential for the implementation of TOD concept in this region by applying several development plans, such as:

1. development of pedestrian facilities to meet the principles of safety, liveliness, and amenities and enliven the pedestrian in this area,
2. alteration of the existing road to be made more pro-pedestrians,
3. widening walkways or sidewalks and adding canopies on the side of arterial roads and local roads so pedestrians could feel more comfortable and safe on foot,
4. vertical development (densification) such as the construction of apartments and other hi-rise building and integrating the development of non-residential sector in this area so that all daily activities could be done within the region and attracted the masses to compact the area,
5. land-use utilization for the development that encouraged the integration of land allotment such as supporting development in the commercial and retail sectors in this region,
6. improvement of LRT Palembang City service and capacity along with the densification of the area, and
7. the provision of perfect integration between each type of public transport that was active in the area so that the public transport range could be more widespread and avoid the use of the same path by different public transport.

However, the scoring system in this study was still subjective and needed to be calibrated so that would be created a more accurate assessments. It indicated that there is a need for a professionals to form indicators that assessed the potential of the area to apply the concept of TOD, not just TOD evaluation indicators.

This research was expected to be utilized to develop the area around the station to be transit-oriented and supported LRT Palembang to become a sustainable public transportation system.

Acknowledgements

The authors are grateful to DRPM Universitas Indonesia for the financial support through PITTA 2018 grants and Mr. Jachrizal Sumabrata who had guided the author in completing this research.

References

1. Institute for Transportation & Development Policy, 2014. *TOD Standard.* v2.1 ed. s.l.:Despacio.
2. Indonesia Republic, 1988. Peraturan Pemerintah Republik Indonesia Nomor 23 Tahun 1988 tentang Perubahan Batas Wilayah Kabupaten Daerah Tingkat II Palembang, dan Kabupaten Daerah Tingkat II Musi Banyuasin dan Kabupaten Daerah Tingkat II Organ Komering Ilir. In: *Lembaran Negara Republik Indonesia Tahun 1988 Nomor 3383.* Jakarta: State Secretariat.
3. Statistic of Palembang Municipality, 2017. *Palembang Muncipality in Figures.* Palembang: Statistic of Palembang Municipality.
4. PT. Tribula Magenta, PT. Waskita Karya, 2017. *Feasibility Study LRT Kota Palembang.* Palembang: Transportation Office of Sumatera Selatan Province.
5. Indonesia Republic, 2015. Peraturan Presiden Republik Indonesia Nomor 116 Tahun 2015 tentang Percepatan Penyelenggaraan Kereta Api Ringan/Light Rail Transit di Provinsi Sumatera Selatan. In: *Lembaran Negara Republik Indonesia Tahun 2015 Nomor 236.* Jakarta: State Secretariat.
6. Caltron, I., 2007. *Histories of Transit-Oriented DEvelopment: Perspectives on the Development of the TOD Concept.* Berkeley: Institute of Urban and Regional Development.
7. Still, T., 2002. Transit-Oriented Development: Reshaping America's Metropolitan Landscape. *On Common Ground,* Issue Winter.
8. Sulistyaningrum, S., 2017. *Pengembangan dan Implementasi Pengukuran Indikator Untuk Menentukan Level Transit Oriented Development (TOD) di Kota Depok Pada Area Buffer Titik Transit Stasiun Kereta Depok Baru, Pondok Cina, dan Universitas Indonesia.* Jakarta: Sekolah Kajian Stratejik dan Global.
9. Galelo, A., Ribeiro, A. & Martinez, L. M., 2014. Measuring nad Evaluating the Impacts of TOD Measures - Searching for Evidence of TOD Characteristics in Azambuja Train Line. *Meeting of the EURO Working Group on Transportation,* 16(111), pp. 899-908.
10. Florida Departement of Transportation, 2009. *Transit Oriented Development Design Guidelines.* s.l.:Florida Departement of Transportation.

AUTHOR INDEX